ADVANCES IN LIQUID CRYSTALS

Volume 2

EDITORIAL ADVISORY BOARD

D. Demus (GDR)
Adriaan de Vries (USA)
G. Durand (France)
George W. Gray (England)
Al Mabis (USA)
Alfred Saupe (USA)
A. Skoulios (France)
Gordon T. Stewart (Scotland)
B. K. Vainshtein (USSR)

Advances in
LIQUID CRYSTALS

VOLUME 2

Edited by Glenn H. Brown

Liquid Crystal Institute
Kent State University
Kent, Ohio

ACADEMIC PRESS   New York San Francisco London   1976
A Subsidiary of Harcourt Brace Jovanovich, Publishers

COPYRIGHT © 1976, BY ACADEMIC PRESS, INC.
ALL RIGHTS RESERVED.
NO PART OF THIS PUBLICATION MAY BE REPRODUCED OR
TRANSMITTED IN ANY FORM OR BY ANY MEANS, ELECTRONIC
OR MECHANICAL, INCLUDING PHOTOCOPY, RECORDING, OR ANY
INFORMATION STORAGE AND RETRIEVAL SYSTEM, WITHOUT
PERMISSION IN WRITING FROM THE PUBLISHER.

ACADEMIC PRESS, INC.
111 Fifth Avenue, New York, New York 10003

United Kingdom Edition published by
ACADEMIC PRESS, INC. (LONDON) LTD.
24/28 Oval Road, London NW1

LIBRARY OF CONGRESS CATALOG CARD NUMBER: 74-17973

ISBN 0-12-025002-0

PRINTED IN THE UNITED STATES OF AMERICA

CONTENTS

LIST OF CONTRIBUTORS	vii
PREFACE	ix
INTRODUCTION TO THE SERIES	x
CONTENTS OF VOLUME 1	xi

Molecular Geometry and the Properties of Nonamphiphilic Liquid Crystals

G. W. Gray

I. Introduction	1
II. General Requirements Relating to Molecular Shape	4
III. General Structural Features Common to Many Mesogens	10
IV. Mesogens Containing Alicyclic Rings	13
V. Mesogens Containing Heterocyclic Rings	17
VI. The Role of Terminal Substituents in Mesogens	23
VII. The Role of Lateral Substituents in Mesogens	39
VIII. The Role of the Central Group in Simple Mesogens	52
IX. Features of Molecular Structure Favoring Formation of Smectic or Nematic/Cholesteric Phases	58
X. Molecular Structural Effects on the Liquid Crystal Properties of Sterol Derivatives	62
References	67

Selective Reflection of Cholesteric Liquid Crystals

W. Elser and R. D. Ennulat

I. Introduction	73
II. Physics of Cholesteric Mesophases	74
III. Chemistry of Cholesteric Liquid Crystals	97
IV. Temperature Dependence of Selective Reflection	125
V. Application	137
References	161

Liquid Crystals and Emulsions

Stig Friberg and Kåre Larsson

I. Prefatory Remarks	173
II. Arrangement of Emulsifier Molecules in the Solid State	174
III. Emulsions with a Crystalline Emulsifier at the Oil/Water Interface	176
IV. The Structure of Liquid Crystalline Phases of Emulsifiers	176
V. Phase Behavior of Different Types of Emulsifier–Water Systems	179
VI. Molecular Arrangement in Surface Films	182
VII. Water/Emulsifier Interaction	184
VIII. Emulsifier/Emulsifier Interaction	186
IX. Hydrocarbon/Emulsifier Interaction	189
X. Emulsion Stability	190
XI. Location of the Liquid Crystal	191
XII. Flocculation and Coalescence	193
XIII. van der Waals Forces	193
References	196

Vibrational Spectroscopy of Liquid Crystals

Bernard J. Bulkin

I. Introduction	199
II. The Vibrational Spectrum as a Source of Information	200
III. Internal Vibrations in Nematic Phases	201
IV. Intermolecular Modes in Nematics and Nematogenic Crystals	213
V. Vibrational Spectra of Smectic Phases	222
VI. Vibrational Spectra of Cholesteric Phases	225
VII. Vibrational Spectra of Lyotropic Mesophases	227
References	229

Equilibrium Theory of Liquid Crystals

J. L. Ericksen

I. Introduction	233
II. Energetics	236
III. Equilibrium Equations	257
IV. Linear Theory	267
V. Nonlinear Problems	270
VI. Linear Problems	289
References	293

SUBJECT INDEX 299

LIST OF CONTRIBUTORS

Numbers in parentheses indicate the pages on which the authors' contributions begin.

BERNARD J. BULKIN* (199), Department of Chemistry, Hunter College of the City University of New York, New York, New York

W. ELSER (73), U. S. Army Electronics Command, Night Vision Laboratory, Fort Belvoir, Virginia

R. D. ENNULAT (73), U. S. Army Electronics Command, Night Vision Laboratory, Fort Belvoir, Virginia

J. L. ERICKSEN (233), The Johns Hopkins University, Baltimore, Maryland

STIG FRIBERG† (173), The Swedish Institute for Surface Chemistry, Drottning Kristinas väg 45, Stockholm, Sweden

G. W. GRAY (1), Department of Chemistry, University of Hull, Hull, England

KÅRE LARSSON (173), Lipid Chemistry Laboratory, University of Göteborg, Rännvägen, Göteborg, Sweden

---

* Present address: Department of Chemistry, Polytechnic Institute of New York, Brooklyn, New York 11201.

† Present address: Department of Chemistry, University of Missouri–Rolla, Rolla, Missouri 65401.

PREFACE

Research studies of the liquid crystalline state of matter continue at an exciting pace. Research on the liquid crystalline state of matter has a history of about 85 years, but only during the past 15 years has the research effort been extensive. The appearance of commercial products incorporating liquid crystals has catalyzed activity in the field.

The five articles in this volume cover studies of both lyotropic and thermotropic liquid crystals. The survey by Professor Gray is concerned with the effects of change in molecular geometry on the properties of liquid crystals formed by nonamphiphilic (thermotropic) compounds. Dr. Elser and Dr. Ennulat address those aspects of the physics and chemistry of cholesteric liquid crystals that are relevant in respect to temperature dependence and temperature sensitivity of the selective reflection of visible light. Dr. Friberg and Dr. Larsson write about various relevant arrangements of emulsifier molecules and the interaction between the different components in emulsions. On this basis, emulsion stability, flocculation, and coalescence are considered. Professor Bulkin discusses the type of information about liquid crystals obtained from vibrational spectroscopy; consideration is directed to both thermotropic and lyotropic liquid crystals. Professor Ericksen covers the continuum theory of liquid crystals as it applies to static equilibrium.

I wish to thank the contributors for their fine cooperation, understanding, and concern, without which this volume could not have been written. My thanks to the members of the Editorial Board for their wise counsel. My sincere thanks to my wife who has helped with the details involved in bringing the manuscript into final form. To all others who have aided in bringing this volume to completion, I owe my appreciation.

GLENN H. BROWN

INTRODUCTION TO THE SERIES

The idea of an *Advances in Liquid Crystals* series of volumes originated a few years ago. At that time the field was beginning to evolve very rapidly but not many people were ready to review selected topics in which they were specialists. The field continues to grow rapidly, and because we can now begin to settle on the "state of the art" in many areas of liquid crystal science, the time now seems appropriate to start the series.

Plans call for coverage of the full spectrum of activity in liquid crystal research. Each chapter in the volumes of the series will be written by an expert(s) in the field and we plan to update a chapter periodically. The period will be determined by the activity in the field. Other topics will appear in subsequent volumes.

It is the wish of the Authors and Editor that the serial publication will become indispensible for those who are working in the field of liquid crystals or are planning to enter the field.

GLENN H. BROWN

CONTENTS OF VOLUME 1

Composition, Properties and Structures of Liquid Crystalline Phases in Systems of Amphiphilic Compounds
    Per Ekwall

Ordering and Structure of Liquid Crystals
    I. Chistyakov

Mesomorphic Properties of Block Copolymers
    A. Skoulios

Plastic Crystals, Liquid Crystals, and the Melting Phenomenon. The Importance of Order
    George W. Smith

Defects in Liquid Crystals
    M. Kléman

Subject Index

MOLECULAR GEOMETRY
AND THE PROPERTIES OF NONAMPHIPHILIC LIQUID CRYSTALS

G. W. Gray

Department of Chemistry
University of Hull
Hull, England

I. Introduction	1
II. General Requirements Relating to Molecular Shape	4
III. General Structural Features Common to Many Mesogens	10
IV. Mesogens Containing Alicyclic Rings	13
V. Mesogens Containing Heterocyclic Rings	17
VI. The Role of Terminal Substituents in Mesogens	23
A. General Considerations	23
B. Terminal Alkyl Chains	25
C. Alternation of Phase Type within Homologous Series	34
VII. The Role of Lateral Substituents in Mesogens	39
A. Molecular Broadening	40
B. Steric Effects	45
C. Branching of Terminal Alkyl Chains	48
VIII. The Role of the Central Group in Simple Mesogens	52
IX. Features of Molecular Structure Favoring Formation of Smectic or Nematic/Cholesteric Phases	58
X. Molecular Structural Effects on the Liquid Crystal Properties of Sterol Derivatives	62
References	67

I. INTRODUCTION

This review is concerned with the effects of change in molecular geometry on the properties of liquid crystals formed by nonamphiphilic compounds. Such liquid crystals have often been called thermotropic, and it is true that most interest in mesophases of this type has centered around those formed by the action of heat on the crystalline states of pure nonamphiphilic mesogens or their mixtures. Increased numbers of nonamphiphilic compounds which exist as thermodynamically stable mesophases at room temperatures are becoming known, but sufficient cooling will cause these to crystallize with greater or lesser ease; the mesophases are then regenerated by heating. Consequently, such mesophases may still be classed as thermotropic.

The term thermotropic also serves to distinguish such liquid crystals from lyotropic liquid crystals which are formed by the action of solvent(s) on suitable compounds (pure materials or mixtures). These compounds are called amphiphiles; the individual molecules contain two distinct regions, one nonpolar (hydrophobic) and the other polar (hydrophilic). These features confer different solubility characteristics on the different molecular regions. Common examples of amphiphiles are metal salts of aliphatic carboxylic acids (soaps) and phospholipids.

In contradistinction, the molecules of nonamphiphiles do not contain ionized, highly polar hydrophilic groupings—they are covalent, and although they may contain strongly dipolar groupings such as —CN, their affinity for ionizing solvents is low.

Difficulties [1] over the terms thermotropic and lyotropic, however, become obvious when one considers the conditions under which amphiphilic and nonamphiphilic mesogens form their mesophases. For example, Aerosol OT (I), a typical *amphiphile*, exists in the pure state at room temperature as a

$$CH_2.CO.OCH_2.CH(C_2H_5)CH_2.CH_2.CH_2.CH_3$$
$$CH(SO_3^-Na^+)CO.O.CH_2.CH(C_2H_5)CH_2.CH_2.CH_2.CH_3$$

(I)

liquid crystal, i.e., no solvent present. The phase is of the type M_2 (classification system of Winsor [2]) and is structurally similar to the M_2 phases formed by many amphiphiles (including Aerosol OT) with solvents. Also, typical anhydrous amphiphiles, such as alkali metal alkanoates and phospholipids, form liquid crystals when heated. Furthermore, the types of mesophase formed by an amphiphile/solvent system are themselves sensitive to temperature as well as composition, i.e., the so-called lyotropic mesophases have a strong dependence on temperature.

The distinction between lyotropic and thermotropic mesophases becomes more diffuse when it is remembered that the so-called thermotropic mesophases of nonamphiphilic mesogens may take up small amounts of typical organic solvents, etc., while retaining liquid crystal properties. Indeed, the use of nonmesogenic additives in thermotropic liquid crystals is important in obtaining room temperature liquid crystals with particular characteristics for use in electrooptic displays. One would not wish, however, to class these mesophases as lyotropic.

It seems preferable therefore not to perpetuate the terms lyotropic and thermotropic, and, instead, to classify *mesogens* in terms of whether they are amphiphiles or nonamphiphiles. Thus, the amphiphile Aerosol OT gives amphiphilic phases irrespective of whether it is pure or mixed with a solvent. Similarly, mesophases formed by nonamphiphiles are nonamphiphilic irrespective of whether they arise from mesogens alone or from mesogens and nonmesogenic additives.

The terms of reference of this review relate to the effects of change in molecular geometry resulting from molecular structural change on the properties of the nonamphiphilic mesophases formed by nonamphiphilic mesogens. For the most part, the review will be concerned with pure, single component nonamphiphilic phases, but its scope may, where necessary, be widened to include such liquid crystals in admixture with one another and with nonmesogens.

The properties of the nonamphiphilic phases with which we will be most concerned are: (i) the nature of the liquid crystal phase or phases formed—nematic, cholesteric, and the various smectic polymorphic forms; (ii) the temperatures at which the mesophases change from one to another (including smectic–smectic changes) or to the amorphous isotropic liquid, cf. p. 43. Since melting temperatures are difficult to correlate with molecular structure, useful conclusions in relation to the thermal ranges of mesophases emerge less readily.

When the structure, and consequently the geometry of a molecule is changed, many molecular parameters are affected and may influence in varying degrees the above liquid crystal properties [3–6]. The effects of structural change on the mesomorphic properties of a compound may therefore be difficult to rationalize, and a great deal remains imperfectly understood. It is therefore the aim of this review to define what relations have been established, even though these may be empirical, and to point out where gaps in knowledge arise.

The motivating reasons behind attempts to obtain a better understanding of the relations between molecular structure and liquid crystal properties through systematic studies of the effects of molecular structural change would seem to be obvious. However, bearing in mind a recent description [7] of such work as of the stamp-collecting variety, it is perhaps necessary to spell them out clearly.

Apart from the challenge to the chemist to understand these phase phenomena in more exact molecular terms, a clearer understanding of the situation allows more accurate predictions about the types of compound likely to give particular types of liquid crystal with the best properties for applications [8–10]. This is important in relation to current needs for (i) stable, low melting cholesterogens and nematogens for electrooptical displays, (ii) mesogens giving thermally stable mesophases of wide thermal ranges for gas–liquid chromatography [11], (iii) cholesterogens giving cholesteric mesophases exhibiting the maximum color sensitivity over required temperature ranges [12, 13]. A better understanding of the relations between molecular structure and the degree and types of smectic polymorphism exhibited is also of help in relation to the synthesis of materials giving various types of smectic phase at temperatures more suitable for physical studies of their molecular organization. For example, the availability of materials giving S_C, S_D, or S_E phases around room temperature would facilitate their study. Finally, the increased numbers of new mesogens made available by such studies increase the chances of discovering whether a wider range of smectic mesophases and of mesophase transitions is possible. Only by so-called stamp-collecting studies was the existence of the S_D phase discovered [14, 15] and the possibility recognized that a S_B phase may transform to the nematic phase [16, 17] or the amorphous isotropic liquid [18, 19] without the formation of an intervening S_A phase.

II. GENERAL REQUIREMENTS RELATING TO MOLECULAR SHAPE

Since the discovery [20] that the nonamphiphile cholesteryl benzoate forms an anisotropic liquid crystal on heating the solid to 150°C and that this phase is thermodynamically stable until 178°C when the amorphous isotropic liquid is formed, an increasing number of nonamphiphilic compounds have become available through the work of organic chemists. Lists of these mesogens were first compiled by Kast [21] in 1960, but, progress has been so fast that more up-to-date lists were soon desirable. These have recently been supplied by Demus et al. [22] who have listed over 5000 mesogens and their main properties, and Verbit [23] has lately computerized information on some 3500 mesogens.

Even a cursory inspection of the molecular structures of a range of nonamphiphilic mesogens makes it obvious that the molecules have a common geometrical feature. All are markedly elongated, and the terms rodlike or preferably lathlike are particularly useful for descriptive purposes. The structures given for 4′-methoxybenzylidene-4-*n*-butylaniline (MBBA) (II), diethyl 4,4′-axozybenzoate (III), 4′-*n*-pentyl-4-cyanobiphenyl (5CB) (IV) and cholesteryl propionate (V) exemplify the above points about molecular shape.

Lathlike molecules have a strong tendency to pack in the solid crystal lattice with the molecular long axes parallel. Early X-ray crystallographic

CH₃O—⟨◯⟩—CH=N—⟨◯⟩—C₄H₉   C₂H₅O₂C—⟨◯⟩—N=N—⟨◯⟩—CO₂C₂H₅
                                                       ↓
                                                       O

(II)   (III)

C₅H₁₁—⟨◯⟩—⟨◯⟩—CN   [cholesteryl structure with C₈H₁₇ and C₂H₅CO.O substituents]

(IV)   (V)

studies [24] of nonamphiphilic mesogens confirmed this, and each of two possibilities can arise:

(i) layer crystal lattices in which the molecules lie with their long axes parallel and their ends in line forming a three-dimensionally organized arrangement of molecular strata in which the long axes may be tilted or orthogonal to the planes;

(ii) crystal lattices in which the molecular long axes are parallel, but in which the molecules are interdigitated; molecular strata do not arise in the three-dimensionally organized system.

These two types of arrangement of lathlike molecules in the precursor solid states of nonamphiphilic mesophases are consistent with the molecular organizations for smectic and nematic (cholesteric) mesophases. In smectic mesophases, there is a statistically parallel arrangement of the lathlike molecules in layers. The layers are free to slide over one another, and only in more highly organized smectic phases (e.g., S_B and S_E) do end-to-end correlations occur over a few molecular lengths [25, 26]. Differing types of intralayer molecular organization and arrangement account satisfactorily for most of the observed types of smectic polymorphs [27, 28]. In nematic mesophases, only a statistically parallel alignment of the molecular long axes occurs, the direction of this axial arrangement (the director) changing in a continuous manner from one part of the nematic phase to another (unless the phase is extensively oriented by external forces) in accordance with the marked fluidity of the phase. The molecules in a nematic phase are free to slide past one another in the direction of their long axes. The cholesteric phase has a nematic type of molecular arrangement with a superimposed helicoidal twist axis at right angles to the long axes of the chiral molecules.

The liquid crystal phases can therefore be envisaged as arising from the parallel arrangements of molecules in the crystal lattices by a stepwise thermal breakdown of the intermolecular forces in the solids. Thus, the layer crystal lattice may first form a smectic phase in which the layers can slide over one another. The smectic phase may then change to the amorphous isotropic liquid direct or via an intermediate nematic mesophase formed by a simple translation of the molecules in the direction of their long axes. Alternatively, the layer lattice could change directly to a nematic phase, without formation of a smectic phase. A nonlayer crystal lattice of parallel, lathlike molecules could only give a nematic phase and thence the amorphous isotropic liquid on heating. Possible sequences of phase transition are therefore

$$C \to N \to I, \quad C \to S \to I, \quad C \to S \to N \to I,$$

where S represents a single smectic phase or a sequence of smectic phases, and N is replaced by Ch (cholesteric) in a system of chiral molecules.

The stepwise decrease in the order of the system on heating is borne out by observed sequences such as

$$C \to S_B \to S_A \to N \to I \quad \text{or} \quad C \to S_B \to S_C \to S_A \to N \to I$$

where S_B involves a hexagonally packed, orthogonal (or tilted [25]) arrangement of the molecules in layers which involve a higher degree of order than that in S_A or S_C layers. In S_C and S_A phases the molecules are tilted and orthogonal, respectively, in relation to the layer planes and the arrangement of the molecular centers in the planes of the layers is random.

For this stepwise breakdown in order, the intermolecular forces must be anisotropic. These forces could be of several types—dipole–dipole, dipole–induced dipole, and induced dipole–induced dipole interactions (dispersion forces). An anisotropy of these forces would therefore arise between molecules having an anisotropy of their polarizability or an anisotropy of their dipolarities. It is clear that anisotropy of molecular geometry or shape is a fundamental requirement for the formation of nonamphiphilic mesophases because this gives rise to an anisotropy of the intermolecular interactions. As we will see, the predominant factor for N–I transition temperatures is the anisotropy of molecular polarizability, i.e., generally speaking, the role of permanent dipoles is not critical in relation to this transition. This is in accord with the Maier–Saupe theory [29]. In so far as S–N and S–I temperatures are concerned, it is likely that permanent dipoles are important.

It is useful to consider the importance of geometric anisotropy from an opposite point of view, by discussing the behavior of nonamphiphilic compounds comprised of molecules which are globular in shape. The anisotropy of polarizability and dipolarity will then approximate to zero, and the intermolecular interactions will be about the same in all directions. Now it is just in these circumstances that the occurrence of plastic crystal mesophases [30, 31] is observed, e.g., with compounds such as CCl_4, $C(CH_3)_4$, cyclohexane, etc. As the temperature of the solid is raised, the approximately spherical units occupying the crystal lattice points (usually cubic) suddenly assume more or less free rotational motion, but the essential point is that the molecules maintain their positions at particular lattice points. The resulting plastic crystal lattice is less rigid, and this is reflected in the softness and volatility of plastic crystals. Only on heating to a higher temperature is the plastic crystal lattice destroyed at a second well-defined transition at which the globular molecules achieve translational freedom and form the amorphous isotropic liquid.

Therefore, at the extremes of molecular shape we have (i) the formation of anisotropic nonamphiphilic liquid crystal mesophases from compounds consisting of anisotropic, lathlike molecules, and (ii) the formation of plastic crystal mesophases from compounds comprised of globular molecules for which the shape anisotropy approximates to zero.

In the anisotropic liquid crystal phases, rotational freedom, e.g., end-over-end rotation, of the lathlike molecules is restricted by the long range parallel organization, although fairly free rotatory motions may occur about the long axes. In plastic crystal phases, the fairly free rotatory motion eliminates long range orientational order, but short-range restrictions on rotational motion affect the type of cubic lattice adopted.

Arguments have recently been put forward [32] for regarding plastic crystals as liquid crystal in type. Certainly they are liquid in that there is a complete absence of long range orientational order, but they are crystalline in the sense that they possess a cubic lattice. As a result, plastic crystal mesophases behave as *optically* isotropic phases, and both amphiphilic and non-amphiphilic mesophases which are optically isotropic and probably possess structures with cubic symmetry made up of freely rotating micelles are known, e.g., smectic D [14]—nonamphiphilic, and viscous isotropic cubic phases such as V_1, V_2, and S_{1c} (in the Winsor system of nomenclature [2])—amphiphilic.

If plastic crystal mesophases are accepted as liquid crystal phases, then it is necessary to define the types of nonamphiphilic molecule conducive to liquid crystal formation as: (i) the near spherical or globular type giving optically isotropic, cubic plastic liquid crystals; (ii) the lathlike type giving anisotropic, optically birefringent liquid crystals.

This classification may eventually have to be widened, because liquid crystal properties are reported as occurring in carbonaceous pitches comprised of flat molecules [33].

For the purposes of this review, we shall confine attention mainly to the lathlike nonamphiphilic mesogens required for the formation of nonamphiphilic mesophases.

In accepting the need for lathlike molecules for the formation of smectic and/or nematic (cholesteric) liquid crystals, we might consider molecules intermediate between the spherical type and the lathlike type, i.e., molecules whose shape is neither globular nor particularly elongated. In these cases, heating the crystal may first cause rotating, tumbling motions, and if the disturbances involved prevent the formation of a cubic plastic crystal mesophase, the amorphous isotropic liquid will be formed direct. Alternatively, if the effect of heat first causes translational motions in the direction of greatest elongation, but the anisotropy of molecular geometry is not enough to preserve parallelism, again the amorphous isotropic liquid will be formed. The majority of organic compounds do possess molecules whose shapes lie intermediate between the upper limits of geometric anisotropy for plastic crystal mesophase formation and the lower limits for liquid crystal mesophase formation. As a consequence, ordinary melting (C–I) takes place.

Recent studies [*34*] by Rooney et al. of the melting behavior of cage hydrocarbons have shown that diamantane (VI), a somewhat elongated cage molecule, is still able to give a plastic crystal phase. From studies of various cage-type compounds, they proposed that a cage hydrocarbon will give a plastic crystal phase if the entropy of transition to the plastic crystal phase (ΔS_{tr}) is less than ca. 55.3 J K^{-1} mol^{-1}. From the effects of elongation of the cage on the entropies of this transition, they calculated that the maximum value of $D_2 - D_1$ (where D_1 and D_2 are the diameters of the molecules perpendicular to the axes about which rotation in the crystal may occur) for a cage hydrocarbon which will still give a plastic crystal phase is ca. 2.8 Å, and that the maximum plastic crystal phase range will be 370° for a spherically shaped hydrocarbon molecule. For adamantane (VII), ΔS_{tr} is 16.2 J K^{-1} mol^{-1}, $D_2 - D_1$ is 0.2 Å, and $T_m - T_r$ is 335° (only 35° less than the maximum predicted). For diamantane (VI), ΔS_{tr} is 27.9 J K^{-1} mol^{-1}, $D_2 - D_1$ is 2.0 Å, and $T_m - T_r$ is 109°.

(VI)   (VII)

Thus, the allowed deviation from sphericity while retaining plastic crystal-forming properties is quite small. The length : breadth ratio for diamantane is ca. 1.4. No such accurate lower limit for the dimensions which lathlike molecules must possess to give anisotropic liquid crystal phases can be obtained, but the length : breadth ratio for 4'-*n*-pentyl-4-cyanobiphenyl (IV) may be quoted as ca. 3.2; the molecules are therefore quite short and the N–I temperature at 35°C is quite low. However, it is doubtful whether such ratios for lathlike mesogens are meaningful, for many compounds with a higher molecular length : breadth ratio than 3.2 do not form liquid crystals on heating, because the melting temperatures of the crystals are too high.

Molecular Geometry

A further point is that many molecules which appear to have elongated structures when drawn in a particular conformation may not retain this shape in a fluid environment (liquid crystalline or amorphous), i.e., the molecules are too flexible and may assume nonlinear conformations by rotations about bonds. Thus, the lathlike analogy extends beyond that of shape to a lathlike rigidity. Therefore in defining a minimum molecular length: breadth ratio for mesogens, this would have to relate only to molecules with the required rigidity for mesophase formation.

Rigidity and anisotropy of polarizability are of course related factors. For example, consider two molecules with structures VIII and IX.

$$X-\text{C}_6\text{H}_4-\text{CH}=\text{CH}-\text{C}_6\text{H}_4-Y \qquad X-\text{C}_6\text{H}_4-\text{CH}_2-\text{CH}_2-\text{C}_6\text{H}_4-Y$$
$$(\text{VIII}) \qquad\qquad (\text{IX})$$

Structure VIII is the more rigid because of restricted rotation about the sp^2 hybridized bonds between the carbons which link the two rings. Structure IX has an elongated conformation when the ring-C–C-ring bonds have a planar zig-zag arrangement, but rotation about the central C–C bond can occur and destroy the lathlike shape. Consequently, the tendency of stilbenes (VIII) to give mesophases is greater than that of 1,2-diphenylethanes. Furthermore, conjugative interactions between the aromatic rings may occur through the —CH=CH— linkage, but not through the —CH$_2$—CH$_2$— linkage. Therefore, in addition to greater rigidity, there is an enhanced anisotropy of molecular polarizability in the stilbene (VIII) because of the greater polarizability of the —CH=CH— unit and the transmission of conjugative effects that it makes possible.

The effects of stiffening otherwise flexible elongated molecules can be illustrated by many examples. For instance, the open chain alkanoic acids are not mesogens, whereas the alka-2,4-dienoic acids with two double bonds are mesogens. In both cases, the molecules form dimers, so that in the unsaturated acids (X), the rigid core part of the molecule is quite extensive (from A to B).

$$R-\underset{A}{|}\text{CH}=\text{CH}-\text{CH}=\text{CH}-\text{C}\begin{smallmatrix}\text{O}----\text{H}-\text{O}\\ \\ \text{O}-\text{H}----\text{O}\end{smallmatrix}\text{C}-\text{CH}=\text{CH}-\text{CH}=\text{CH}\underset{B}{|}-R$$
$$(\text{X})$$

III. GENERAL STRUCTURAL FEATURES COMMON TO MANY MESOGENS [3-6, 35]

The need for rigidity of the molecules of nonamphiphilic mesogens is the reason few aliphatic compounds are mesogenic and the majority of nonamphiphilic mesogens contain aromatic rings. These rings are polarizable, planar, and rigid, and by suitable positioning of substituents and linking units, stiff lath-shaped molecules are readily built up.

A common pattern for nonamphiphilic mesogens is two *p*-phenylene rings linked via an unsaturated linkage represented in XI as —A—B—. The nature and the role of the substituents X and Y are discussed later.

$$X-\phenyl-A=B-\phenyl-Y$$

(XI)

Examples of the linking units —A—B— are given below, and more will be said about their importance in relation to low melting mesogens for technological applications.

$$-(CH=CH)_n-;\ -C\equiv C-;\ -CH=N-;\ -N=N-;\ -\underset{\downarrow O}{N}=N-;\ -CH=\underset{\downarrow O}{N}-;\ -CH=N-N=CH-$$

To this list may be added

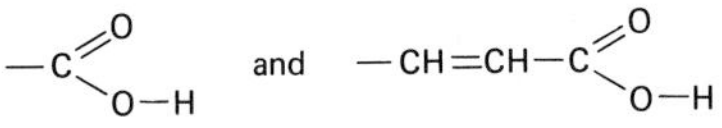

$$-C\overset{O}{\underset{O-H}{\diagdown}}\quad \text{and} \quad -CH=CH-C\overset{O}{\underset{O-H}{\diagdown}}$$

both of which give cyclic dimers, in which rotational motion is restricted.

Linking units such as —CO.O—, in aryl benzoate esters, do not involve a double bond actually linking the two rings together, but contributions from resonance structures such as

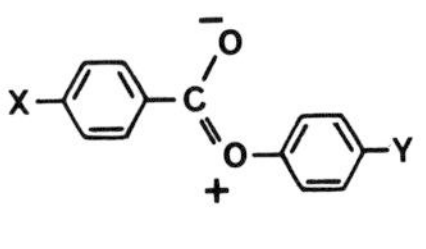

$$X-\phenyl-\overset{\overset{-}{O}}{\underset{\underset{+}{O}}{C}}-\phenyl-Y$$

can enhance the rigidity of such molecules. This also applies to the linking unit —CH=CH—CO.O—, in cinnamate esters.

Molecular Geometry

With the linking units —CH_2—CH_2— and —O—CH_2—CH_2—O—, the CH_2—CH_2 and O—CH_2 linkages cannot adopt any double bond character through resonance interactions, and do not contribute to molecular rigidity. As a consequence, these linking units are less commonly encountered, and when mesogens do arise that contain them as linking functions between two p-phenylene rings, the mesophases have low thermal stabilities.

As would be expected, the more extended the conjugated central unit, the more lathlike the molecule will be. As a consequence, replacement of —CH=N— by —CH=N—N=CH— or of —CO.O— by —CH=CH—CO.O— enhances the mesophase-forming properties, e.g., higher mesophase–amorphous isotropic liquid transition temperatures are observed.

For the same reasons, if we move from a two ring mesogen with one linking unit to a three ring mesogen with two linking units, mesophase thermal stabilities are greatly enhanced. For example, XII has an N–I (monotropic) of 121°C, whereas the N–I of XIII is 297.5°C.

C_4H_9O—⟨⟩—CH=N—⟨⟩—OC_4H_9 C_4H_9O—⟨⟩—CH=N—⟨⟩—N=CH—⟨⟩—OC_4H_9

(XII)   (XIII)

When three rings are involved, and two are linked by a unit which preserves conjugative interactions and molecular rigidity, the second linking unit can be more flexible, e.g., —CH_2—CH_2—, —O—CH_2—CH_2—O—, —(CH_2)$_4$—, —CH_2—NH—, —NH—, —OC.O—(CH_2)$_2$O.CO—, etc. This implies that if a considerable proportion of a lathlike molecule is rigid and packs parallel to similar portions of neighboring molecules, more flexible parts of the molecules may be constrained to lie in line with more rigid parts. This is relevant in connection with flexible alkyl terminal substituents.

With three ring mesogens, a combination of a rigid and a flexible linking unit gives less thermally stable mesophases than those that occur with three ring mesogens with two rigid linking units.

For the same reasons, replacement of a single p-phenylene ring by the 4,4'-biphenylene ring system gives a more thermally stable mesophase. The biphenyl system is fairly rigid, and if the individual biphenyl rings are near planar, conjugative interactions in the molecule will be enhanced. Thus, XIV has a monotropic N–I of 99°C and XV has an N–I of 318°C.

CH_3O—⟨⟩—CH=N—⟨⟩—OCH_3 CH_3O—⟨⟩—CH=N—⟨⟩—⟨⟩—OCH_3

(XIV)   (XV)

Indeed, the biphenyl ring system has such strong mesophase-promoting tendencies, that *simple* 4,4'-disubstituted biphenyls can be mesogens. Examples of current interest for display devices are the 4'-*n*-alkyl- and 4'-*n*-alkoxy-4-cyanobiphenyls [*16, 36, 37*]. The 4'-*n*-pentyl-compound (5CB) (IV) has C–N, 22.5°C; N–I, 35°C and the nematic melt supercools for long periods down to 4°C. The 4'-*n*-octyl-compound gives a room temperature smectic A mesophase. Materials such as these can be regarded as two ring mesogens with the linking unit replaced by the ring–ring bond.

$$C_5H_{11}\text{–}\text{–}\text{–}\text{–CN}$$

(XVI)

4,4''-Disubstituted *p*-terphenyls are even more pronounced mesogens [*38, 39*] and XVI [*40, 41*], the terphenyl analog of 5CB, has C–N, 130°C; N–I, 239°C. If the number of rings is increased, still more thermally stable mesophases are produced, and derivatives which are not mesomorphic in the biphenyl series are mesomorphic in the *p*-terphenyl or *p*-quaterphenyl series [*39*]; *p*-pentaphenyl with five rings and no terminal substituents forms a liquid crystal.

Dewar and Goldberg [*42*] illustrated the importance of the aromatic character of benzenoid systems in permitting conjugation in mesogens, and showed that under suitable circumstances, the enhanced anisotropy of polarizability led to higher transition temperatures, i.e., the *p*-phenylene ring promotes liquid crystal formation not only because of shape and rigidity, but also by permitting conjugative interactions between substituents. For example, the amounts by which the N–I temperatures of the quinol esters (XVII) exceeded those of the corresponding terephthalic esters (XVIII) were greatest when the

$$X\text{–}\text{–C(=O)–O–}\text{–O–C(=O)–}\text{–X} \qquad X\text{–}\text{–O–C(=O)–}\text{–C(=O)–O–}\text{–X}$$

(XVII)   (XVIII)

substituent X supplied electrons readily by conjugation with the ring. For example, when X = CH$_3$O, conjugation can extend through to the —CO.O— function in the quinol esters, but this is not so in the terephthalic esters where the ester function is —O.OC—.

When X = NO$_2$ (strongly electron withdrawing) the difference in the N–I temperatures was not that expected, and this required the use of more subtle

arguments. However, in general, the results confirmed a dual role for the
p-phenylene group in stabilizing mesophases—a geometrical and an electronic
function. The latter will obviously vary in magnitude depending on the terminal
substituents, etc.

De Jeu, van der Veen *et al.* [43] advanced similar ideas to explain the
enhanced N–I transition temperatures for compounds containing terminal
alkoxy groups compared with terminal alkyl groups of the same length. Two
examples are given in Table I. According to Maier–Saupe theory, the dipole

TABLE I

COMPARISON OF N–I TEMPERATURES FOR ALKYL AND ALKOXY SUBSTITUTED MESOGENS

	Substituent	N–I (°C)	Substituent	N–I (°C)
p-Substituted benzoic acids	*n*-C_3H_7O	154	*n*-C_4H_9	113
	n-C_4H_9O	160	*n*-C_5H_{11}	126.5
4′-Substituted 4-cyanobiphenyls	*n*-C_4H_9O	75.5	*n*-C_5H_{11}	35
	n-$C_5H_{11}O$	67.5	*n*-C_6H_{13}	27

of the alkoxy group should not be responsible, and de Jeu points out that
"the non-bonded electrons of the oxygen are easily coupled to the π-electron
system. This favors the polarizability along the long axis more than would
an alkyl group of the same length." In these words, de Jeu describes what we
call greater conjugative interactions between an alkoxy group and the ring
than between an alkyl group and the ring. De Jeu *et al.* quote calculated
values for the molecular Kerr constants for *p*-substituted toluenes and
p-substituted anisoles in support of this proposal. They also point out that
the difference between the total dipole moment (μ) for a substituted azobenzene
containing an alkyl and an alkoxy group and that for the dialkyl compound
(O) is greater than the difference between the total dipole moments for the
dialkoxy ($\mu/\sqrt{2}$) and alkylalkoxy (μ) compounds. This difference is not
reflected in the N–I transition temperatures and confirms that permanent
dipoles are not important in relation to nematic thermal stability.

IV. MESOGENS CONTAINING ALICYCLIC RINGS

Dewar *et al.* [42, 44, 45] also sought to confirm the dual role of *p*-phenylene
rings by replacement of one, two, or all the *p*-phenylene rings in compounds
of the type XIX by 1,4-bicyclo[2,2,2]octylene rings (XX). This nonaromatic

RO—⟨1⟩—C(=O)—O—⟨2⟩—O—C(=O)—⟨3⟩—OR

(XIX)

(XX)

ring system was chosen because it has a similar width to a *p*-phenylene ring and the 1,4-bonds are collinear; it differs only by being slightly thicker. Therefore if geometry alone is important, replacement of *p*-phenylene by 1,4-bicyclo[2,2,2]octylene rings should have little effect. However, replacement of ring 2 reduced the N–I temperature by 28° (R = CH_3). Replacement of terminal ring 1 or 3 lowered the N–I temperature by 76°, and replacement of rings 1 and 3 (R = CH_3) gave a nonmesogen (extrapolated N–I, ca. 130°C, a decrease of 167°). The effect of replacing two rings is therefore approximately additive. Replacement of all three rings also gave a nonmesogen for which an extrapolated N–I temperature could not be obtained. These large effects were accounted for in terms of the polarizable electrons of the aromatic ring system and the involvement of π-electrons in conjugation with substituents. Dewar *et al.* say "In either case the terminal rings ... would be expected to play a major part, for enhanced ... polarizability at the ends of a linear molecule should have a greater effect in making intermolecular forces anisotropic than would a corresponding change in the middle." In addition, replacement of rings 1 and/or 3 should interfere with conjugative interactions between RO— and —CO.O—, whereas replacement of ring 2 should have a smaller effect, as this ring is subject to equal and opposite effects from two —O.OC— groups.

Dewar *et al.* nonetheless stressed that the geometric role of the *p*-phenylene group is very important because replacement of *p*-phenylene ring 2 by *trans*-1,4-cyclohexylene gave (R = CH_3) a monotropic N–I of 195°C, i.e., a decrease of 75° from the 1,4-bicyclo[2,2,2]octylene analog and of 103° from the *p*-phenylene analog. This was attributed to the greater flexibility of the 1,4-cyclohexylene ring, and the conclusion was that a rigid geometry is essential (cf. also p. 10).

However, the situation is less simple, for Schubert *et al.* [46] compared the *trans*-equatorial 4-*n*-alkylcyclohexane-1-carboxylic acids with the corresponding 4-*n*-alkylbenzoic acids. Conjugation between the alkyl and carboxyl groups should not be important in the benzoic acids, and differences in the N–I temperatures should (cf. the results of Dewer *et al.*) be largely concerned with geometry and rigidity. Since two cyclohexylene rings are involved in the dimers (XXI), appreciably lower N–I temperatures would be anticipated. In fact, the N–I temperature for XXI is only 19°–20° below that of the analogous benzoic acid. The effects of the puckered rings on molecular thickness and flexibility do not therefore seem to play a great role in this case.

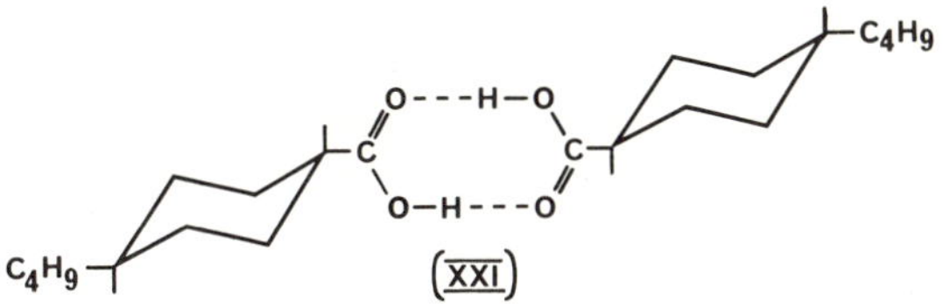

Molecular Geometry

(XXI)

However, in their studies of *p*-terphenyl compounds, Schubert and Dehne [39] also examined the effects of reducing certain of the rings. For example, XXII has N–I, ca. 315°C, whereas, with the middle ring reduced to cyclohexylene, XXIII has N–I ca. 239°C. This decrease of 76° is more in line with Dewar's result on introducing a cyclohexylene ring centrally in XIX.

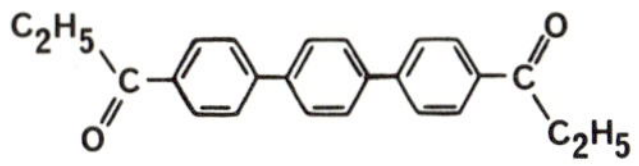

(XXII) 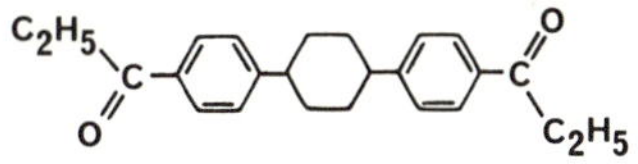   (XXIII)

These results can only be reconciled by concluding that the flexibility of a cyclohexylene ring placed centrally in a molecule is very critical, whereas more terminally (XXI), this has a less serious effect. An analogy is provided by flexible alkyl chains which function satisfactorily as terminal groups in lathlike mesogens, whereas a flexible methylene chain located centrally as a linking unit militates strongly against mesophase formation. However, it must also be concluded from the work on three-ring compounds that interference with conjugative interactions in the end rings has more serious consequences on anisotropy of polarizability and the intermolecular forces than interference with conjugation in the central ring.

The variable effects of replacement of *p*-phenylene rings suggest that they do function in mesogens in more than one way, and that the roles played by the geometric and conjugative factors vary from system to system.

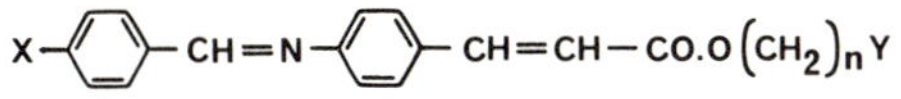

$$X-\bigcirc-CH=N-\bigcirc-CH=CH-CO.O(CH_2)_nY$$

(XXIV)

Studies [47, 48] have also been made of esters of the type XXIV where $X = -CN, -NO_2$ or Ph; $n = 0$ and 2; $Y = $ Ph or cyclohexyl.

TABLE II

Transition Temperatures (°C) for Compounds of the Type XXIV

		Y = phenyl		Y = cyclohexyl	
		C–N	N–I	C–N	N–I
X = CN	$n = 0$	167	277.5	165	(120.5)[a]
	$n = 2$	144.5	186	116	136
X = NO$_2$	$n = 0$	183.5	238	151	(99.5)
	$n = 2$	159	178.5	128.5	(115)

		X = phenyl				
		C–S$_A$ or S$_B$	S$_E$–S$_B$	S$_B$–S$_A$	S$_A$–N	S$_A$ or N–I
Y = phenyl	$n = 0$	188	—	—	190	266
	$n = 2$	168.5	—	(152.5)	179	217.5
Y = cyclohexyl	$n = 0$	172	—	(159)	—	191
	$n = 2$	114	(95.5)	158	—	187

[a] Temperatures in parentheses are for monotropic transitions.

Data for these compounds are given in Table II. Results for $n = 1$ are not included, because, as discussed later, an odd number of methylene groups forces the terminal ring Y out of line with the rest of the molecule, with serious consequences on the N–I temperatures.

The results for X = CN and NO$_2$ show that a terminal cyclohexyl ring strongly disfavors nematic phase stability. The difference is smaller when the change phenyl to cyclohexyl is made at the end of the flexible —CH$_2$—CH$_2$— chain.

When X = Ph, a cyclohexyl ring again disfavors nematic properties; neither ester exhibits a nematic phase. The smectic thermal stabilities of the cyclohexyl and phenyl analogs are quite close, however, although in all three cases where comparisons can be made, the cyclohexyl compounds give slightly higher transition temperatures (191°C and 190°C; 187°C and 179°C for S$_A$; 158°C and 152.5°C for S$_B$). Thus, a terminal cyclohexyl ring disfavors nematic phases strongly, but may slightly favor S$_A$ and S$_B$ phases, relative to a terminal phenyl ring.

Alicyclic rings do, of course, feature in several well-known mesogens recorded in Kast's lists [21], but in these systems, the ring is subject to certain constraints which diminish its flexibility. For example, in the derivative (XXV)

Molecular Geometry

$$CH_3O-\bigcirc-CH=\underset{O}{C}-CH=\bigcirc-OCH_3$$

(XXV)

of cyclopentanone, N–I, 195°C, and the corresponding derivative of cyclohexanone, N–I, 173°C, three of the ring carbons are sp² hybridized. Also, in steryl esters, the flexibility of individual 6- and 5-membered rings is restricted by the fused nature of the alicyclic skeleton, and frequently, too, by the presence of double bonds in the rings, e.g., in the 5,6-position of cholesterol.

V. MESOGENS CONTAINING HETEROCYCLIC RINGS

Kast's lists [21] refer to only a few pyridazinyl derivatives (XXVI), but later Schubert et al. [49, 50] examined a range of related compounds (XXVII) containing different heterocyclic rings X as the central ring and having either

Alkyl—⟨◯⟩—⟨◯⟩—Alkyl   R—⟨◯⟩—X—⟨◯⟩—R
          N=N

(XXVI)   (XXVII)

two alkyl or alkoxy groups in the terminal positions. Typical results are shown in Figs. 1 and 2; they relate to the heterocyclic diethyl and di-n-pentyl compounds and to the corresponding p-terphenyl compounds (X = p-phenylene).

The low melting points of the pyridyl and pyrimidinyl derivatives—particularly in the di-n-pentyl series—are worthy of note. With regard to mesophase thermal stabilities, the results are difficult to assess because of the nonmesomorphic properties of 4,4″-diethyl-p-terphenyl, the purely smectic properties of 4,4″-di-n-pentyl-p-terphenyl and the variations in phase type within each series. From the data in Figs. 1 and 2 the order of effect of the central ring in promoting mesophase thermal stability would however appear to be as follows.

Smectic:

⟨◯⟩ > ⟨◯⟩ > ⟨◯⟩ > ⟨◯⟩ > ⟨◯⟩ > (⟨◯⟩)
N=N   N=   N=   N   N=N

Pyridazinyl Phenyl Pyridyl Pyrimidinyl Pyrazinyl (Tetrazinyl)*

* Purely nematic.

Nematic:

Pyridazinyl > Pyridyl > Pyrazinyl > Pyrimidinyl > Tetrazinyl

The position of phenyl in the nematic order cannot be decided.

The two orders are similar and may derive from the influence of the different central rings on conjugative interactions and molecular breadth. Conjugative interactions with the terminal phenylene rings should be greatest when they are linked to a position adjacent to a heteronitrogen in the central ring. Thus,

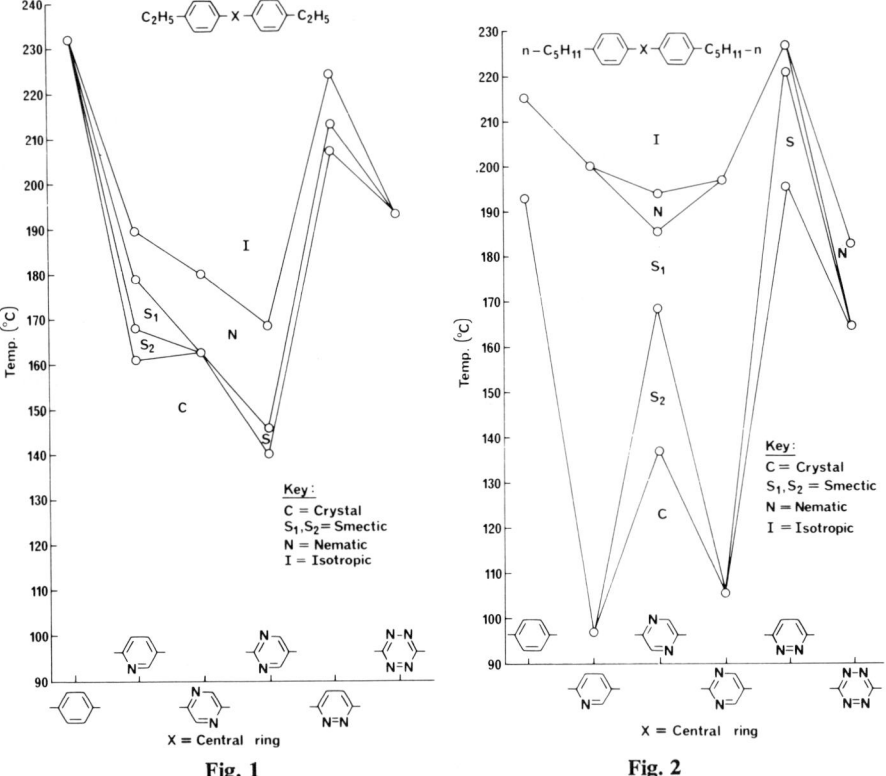

Fig. 1. Graphical presentation of the liquid crystal transition temperatures for 4,4″-diethyl-*p*-terphenyl and various diethyl substituted heterocycles of the structure indicated in the figure.

Fig. 2. Graphical presentation of the liquid crystal transition temperatures for 4,4″-di-*n*-pentyl-*p*-terphenyl and various di-*n*-pentyl-substituted heterocycles of the structure indicated in the figure.

Molecular Geometry

both terminal phenylene rings may conjugate with a central pyridazinyl, pyrazinyl, or tetrazinyl ring. However, the molecular broadening effects (see p. 40) arising from the orbitals of the heteronitrogens will be greatest for pyrazinyl, pyrimidinyl, and tetrazinyl. A combination of these effects would explain the nematic order. The inversion of the positions of pyrimidinyl and pyrazinyl in the smectic order might indicate that another factor is at work here, namely, the permanent dipole associated with a heteronitrogen. It would not appear that the ring nitrogens play any significant role in causing enhanced twisting of the molecules about the interannular bonds.

Information [51] relating to the effect on nematic thermal stability of replacement of a *p*-phenylene ring by a pyridyl ring in the central core part of a mesogen is summarized in Table III. Certain results quoted by Dewar *et al.* [45] are also relevant.

TABLE III

Comparison of Mesomorphic Transition Temperatures for Some Mesogens Containing Pyridyl and Phenyl Rings

No.	Compound	Constants (°C)	Constants for phenyl analog (°C)
1	$C_5H_{11}O$—pyridyl—CO.OH	C–I, 125	C–N, 124; N–I, 152
2	$C_5H_{11}O$—pyridyl—CO.O—phenyl—OC_4H_9	C–I, 91	C–N, 67; N–I, 82
3	CH_3O—pyridyl—N=CH—phenyl—$O.CO.OC_6H_{13}$	C–I, 64	C–N, 71; N–I, 82
4	CH_3O—pyridyl—N=CH—phenyl—CH=N—pyridyl—OCH_3	C–N, 190; N–I, 267	C–N, 224; N–I, 330
5	CH_3—pyridyl—N=CH—phenyl—CH=N—pyridyl—CH_3	C–N, 215; N–I, 258	C–N, 188; N–I, 283
6	CH_3O—phenyl—CO.O—pyridyl—CH=CH—phenyl	C–I, 53	C–N, 182; N–I, 225
7	$C_8H_{17}O$—phenyl—CO.O—pyridyl—CH=CH—phenyl	C–S_A, 62; S_A–I, 64	C–N, 125; N–I, 183

Brief comments only are justified on the consistent decreases in nematogenic tendencies for the pyridyl compounds since a limited range of materials is under consideration. The C–I temperature for compound 2 is higher than the N–I temperature of the phenyl analog and no real comment can be made on the absence of a nematic phase. Reference to compounds 3 and 4 suggests that location of the heteronitrogen adjacent to the methoxy group and in the 3-position relative to the N of the Schiff's base linkage decreases conjugation between this linkage and the alkoxy group and inhibits nematic phase formation (reduction in anisotropy of polarizability). In support of this, the decrease in N–I temperature from the phenyl analog to compound 5 is lower when the terminal methyl rather than a methoxy group is involved. Similar inhibition of conjugation between substituents may apply in compounds 1 and 2, and in 1 an additional feature preventing mesophase formation may be intermolecular hydrogen bonding between the —CO.OH group and the heteronitrogen so that lathlike dimers do not arise. The biggest decreases in N–I temperatures arise for stilbazoles 6 and 7, and could be consistent with a steric effect between the heteronitrogen and the olefinic unit resulting in a twisted molecule (see p. 45). This, coupled with inhibition of conjugation as for compounds 3 and 4, may account simply for the absence of nematic phases at quite low temperatures. The purely smectic properties of compound 7, where the same arguments apply, are particularly interesting. Champa [52] recorded a very low N–I temperature for stilbazole (XXVIII) compared with that of a related stilbene.

CH_3O—⟨⟩—CH=CH—⟨N⟩—C_2H_5 RO—⟨N⟩—N=CH—⟨⟩—OR'

(XXVIII)   (XXIX)

The results for compounds 1–5 may be compared with those of Champa [53] and of Oh [54] who investigated compounds of the type XXIX and XXX, as part of a search for lower melting analogs of MBBA and EBBA. Generally,

R—⟨N⟩—N=CH—⟨⟩—OR' $C_8H_{17}O$—⟨⟩—CH=N—⟨N⟩—CH_3

(XXX)   (XXXI)

melting points were lower for the heterocyclic compounds, but consistent with the results of Nash and Gray [51], N–I temperatures were also lower, sometimes by as much as 50°. The pyridyl analog of MBBA has a monotropic N–I at 6.4°C (ΔT_{N-I} ca. 40°); in this case the compound does melt higher by about 20°. The pyridyl analog of EBBA is low melting (27°–28°C), but gives two smectic phases; S_2–S_1, 39°C; S_1–N, 49–50°C; N–I, 57.6°C. EBBA (C–N, 35.3°C; N–I, 79.0°C) has no smectic phase [55].

In the dialkoxy series (XXIX), the tendency of the pyridyl compounds to give smectic phases was again marked, and the enhanced smectic tendencies of pyridyl compounds could be connected with the dipole moment associated with the heteroatom.

Champa also examined four compounds in which the heteronitrogen was adjacent to the Schiff's base linkage; of these, only one (XXXI) was a mesogen with a monotropic nematic phase, C–I, 67°–68°C; N–I, 64°C.

Studies of the effects of terminal heterocyclic rings were made by Young *et al.* [56] for compounds of the type XXXII, where R′ = CH_3 and R = a range

$$R'O-\bigcirc-\bigcirc-N=CH-R$$

(XXXII)

of heterocyclic rings, and later by Nash and Gray [51], for the homologs with R′ = *n*-octyl, so that information was obtained on smectic thermal stabilities. The orders obtained for nematic and smectic (S_A) thermal stabilities were identical (see below). Inversion of the position of 4- and 3-pyridyl occurred

$$R = \;-\bigcirc_{=N} \;>\; -\bigcirc_N \;>\; -\bigcirc \;\gg\; -\bigcirc_{N=}$$

when the S_B and S_E phases of the *n*-octyl compounds were considered. Only the *n*-octyl compound with R = 2-pyridyl was nematic as well as smectic.

The 2-pyridyl ring may increase the angle of twist at the Schiff's base linkage through repulsive interactions between the heteronitrogen and the N = CH linkage and this, together with the dipole of the heteronitrogen, may decrease the anisotropy of the molecular polarizability relative to the phenyl analog. The lower (ca. 14°) nematic thermal stability of the 4-pyridyl compound (R′ = CH_3) compared with the 3-pyridyl analog may be a result of repulsion between the terminal lone pair of the 4-pyridyl nitrogen and either the oxygen of a CH_3O group or the lone pair of another 4-pyridyl ring. The difference in smectic thermal stability was less (6°–7°) for the *n*-octyl compounds in which the longer chain would shield the oxygen more efficiently. To check this, the compounds XXXIII were prepared and the order was

$$R-CH=N-\bigcirc-\bigcirc-N=CH-R$$

(XXXIII)

R = [3-pyridyl] > [phenyl] > [4-pyridyl] & [2-pyridyl]

neither of the last two compounds being mesomorphic. The decrease in N–I temperature from 3- to 4-pyridyl compound was now >68°, and the positions of 4-pyridyl and phenyl were inverted, in keeping with the large repulsive interactions expected between two 4-pyridyl rings placed end to end.

The high nematic thermal stabilities of the 3-pyridyl compounds suggest that the diminution in conjugation between a terminal phenyl ring and the CH of the Schiff's base linkage that must occur when this ring is replaced by 3-pyridyl are offset by the enhanced axial polarizabilities resulting from the 3-heteronitrogen.

Studies of the system XXXIV, where R' = CH_3 or n-octyl and R = a range

R'O–[phenyl]–CO.O–[phenyl]–CH=CH–R   R'O–[phenyl]–CO.O–[phenyl]–CH=C(R")–R

(XXXIV)   (XXXV)

of heterocyclic rings or phenyl, gave the order of nematic phase thermal stability as

[2-pyridyl] ≈ [pyridazine N=N] > [phenyl] > [pyrimidine] > [4-pyridyl] ≫ [3-pyridyl]

Even with R' = n-octyl, no smectic phases were observed. A feature of this order is the low position occupied by 3-pyridyl, compared with that in the order for the system XXXV, where R' = CH_3 or n-octyl and R" = CN. The order for R was

[3-pyridyl] > [phenyl]

as for the Schiff's bases.

If in the *planar* stilbenes (XXXIV), conjugation from the phenyl group R through to the O of the ester function is strong, the 3-pyridyl ring will tend to reduce this, possibly lowering nematic thermal stability more than the lone pair of the heteronitrogen can raise the N–I temperature by contributing to axial polarizability. In the cyanostilbenes (XXXV), the molecules will almost certainly be twisted, and conjugation with phenyl group R may be unimportant. Replacement by R = 3-pyridyl may not significantly affect the conjugation

Molecular Geometry 23

therefore, and a net axial polarizability increase may result from the lone pair of the heteronitrogen. The position of the diazinyl ring in the above order is difficult to explain.

Young *et al.* and Nash and Gray also used terminal 5-membered heterocyclic rings as the groups R in Schiff's bases of the type XXXII. The groups R were

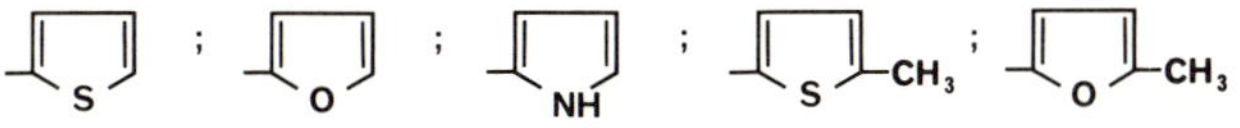

$$\underset{S}{\bigcirc} \; ; \; \underset{O}{\bigcirc} \; ; \; \underset{NH}{\bigcirc} \; ; \; \underset{S}{\bigcirc}-CH_3 \; ; \; \underset{O}{\bigcirc}-CH_3$$

Of these, only the pyrrolyl and 5-methylthienyl groups consistently promoted N–I or S–I transition temperatures relative to R = phenyl. However, the results obtained were too few to permit general conclusions to be reached.

The studies described represent all the *systematic* work conducted on heterocyclic mesogens; other heterocycles have been found to be compatible with mesomorphism—e.g., compounds XXXVI and XXXVII, but the numbers of such materials are limited and comparisons would be premature.

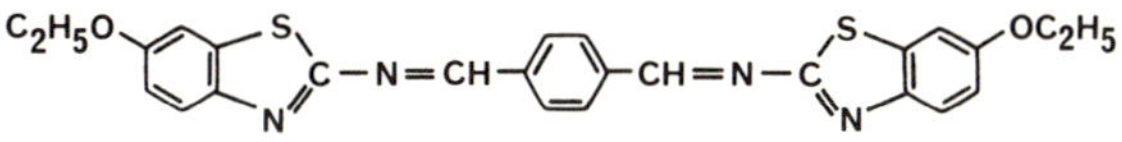

(XXXVI)

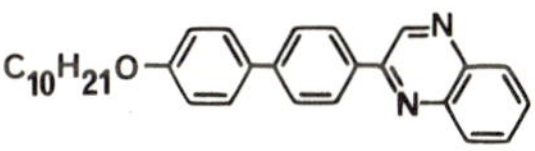

(XXXVII)

VI. THE ROLE OF TERMINAL SUBSTITUENTS IN MESOGENS

A. General Considerations

By examining a range of pure mesogens (XXXVIII), it was found [3, 6] that any substituent X which was not too bulky, i.e., did not increase the molecular breadth (see p. 40) and which extended the major molecular axis

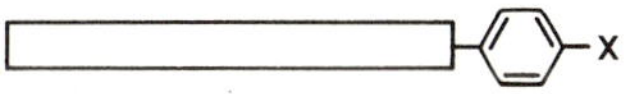

(XXXVIII)

of the unsubstituted parent mesogen enhanced the N–I temperature. Complementary studies [57] made using mixtures of mesogens and of nonmesogens to determine the slopes of the N–I transition lines gave similar results; i.e., of the substituents selected, all were more efficient than a terminal hydrogen in promoting nematic properties.

From both studies, orders of terminal group efficiency in promoting nematic properties were obtained. The orders differed in points of detail from one molecular system to another and dependent upon whether mixtures or pure mesogens were being examined, but the general conclusion reached was that any terminal substituent conforming to the requirements mentioned above enhances the anisotropy of polarizability and raises the N–I temperature. The polarizability along the long molecular axis should be increased most if X can conjugate with the aromatic parts of the molecule, or to quote de Jeu [43], "if the *para*-substituent can easily be imbedded in the conjugated system" (see van der Veen [57a]).

An average terminal group efficiency order which has been quoted is Ph > $NH.CO.CH_3$ > CN > OCH_3 > NO_2 > Cl > Br > $N(CH_3)_2$ > CH_3 > F > H. Higher N–I temperatures therefore arise when H is replaced terminally by any of these substituents and the effect is (i) greatest with groups which contribute strongly to axial polarizability and/or are capable of strong conjugative interactions, and (ii) smallest with small substituents (CH_3, F), substituents which conjugate weakly (F, Br) or a substituent such as $N(CH_3)_2$ with an off-axial dipole moment.

Although cholesteric terminal group efficiency orders have been established ([6] and ref. to Vora [58]) for fewer systems, the cholesteric order is apparently the same as the nematic order for the same group of substituents. This is as expected since Ch–I transition temperatures for optically active isomers are the same as the N–I temperature for the racemic modification.

The situation [6] for smectic terminal group efficiency orders is less clear; considerable variations occur for the few systems investigated, and studies of mixtures do not give any information, because transition lines across phase diagrams deviate from linearity. Results have been discussed elsewhere [6] and it has been noted that in certain cases replacement of H by OCH_3, CN, or NO_2 can *decrease* smectic A thermal stability. This is not always true, however. 4'-*n*-Octyloxy-4-cyanobiphenyl [16, 37, 40], has C–S_A, 54.5°C; S_A–N, 67°C; N–I, 80°C, but 4'-*n*-octyloxybiphenyl is not a mesogen and gives a direct C–I transition at 70°C, the melt supercooling until 64°C with no mesophase formation. Therefore, for both smectic A and nematic thermal stabilities, CN > H in the order. Reasons for the promotion of smectic A order by CN in this series and not in others are not clear. However, it is noted that the smectic phases for the 4'-*n*-alkoxy- (and -alkyl-) 4-cyanobiphenyls, although S_A in type by miscibility, have layer spacings of 1.5–1.6 molecular lengths [*16, 59, 60*]. Some type of interdigitated bilayer occurs in these smectic phases, and it is possible that this is also the case for smectic phases formed by other cyano-substituted mesogens. If this bilayer arrangement is a consequence of the presence of a terminal CN group, it is possible that enhancement of the smectic thermal stability relative to that of the parent system with a terminal hydrogen is a secondary consequence of this structure.

However, OCH_3 too has been found to lie either above or below H in the orders for two systems (both S_A) that have been studied [*6*]. Recent results for ω-phenylalkyl and *p*-substituted-ω-phenylalkyl 4-(*p*-phenylbenzylideneamino)cinnamates are also relevant [*16, 18*]. Such esters are discussed later (see p. 34), but when the number of CH_2 units in the group [$Ar(CH_2)_n$-] is even, the molecule is lathlike and any *p*-substituent acts as a true terminal function. When *n* is odd, a different situation arises (see p. 36). Such esters give S_A, S_B, and S_E phases, and information is therefore obtained about the effects of *p*-CH_3 and *p*-Cl substituents (relative to hydrogen) on the thermal stabilities of these phases. It was found that

i. S_E thermal stabilities are little affected (either increased or decreased slightly) by replacement of hydrogen by *p*-CH_3 or *p*-Cl;
ii. S_B thermal stability is diminished slightly by a *p*-CH_3 group;
iii. the order of decreasing S_A thermal stability is H > CH_3 ≫ Cl.

For the S_A phases, CH_3 and particularly Cl substituents therefore once more *diminish* the thermal stability, whereas S_B and S_E phases are less affected. We should note, however, that for these esters, the *p*-substituent is introduced into the terminal position of a *flexibly* attached end group and not into that of the *rigid* core part of the molecule; this could be significant.

The position really is that too few systems have been examined to be sure of *either* the extent of variation in smectic terminal group efficiency orders for a given type of smectic phase over a number of molecular systems *or* how the orders are affected if the type of smectic phase varies.

B. Terminal Alkyl Chains

A common terminal substituent is the *n*-alkyl group, e.g., attached directly to the end ring or via an oxygen. Studies [*3, 6*] of such homologous series have shown that temperatures for mesophase–mesophase and mesophase–amorphous isotropic liquid transitions show regular trends as the series are ascended and the $CH_3(CH_2)_n$ chain is extended. In several cases for which irregularities were reported, further investigation, with more careful purification of the compounds, confirmed that regular trends in fact occur. These trends are represented by plotting the transition temperatures against the number of carbons in the alkyl or alkoxy group. The various types of smooth curve relations for a large number of different series are well known and have been exemplified elsewhere [*3, 6*]. We shall concern ourselves only with recent developments.

There is always a regular alternation of N–I temperatures as a series is ascended, such that for *n*-alkyl aromatics, odd members have the higher and even members the lower values. In alkoxy substituted series, the oxygen is equivalent to a CH_2 group and the reverse situation is found. A common behavior for the two separate curves for a series is that both either fall or rise initially and then level out, the alternation becoming less as the series is ascended. Alternatively, the upper curve may fall and level out, while the lower curve rises only slightly or stays almost level as the series is ascended. Figures 3, 4, and 5 illustrate these possibilities. The falling types of curve were once the commonest and are associated with series whose members have relatively high N–I temperatures. Indeed, the average slope for a given curve is greater the higher the N–I temperatures involved. The quest for low melting mesogens has however led to the study of more series whose members have relatively low N–I temperatures, and has increased the number of examples for which the curves rise initially and then level out.

Attempts have been made to explain the shapes of these types of N–I curve. For example, de Jeu, van der Veen, and Goossens [*61*] consider the alternation of N–I temperatures in the context of the effects on anisotropy of molecular polarizability arising from increases in chain length on passing from

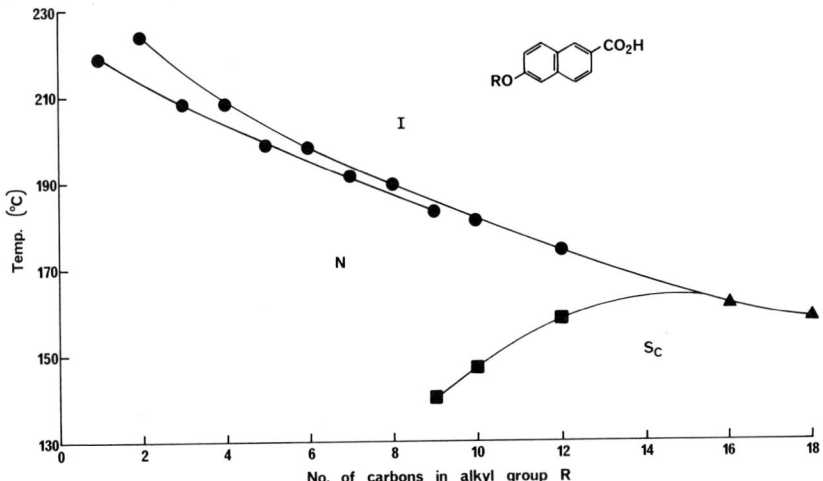

Fig. 3. Plot of the liquid crystal transition temperatures against the length of the *n*-alkyl chain for the 6-*n*-alkoxy-2-naphthoic acids (● = N–I; ▲ = S_C–I; ■ = S_C–N); both N–I lines fall.

Molecular Geometry

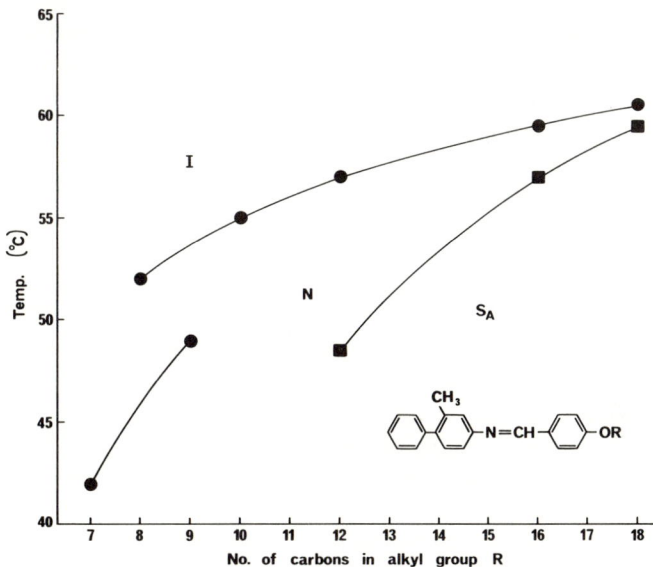

Fig. 4. Plot of the liquid crystal transition temperatures against the length of the *n*-alkyl chain for the 4-(4'-*n*-alkoxybenzylideneamino)-2-methylbiphenyls (● = N–I; ■ = S$_A$–N); both N–I lines rise.

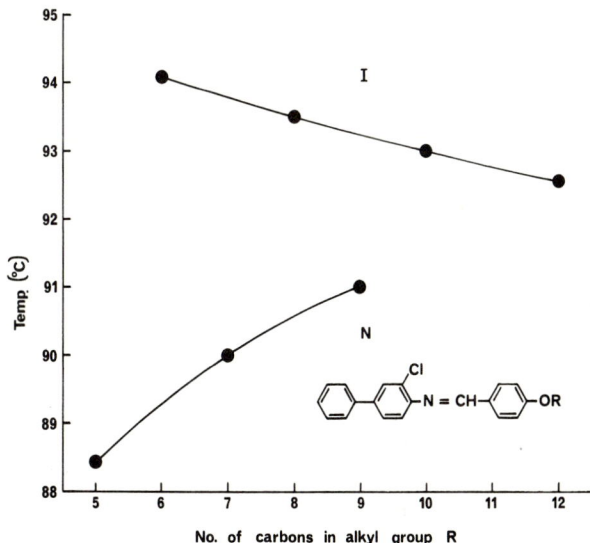

Fig. 5. Plot of the liquid crystal transition temperatures against the length of the *n*-alkyl chain for the 4-(4'-*n*-alkoxybenzylideneamino)-3-chlorobiphenyls (● = N–I); one rising and one falling N–I line.

even to odd or odd to even numbers of carbons. Assuming a rigid, extended, planar zig-zag chain [4], the axial polarizability is increased about twice as much as the polarizability at right angles to the axis on passing from an even to an odd carbon chain. On passing from an odd to an even carbon chain, the polarizabilities along and perpendicular to the long axis are affected almost equally. The anisotropy of molecular polarizability is therefore greater for compounds with odd alkyl chains, and the N–I temperatures for the methyl, propyl, etc., homologs lie on a curve above that for the ethyl, butyl, etc., homologs. The same situation arises for n-alkyl esters, and the opposite for n-alkyl ethers.

Arguments that an alkyl chain cannot adopt a single, extended zig-zag conformation in a fluid nematic are justified, but to explain the alternation, some *preference* for such a conformation must be accepted, despite the rotational and flexing motions which must occur with alkyl chains in a real situation. In fact, motions of the chain explain the damping of the N–I alternation as the chain is lengthened. The increased flexibility of long chains will progressively decrease the differences in the anisotropies of molecular polarizability between odd and even alkyl derivatives with shorter and more rigid chains.

As de Jeu et al. [61] point out, the above arguments assume that the anisotropy of molecular polarizability *increases* as the series is ascended and lead to the conclusion that N–I temperatures should *rise* as the series are ascended, leveling off at some limiting value for long chains. However, this is the case only for series involving relatively low N–I temperatures.

In the Maier and Saupe theory [29], attractive dispersion forces alone are taken into account when considering the free energy difference (ΔF) between the nematic and isotropic phases. However, measurements of order parameters at various pressures and at constant volume show that the entropy contribution to ΔF due to steric intermolecular interactions (repulsive forces) is equally important and the simplest expression for ΔF which considers both contributions is

$$\Delta F = -AS^2 - BTS^2 - T\Sigma(S)$$

where S is the order parameter, $-AS^2$ is the internal energy stemming from dispersion forces (A being approximately proportional to the anisotropy of molecular polarizability), BS^2 is the packing entropy due to excluded volume effects that favor molecular alignment, and Σ is the orientational entropy. The condition $\Delta F = 0$ for a phase transition at constant volume leads to the expression for the N–I transition temperature

$$T_c = \frac{2A}{4.54k - 2B}$$

De Jeu et al. [*61*] consider that B is in some way proportional to the effective molecular length/breadth ratio. They quote Stenschke [*62*] who argues that the reduction in the effective length/breadth ratio of a molecule due to bending of alkyl chains is proportional to molecular length and to temperature, and conclude that as the alkyl chain lengthens, so B will assume smaller values. The effect will be small at lower temperatures, and so variations in A will determine the change in T_c with chain length. When higher T_c values are involved, chain flexing will be more important, and the decrease in B may predominate, leading to a fall in T_c as the chain lengthens—see also Kaplan [*63*].

Important as these attempts are to explain the shapes of N–I curves, they do not provide a complete answer. For example, the N–I temperatures for the 2-(*p-n*-alkoxybenzylideneamino)fluorenones [*3, 64*] do not conform to the trends discussed above. After exhaustive purification of the members, the plot for the series was as shown in Fig. 6. Alternation still occurs, but both

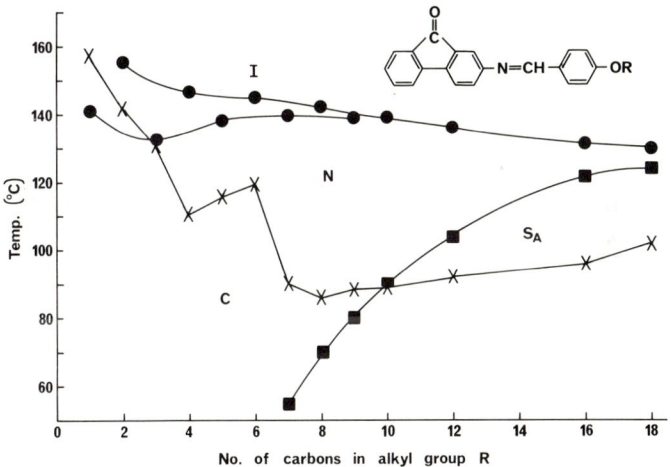

Fig. 6. Plot of the liquid crystal transition temperatures against the length of the *n*-alkyl chain for the 2-(4′-*n*-alkoxybenzylideneamino)fluorenones (● = N–I; ■ = S_A–N; × = C–I, N, or S_A).

N–I curves fall initially to shallow minima, then rise slightly and fall again. A regular effect still occurs, but of a more complex kind than for the commoner cases treated theoretically. Curves involving minima and maxima are also obtained for esters of 4,4′-azoxyphenol [*65*] and for 4-(*p-n*-alkoxybenzylideneamino)acetophenones [*66*]. Also one may consider the plots obtained for the 4′-*n*-alkyl-4-cyanobiphenyls (Fig. 7) and 4′-*n*-alkoxy-4-cyanobiphenyls (Fig. 8)

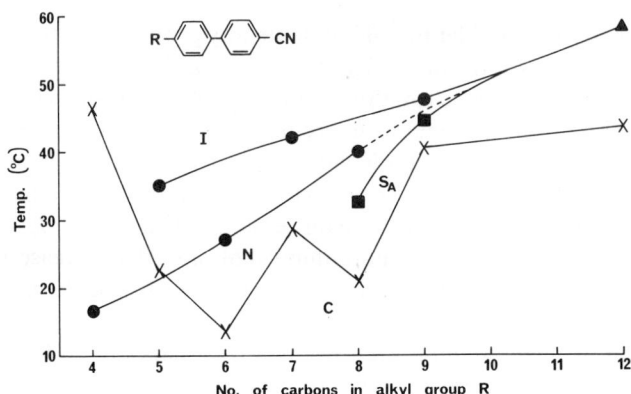

Fig. 7. Plot of the liquid crystal transition temperatures against the length of the *n*-alkyl chain for the 4′-*n*-alkyl-4-cyanobiphenyls (● = N–I; ▲ = S_A–I; ■ = S_A–N; × = C–I, N, or S_A).

[*16, 37*]. For the alkoxy series, both curves commence fairly level and then rise more steeply, the curves having an upward concavity. Although the series is still incomplete, the path of the N–I curve through the odd carbon chain members then assumes a convexity upward, and so must that for the even members if it is to pass through the S–I temperature for the dodecyloxy compound. Similar changes in the concavity and convexity of the N–I curves

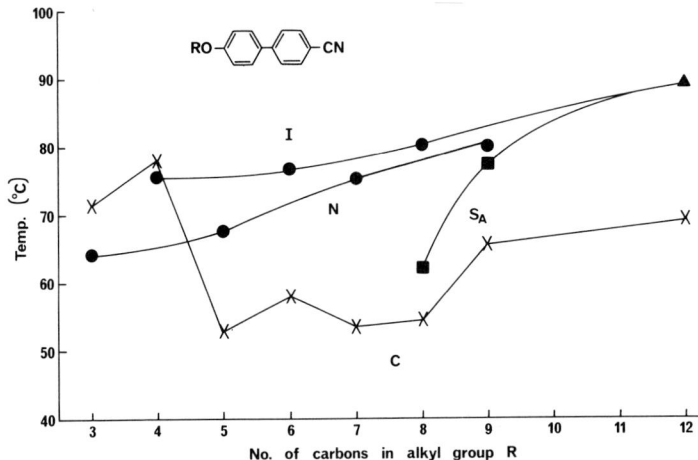

Fig. 8. Plot of the liquid crystal transition temperatures against the length of the *n*-alkyl chain for the 4′-*n*-alkoxy-4-cyanobiphenyls (● = N–I; ▲ = S_A–I; ■ = S_A–N; × = C–I, N, or S_A).

Molecular Geometry

are seen in Fig. 7 for the *n*-alkyl series for which preliminary results indicate that the virtual N–I temperatures of the methyl and ethyl compounds are *higher* than those of the propyl and butyl compounds, respectively (cf. Fig. 7).*
All the above effects are subtleties for which only tentative explanations can be given, but they are important effects, for similar features arise with series of cholesterogens.

For example, the plot of the Ch–I temperatures for the cholesteryl *n*-alkanoates exhibits unusual features [6]. The alternation gives two curves: one rises from acetate to butyrate and then falls to stearate; the other falls from propionate to nonanoate, but if the formate is regarded as a homolog with an even number of carbons (zero) in the alkyl portion of the acyloxy function, this curve too rises to a maximum (at the propionate) and then falls. These curves then become similar in shape to the portions of the N–I curves for the series shown in Fig. 6 beyond C_3 and C_4, respectively.

Most homologous series of cholesterogens are derived from sterols. Purification problems are probably more acute for the members of such series, and reported deviations from smooth curve relations have been most frequent for such series [6, 68, 69]. An additional problem is that when the cholesteric phase first forms from the amorphous liquid on cooling a thin film between glass surfaces, it escapes detection unless *very* careful microscopic observations are made, because it is almost optically extinct in transmitted light [70]. In reflected light, the texture is, however, clearly seen to consist of platelets [70–72]. The temperature range for the almost extinct texture, before it assumes the birefringent "focal-conic" texture, varies from homolog to homolog. Therefore, if Ch–I temperatures are measured by optical microscopy, on cooling cycles, using transmitted light, low values can be obtained. While such problems remain as possible relevant factors, it is not justifiable to consider series of cholesterogens derived from sterols as exceptions to the idea of smooth curve relations for the mesomorphic transition temperatures of homologous series. In fact, where deviations from smooth curve trends are reported, the deviations are not such that the temperatures are randomly scattered. The deviations are usually small, and the curves approximate in shape to those that would be anticipated by analogy with the cholesteryl *n*-alkanoates or series of nematogens.

Finally, we consider transitions involving smectic phases [6]. When the change is direct from smectic (usually S_A or S_C) to amorphous liquid, then, as the series are ascended, diminishing alternation of the kind found for N–I and Ch–I transitions occurs and either (i) both curves rise to early maxima

---

* *Note added in proof*. The preliminary results have been confirmed and a similar situation in fact occurs in the alkoxy series (Fig. 8) for the methoxy and ethoxy derivatives. Both series are therefore similar in behavior to that of the 2-(*p-n*-alkoxybenzylideneamino)fluorenones (Fig. 6). A full account of these results has been published recently [67].

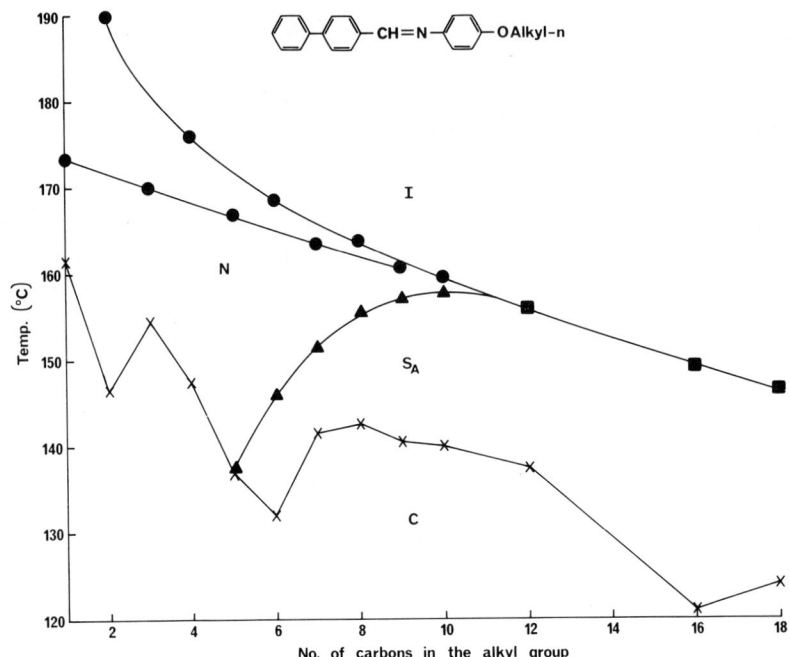

Fig. 9. Plot of the liquid crystal transition temperatures against the length of the *n*-alkyl chain for the 4-(4′-*n*-alkoxybenzylideneamino) biphenyls (● = N–I; ▲ = S_A–N; ■ = S_A–I; × = C–N or S_A).

and then fall, (ii) both curves rise or fall progressively over the series as far as they have been studied, or (iii) one curve rises to an early maximum and then falls, while the other falls continuously over the series.

As such, S–I transitions closely resemble N–I or Ch–I transitions in their temperature trends over homologous series.

At one time, S–N transitions followed a common pattern for all series examined—the S phases all being S_A or S_C. No obvious alternation of the temperatures occurred, and all points for a series lay on one curve which rose steeply at first, then leveled out and became coincident with the line* drawn through the N–I points for the lower homologs. Figure 9 shows a typical example. In some series, the merging of the S–N and N–I curves occurred only after the S–N curve had reached a maximum and fallen again slightly. In series of steryl derivatives, the S–Ch curves do not merge, however, with the Ch–I curves for the highest homologs studied.

More recently, studies of homologous series, particularly those giving

---

* Usually the S–N and N–I curves merge at a stage in the series at which the alternation of N–I temperatures is small, so that a single mean N–I curve can be considered.

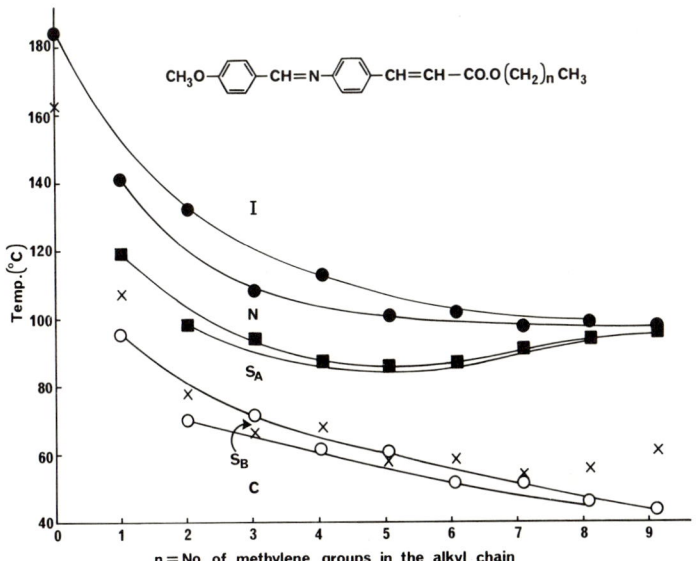

Fig. 10. Plot of the liquid crystal transition temperatures against the number of methylene groups (n) in the ester alkyl chain for the n-alkyl 4-(p-methoxybenzylideneamino)cinnamates (● = N–I; ■ = S_A–N; ○ = S_B–S_A; × = C–N, S_A, or S_B).

S–N transitions over many members, have revealed [6, 47, 48, 70, 73] that alternation does occur and that the shapes of the curves can be complex—see the two S_A–N curves in Fig. 10. To explain these shapes is obviously difficult. Figure 10 also shows that the S_B–S_A temperatures alternate, but this is more clearly shown in Fig. 11 where the shapes of the curves are simpler and a more extended range of S_E–S_B and S_B–S_A transitions occurs. It is noted that: (i) the alternation of the S_A–N and N–I temperatures is of the same sense and is perpetuated in the S_A–I temperatures; (ii) the alternation of the S_B–S_A and S_E–S_B temperatures is of the same sense, but opposite to that of the S_A–N, N–I, and S_A–I temperatures.

This opposite sense of alternation of thermal stability for S_A phases compared with S_B and S_E phases has been confirmed in other instances [18], i.e., factors of molecular structure that elevate S_A thermal stability decrease S_B and S_E thermal stability. It can be inferred then that enhancement of the anisotropy of molecular polarizability stabilizes S_A and N phases, but destabilizes S_B and S_E phases, which could be stabilized more by enhancement of the polarizability *across* the molecular long axis. S_B and S_E phases are more highly ordered with respect to the molecular organization within the smectic lamellae, and lateral attractions could be of greater importance in determining the sense of alternation.

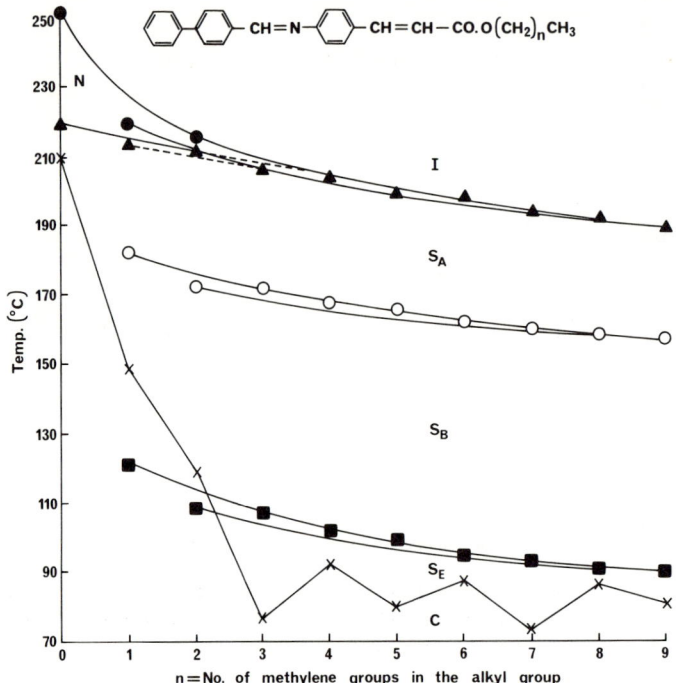

Fig. 11. Plot of the liquid crystal transition temperatures against the number of methylene groups (n) in the ester alkyl chain for the n-alkyl 4-(p-phenylbenzylideneamino)cinnamates (● = N–I; ▲ = S_A–N or I; ○ = S_B–S_A; ■ = S_E–S_B; × = C–S_A, S_B, or S_E).

C. Alternation of Phase Type within Homologous Series

Extremely large alternations of N–I temperatures for homologous mesogens were demonstrated [47, 48] for the ω-phenylalkyl 4-(p-cyano- and -nitro-benzylideneamino)cinnamates (XXXIX). When n is 0 or even, the

$$O_2N \text{ or } NC-\bigcirc-CH=N-\bigcirc-CH=CH-CO.O(CH_2)_n-\bigcirc$$

(XXXIX)

terminal phenyl ring lies along the major molecular axis and enhances the anisotropy of molecular polarizability. However, for odd n values, the phenyl group protrudes beyond the perimeter defining the rest of the molecule, as shown in Fig. 12, which illustrates the L-shape of the molecule. For odd n values, the terminal ring therefore diminishes the anisotropy of molecular polarizability; also, in a nematic phase where the molecules must be free to slide past one another, a greater axial separation of the L-shaped molecules

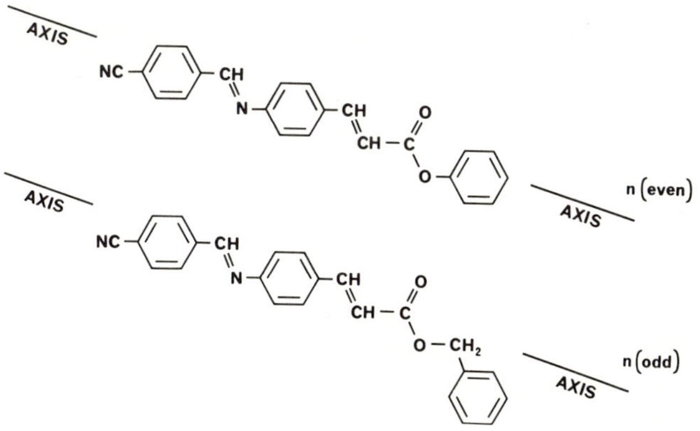

Molecular Geometry

Fig. 12. Stereochemistry of molecules of ω-phenylalkyl 4-(p-cyanobenzylideneamino)cinnamates with even ($n = 0, 2, ...$) and odd ($n = 1, 3, ...$) numbers of methylene groups in the ω-phenylalkyl group.

will be involved. These arguments again assume a preferred planar, zig-zag conformation of the chain, and this is supported by the extremely large alternation between the N–I temperatures for esters with even and odd values of n, as shown in Fig. 13 where the nematic phases alternate between being enantiotropic and monotropic.

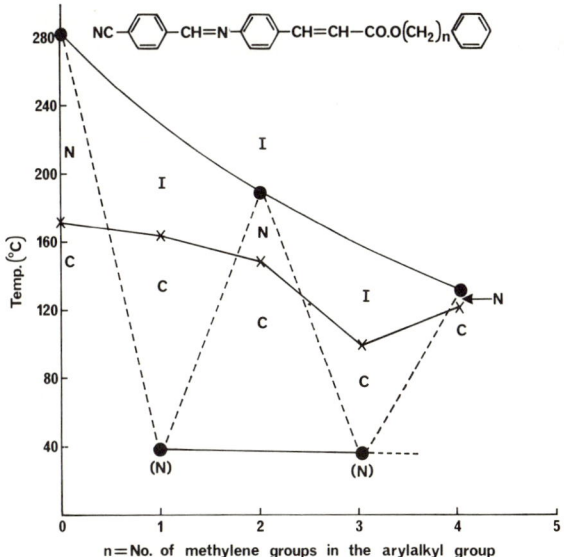

Fig. 13. Plot of liquid crystal transition temperatures against the number of methylene groups (n) in the arylalkyl group for ω-phenylalkyl 4-(p-cyanobenzylideneamino)cinnamates (● = N–I; × = C–N or I).

Similar large alternations of Ch–I temperatures were observed [74–76] for series of cholesteryl, cholestanyl, and S-cholesteryl ω-phenylalkanoates, and the authors proposed similar explanations for the effect—see also [77].

More recently, even higher alternations of N–I temperatures were obtained [16] with esters of 4-(p-cyanobenzylideneamino)cinnamic acid involving p-substituted ω-phenylalkyl groups. Figure 12 shows that such a p-substituent would lie approximately along the molecular long axis when n is even, but when n is odd, it further exaggerates the off-axial effect of the phenyl ring. The alternation in anisotropy of shape and molecular polarizability is therefore enhanced, and this is reflected in larger alternations between the N–I temperatures. For the 4-chloro esters, the N–I temperatures are $n = 0$, 338°C; $n = 1$, $< -30°C$; $n = 2$, 217.8°C; $n = 3$, 13°C—a very dramatic alternation effect.

As expected, the N–I temperatures for say the 4-methyl- and 4-chloro-ω-phenylalkyl esters are higher than those for the corresponding unsubstituted ω-phenylalkyl esters when n is even. However, for each *odd* value of n, a decrease in N–I temperature, relative to the corresponding unsubstituted ω-phenylalkyl ester, arises from what might be construed to be the introduction of a terminal p-substituent (X) on the basis of inspection of a simple linear formula $(NC.C_6H_4.CH{=}N.C_6H_4.CH{=}CH.CO.O(CH_2)_3C_6H_4.X)$. Only when the molecular geometry is taken into account (see Fig. 12) is it appreciated that the substituent X plays a different role than that of terminal substituents discussed previously. Examples of this kind warn against reaching hasty conclusions concerning the effects of molecular structure on liquid crystal properties.

Similar effects have been observed [47, 48] in series of ω-phenylalkyl esters of 4-(p-substituted-benzylideneamino)cinnamic acids carrying substituents other than cyano or nitro; high alternations of N–I temperatures occur when CH_3O, Ph, or $CH_3CO.O$ substituents are used. Smectogenic tendencies are quite high for these series, and it was also shown that the thermal stabilities of the most thermally stable S_A phases were not greatly affected by the alternation in shape and anisotropy of molecular polarizability when n changes from even to odd. A possible explanation was proposed in terms of L-shaped molecules packing almost as closely as lath-shaped molecules (even n values) in smectic layers. The interesting effect therefore arose of an alternation between compounds giving smectic and nematic phases and those giving *only* smectic phases as the series were ascended. That is, an alternation arose of phase type exhibited, because nematic properties are extinguished for odd n values. Figure 14 illustrates some results: $n = 0$, enantiotropic S_A and N; $n = 1$, monotropic S_A and S_B; $n = 2$, enantiotropic S_A and N and monotropic S_B; $n = 3$, enantiotropic S_A and S_B; $n = 4$, enantiotropic S_A, S_B, and N.

These results have been confirmed [18] by studies of ω-phenylalkyl

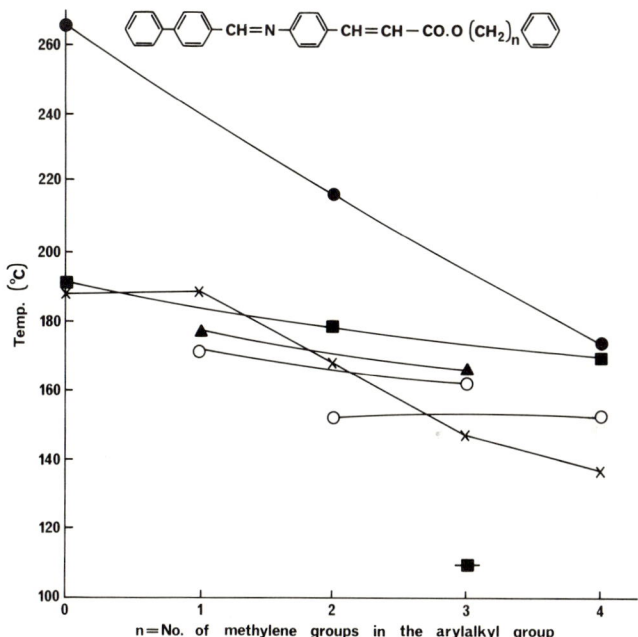

Fig. 14. Plot of liquid crystal transition temperatures against the number of methylene groups (*n*) in the arylalkyl group for ω-phenylalkyl 4-(*p*-phenylbenzylideneamino)cinnamates (● = N–I; ■ = S_A–N; ▲ = S_A–I; ○ = S_B–S_A; ▆ = S_E–S_B; × = C–I, S_A, or S_B).

4-(*p*-phenyl-, -acetoxy- and -methoxy-benzylideneamino)cinnamates in which the phenyl ring of the ester function itself carries a *m*- or a *p*-substituent (CH_3 or Cl). In one series, a *fortuitous* distribution of the C-mesophase temperatures eliminated even monotropic smectic phases for even *n* values, but for odd *n* values, smectic phases were seen. Since *no* nematic phases occurred for odd *n* values, an alternation from purely nematic to purely smectic occurred along the series (Fig. 15).

The results in Fig. 14 show how S_A–N and S_A–I temperatures alternate slightly along the series and with the same sense as the much larger alternation for the nematic phases which become "monotropic" with respect to the smectic phases for odd *n* values. However, the S_B–S_A temperatures alternate in the opposite sense (see also p. 33). This has also been found for substituted ω-phenylalkyl esters, for which the S_B–S_E temperatures, too, alternate in the opposite sense to the S_A–N or –I temperatures. Once more we conclude that deviations from a strict lathlike shape slightly *diminish* S_A thermal stabilities, whereas for more highly ordered S_B and S_E phases, the deviations from lathlike shape *enhance* their thermal stabilities. In describing possible packing arrangements for L-shaped molecules in smectic layers, it was noted [*18*] that if some

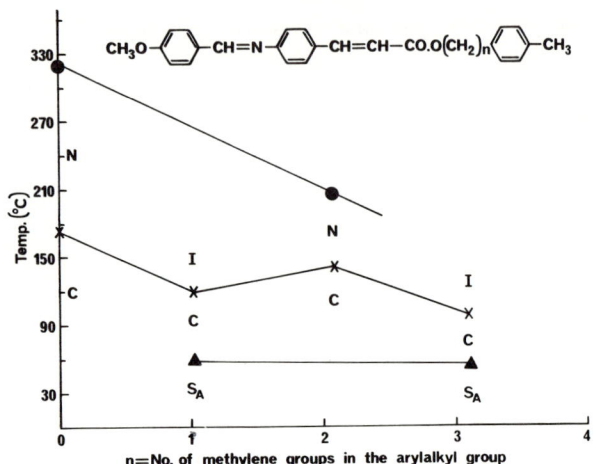

Fig. 15. Plot of liquid crystal transition temperatures against the number of methylene groups (n) in the arylalkyl group for ω-4′-methylphenylalkyl 4-(p-methoxybenzylideneamino)-cinnamates (● = N–I; ▲ = S_A–I; × = C–N or I).

degree of rotational motion about the molecular long axes is possible in S_A phases (offering another explanation of the lower S_A thermal stabilities for odd n values), this must be quenched or co-operative in S_B and S_E phases which give the opposite trends in thermal stability with n (see Doucet et al. [25]).

When the opposite alternation of S_A and S_B thermal stabilities is quite high, e.g., in the series with structure XL, an interesting effect occurs (Fig. 16).

The thermal range of the S_A phase alternates from wide to zero to wide to small as n changes from 0 to 3. When the CH_3 group in XL is changed to Cl, the S_A range is narrow when $n = 1$, and zero when $n = 3$. Miscibility and microscopic studies of the compounds having zero S_A range showed that the S_B phase forms from the isotropic liquid via a transitional S_A phase. Similar S_{AB}–I transitions were observed for compounds with structure XL with 4-CH_3 or 4-Cl substituents. Similar microscopic textures to those published by Coates and Gray [18] have been observed by Billard [19] at the I–S transition for compound XLI, but he concludes that a direct S_B–I transition occurs. This difference is, however, one of degree, not kind, and such compounds giving transitions from the S_B phase to the amorphous liquid, with or

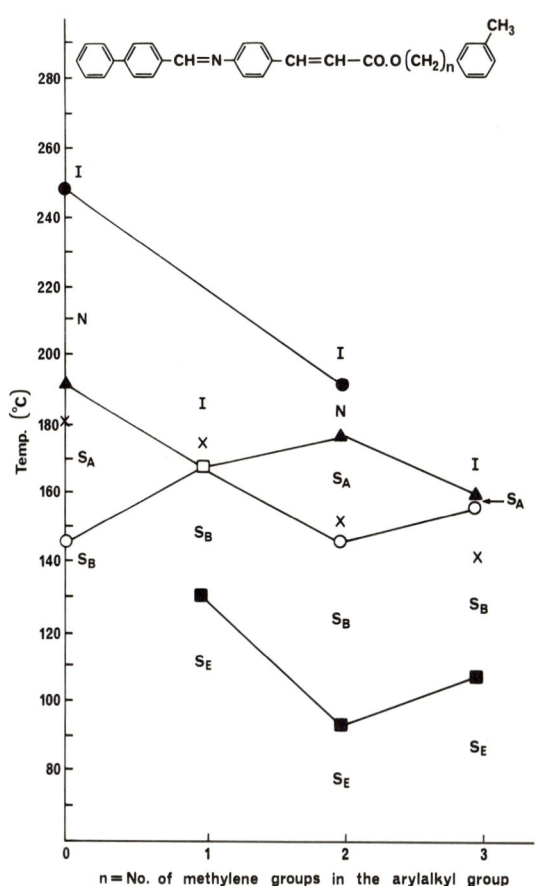

Fig. 16. Plot of liquid crystal transition temperatures against the number of methylene groups (n) in the arylalkyl group for ω-3'-methylphenylalkyl 4-(p-phenylbenzylideneamino)cinnamates (● = N–I; ▲ = S_A–N or I; ○ = S_B–S_A; □ = S_{AB}–I; ■ = S_E–S_B; × = C–I, S_A, or S_B).

without a transitional S_A phase, provide examples of a new mesophase transition which complements the transition S_B–N which has been observed for three compounds [17, 41].

VII. THE ROLE OF LATERAL SUBSTITUENTS IN MESOGENS [3, 6]

The effects of lateral substituents in the lathlike molecules of mesogens are considered under different headings, because the molecular geometry is differently affected depending on the molecule involved and the position of the substituent.

A. Molecular Broadening

The broadening effect of a lateral substituent is most simply demonstrated when this is introduced into the central core part of the molecule so that no steric effects arise and the substituent exerts its full broadening influence. Experimental results have been obtained for many different kinds of mesogen, but the most fully investigated system was the 4'-*n*-alkoxybiphenyl-4-carboxylic acids [3, 6]; these acids give nematic and smectic C phases, and the effects of a range of 3'-substituents of different sizes and dipolarities on the mesophase thermal stabilities were established, e.g., structure XLII with X = F, Cl, Br, I, NO_2, and CH_3.

$$\text{AlkylO} - \underset{\text{X}}{\text{C}_6\text{H}_3} - \text{C}_6\text{H}_4 - \text{CO}_2\text{H}$$

(XLII)

The substituent (i) increases the polarizability across the long axis and reduces the anisotropy of molecular polarizability and (ii) increases the axial separation, so decreasing intermolecular attractions. As expected, decreases in the N–I temperatures accompany replacement of the 3'-H by another substituent. For system XLII, the N–I temperatures fell approximately in proportion to the size of the substituent X, irrespective of its dipolarity. The ineffectiveness of substituent dipoles is in accord with the Maier–Saupe theory [29]. A dependence of the N–I temperature on substituent size is therefore expected, since this determines both the axial molecular separation and the contribution to the polarizability at right angles to the long molecular axis, which together govern the magnitude and anisotropy of the dispersion forces.

The picture was less simple for the smectic C thermal stabilities. If the substituent X was strongly dipolar, it had a smaller effect on S–N or S–I temperatures compared with that of a less dipolar substituent of similar size. Thus, dipole moments contribute to the lamellar smectic order and prevent the occurrence of simple relationships between substituent size and smectic transition temperatures; CH_3 and Cl substituents have similar effects in decreasing N–I temperatures, but the CH_3 group reduces S–N or S–I temperatures more than the Cl substituent. Therefore the relative effect of a substituent on nematic and smectic thermal stabilities depends on the size and dipolarity of the substituent relative to H. As discussed elsewhere [3, 6], either the smectic or the nematic thermal stability may be the more affected.

Other systems were studied, but none allowed the effects to be established

Molecular Geometry

over such a range of substituents. However, these studies confirmed the general effects of lateral substituents in a range of different mesogens and drew attention to the fact that

i. a given substituent decreases N–I transition temperatures more if the molecule is short—presumably the anisotropy of molecular polarizability is more seriously affected;

ii. a given substituent decreases S–N or S–I temperatures more if the molecule is short, but dependence on molecular length is less critical than for nematic thermal stabilities. This could be consistent with the effect of substituent dipoles in counteracting their size effects.

Exceptions to these general rules relating to lateral substituents were recognized in the 5-substituted 6-*n*-alkoxy-2-naphthoic acids, which provide a classical example [*3, 6, 78*] in which a substituent does not exert its full breadth increasing effect because it is completely (partially) contained within the perimeter of the rest of the molecule. For 5-chloro and 5-bromo substituents, for example, the following is observed.

i. Small but definite increases in the N–I temperatures arise. Since these occur despite some decrease in anisotropy of molecular polarizability, the filling up of the gap by the 5-substituent must enhance intermolecular cohesions by improving molecular contacts.

ii. The smectic C thermal stabilities are increased more than the nematic, i.e., in the absence of breadth-increasing effects, dipoles play a role in enhancing smectic thermal stability.

Other examples of a less marked kind are given by 4-(*p-n*-alkoxybenzylidene-amino)-3′-substituted biphenyls, and probably by 3-substituted *trans*-4-*n*-alkoxycinnamic acids [*6*].

Recent examples of lateral substitution enhancing N–I temperatures are provided by 2-hydroxy Schiff's bases (XLIII) [*79*]. The hydroxy group does

(XLIII)   (XLIV)

not exert its full breadth effect, because it is held in the cis-position through hydrogen bonding. This will reduce rotational motions within the molecule and minimize any twist from noncoplanarity of the kind shown to occur for certain Schiff's bases [*80, 81*]. These effects obviously combine to give enhanced nematic thermal stabilities—and an enhanced stability of the Schiff's base linkage to cleavage.

Similar results have been obtained for analogous 2-hydroxy-azo compounds, but van der Veen and Hegge [82] explained the generally higher N–I temperatures, compared with those for the parent compounds without the hydroxyl group, solely in terms of the enhanced axial molecular polarizability resulting from delocalization of π-electrons due to intramolecular hydrogen bonding. The corresponding azoxy compounds XLIV are reported [82] as having *similar* N–I temperatures to the nonhydroxy compounds. As shown by XLIV, hydrogen bonding now occurs between the two oxygens, and the hydroxy group lies trans to the azoxy linking. The hydroxy group therefore exerts a larger breadth increase than in the Schiff's bases or azo compounds. However, relative to the parent azoxy compounds, which possess the protruding azoxy oxygen, the breadth increase is probably not large. The various effects apparently approximately cancel for this system.

Recent examples of substituents decreasing mesophase thermal stabilities have sprung mainly from attempts to use lateral substituents to lower C-mesophase temperatures and obtain nematic mesophases at temperatures suitable for applications. For example, 4′-*n*-alkoxybenzylidene-2-methyl-4-*n*-alkylanilines (XLV) have been compared [83] with the unmethylated compounds. The N–I temperatures were lowered by the 2-methylation, but, as

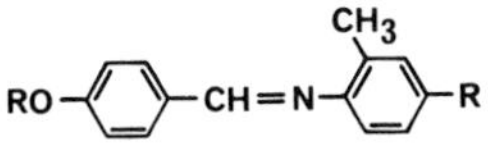

$$RO--CH=N--R \quad (CH_3)$$

(XLV)

expected, this has an even larger effect on smectic thermal stabilities, and none of the compounds XLV exhibited smectic mesophases. Arora *et al.* [84] have studied 4′-alkoxybenzoate esters of 2-methylhydroquinone and found decreases in both melting and clearing temperatures resulting from the CH_3 group.

Van Meter and Klanderman [10, 85] prepared a range of chlorinated derivatives of terminally disubstituted phenyl *p*-benzoyloxybenzoates (XLVI). The results in Table IV show the substantial reductions in N–I temperatures that accompany monochlorination; in one case, the chloro substituent is very effective in lowering the C–N temperature. The results in the second half of

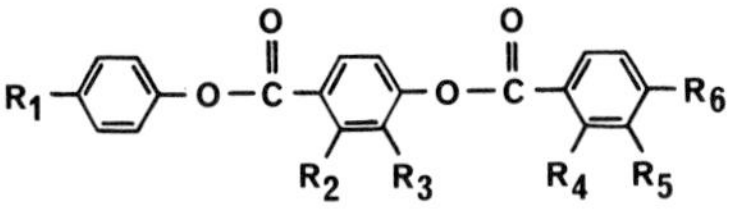

$$R_1--O-C(=O)-(R_2,R_3)-O-C(=O)-(R_4,R_5)-R_6$$

(XLVI)

TABLE IV

Transition Temperatures for Esters of the Type XLVI

R_1	R_2	R_3	R_4	R_5	R_6	C–N (°C)	N–I (°C)
$n\text{-}C_5H_{11}$	H	H	H	H	$n\text{-}C_5H_{11}$	78	179.5
$n\text{-}C_5H_{11}$	H	Cl	H	H	$n\text{-}C_5H_{11}$	67	130
$n\text{-}C_5H_{11}$	Cl	H	H	H	$n\text{-}C_5H_{11}$	39	122
$n\text{-}C_5H_{11}$	H	H	H	H	Cl	—	212
$n\text{-}C_5H_{11}$	H	Cl	H	H	Cl	—	155
$n\text{-}C_5H_{11}$	H	H	Cl	H	Cl	—	153.5
$n\text{-}C_5H_{11}$	H	Cl	Cl	H	Cl	—	89

Table IV show that dichlorination has an additive effect in reducing the N–I temperatures, and this agrees with earlier observations [3, 6, 86] of additive effects on mesophase transition temperatures accompanying mono-, di-, and tri-substitution of the biphenyl nuclei of 4,4'-di-(*p-n*-alkoxybenzylidene-amino)benzidines.

Young, Haller, and Green [87] carried out similar studies of esters of type (XLVII), where A, B, Y, and Z were entirely CH_3 groups or H, or combinations of CH_3 groups and hydrogens. The decreases in N–I temperatures

(XLVII)

accompanying introduction of one CH_3 group were considerable and comparable with the effect of a single Cl substituent in (XLVI). An approximate additive effect on dimethylation was observed (see also Gunjima *et al.* [88]).

Like Dewar *et al.* [45], Young *et al.* [87] note that the decreases in N–I temperatures on methylation are not always accompanied by decreases in the enthalpies and entropies of the N–I transitions, and that *increases* in ΔS occur on substitution in the central ring. Although the enthalpy differences being considered by these authors were considerably *greater*—the total spread of ΔH values was from 0.25 to 0.51 kcal mol^{-1}—than those noted by Dewar *et al.*, these authors, while pointing out the relevance of entropy and enthalpy changes in relation to N–I temperatures, sound the cautionary note that "there are a large number of degrees of freedom, both internal and external, that contribute to the entropy, and subtle changes in the restrictions that these experience in the two phases may result in a non-systematic variation in ΔS_{NI} with substitution." This seems a fair statement of the present situation. At least we can see regular effects on mesophase transition temperatures which can be correlated with molecular structural changes, and it seems worthwhile to extend such correlations. It seems unwise, on the other hand, to oversimplify the situation with regard to enthalpy and entropy changes, and premature to base strong arguments on trends which are observed in these quantities, particularly when the low enthalpies of N–I transitions are under consideration and experimental error could account for the effects.

Finally we refer again to ω-phenylalkyl 4-(p-substituted-benzylidene-amino)cinnamates [*16, 18*]. The situation for esters carrying substituents in the p-position of the ω-phenylalkyl group has been discussed (see Fig. 12). When m-substituents are used and the number of CH_2 units is odd, it can be seen (Fig. 17) that of two extreme conformational positions for the end ring,

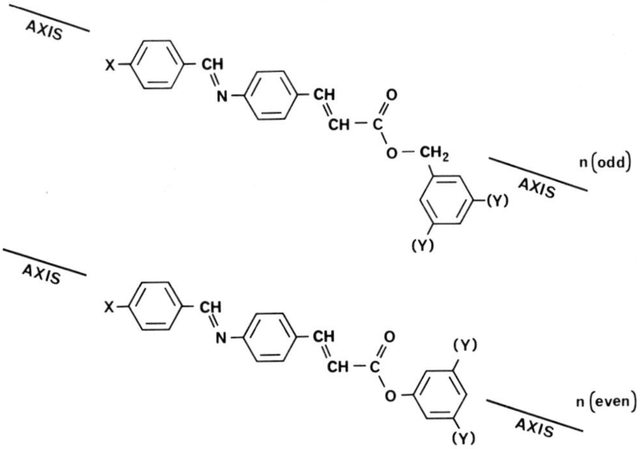

Fig. 17. Stereochemistry of molecules of ω-m-substituted-phenylalkyl 4-(p-X-benzylidene-amino)cinnamates with odd ($n = 1, 3, ...$) and even ($n = 0, 2, ...$) numbers of methylene units in the ω-arylalkyl group.

in one the substituent broadens the molecule (decreasing N–I), whereas in the other it lengthens it (enhancing N–I). When n is even (Fig. 17), the molecule is broadened by Y irrespective of which extreme conformer is considered as a result of rotation about the ring–O or ring–CH_2 bonds. When X = CN [*16*] and n is odd (1 and 3), m-CH_3 or m-Cl substituents diminish the N–I temperature in three cases and increase it in one. This suggests that the amounts of different rotational conformers differ with substituent type and methylene chain length, and that continuous rotation about the CH_2–ring bond does not occur in the nematic melt. When X = CN and n is even (0 and 2), m-CH$_3$ or m-Cl substituents markedly lower the N–I temperature in *all* four cases. However, for $n = 0$, the effect of m-CH$_3$ is considerably *greater* than that of m-Cl. This is unusual, and may again imply different amounts of different conformers when the substituent type and length of the methylene chain vary.

When X = Ph [*18*], no nematic phases are found when n is odd,* but the effects of m-substituents on the thermal stabilities of the different *smectic* phases for $n = 0$ to 3 are (i) S$_E$ thermal stabilities are either increased or decreased, but only by a few degrees (largest increase = 6.1°), by m-CH$_3$ or m-Cl; (ii) S$_B$ thermal stabilities are consistently decreased by m-CH$_3$ or m-Cl, the largest effect being ca. 10°; (iii) S$_A$ thermal stabilities are consistently decreased by m-CH$_3$ or m-Cl; when $n = 2$ or 3 the decreases are small, but rise to 10 to 13° when $n = 1$. However, when $n = 0$, m-CH$_3$ slightly *raises* the thermal stability, whereas m-Cl decreases it by 26.6°. These substituents are similar in size, but differ in dipolarity. Since $n = 0$, the proximity of the substituent to the ester carbonyl group is close and different dipolar interactions may result in different contributions from the possible extreme rotational conformers (see Fig. 17). If this is the case, a restriction of rotation about the O–ring bond is once more implied, in this case in a S$_A$ phase.

These results on the benzylidene aminocinnamate esters demonstrate not only that we have much to learn about the effects of molecular broadening on the thermal stabilities of different smectic polymorphic forms,† but also that all aspects of molecular geometry must be considered before being certain that a substituent will act in a lateral, molecular-broadening capacity on the thermal stability of a mesophase.

B. Steric Effects

Systematic studies were first made [*3, 6, 86*] of steric effects on liquid crystal thermal stabilities by introducing lateral substituents which both broaden the mesogen molecule and twist it. The systems studied were Schiff's bases derived from sterically affected 2-substituted biphenyls (XLVIII–L).

Subsequently, it was shown [*80, 81*] that Schiff's bases involve an inherent molecular twist such that the plane of the *N*-aryl group makes an angle of 40°–60° with the rest of the molecule. It can probably be assumed that a similar twist occurs about the Schiff's base linkage in (XLVIII–L). However, if this twist is constant for all these Schiff's bases, the effect studied was the influence of the extra twist imposed on the biphenyl nucleus by the 2-substituent

---

* When n is even, N–I temperatures are again diminished by m-CH$_3$ or m-Cl substituents, and again m-CH$_3$ has a larger effect than m-Cl, when $n = 0$.

† Early results established for carboxylic acids are mainly relevant to S$_C$ phases [*89*].

AlkylO—⟨⟩—CH=N—⟨⟩(X)—⟨⟩

(XLVIII)

AlkylO—⟨⟩—CH=N—⟨⟩(X)—⟨⟩

(XLIX)

AlkylO—⟨⟩—CH=N—⟨⟩(X)—⟨⟩—N=CH—⟨⟩—OAlkyl

(L)

X which causes rotation about the interannular bond. The implicit assumption existed initially that when X = H the system did not involve a large interplanar angle for the biphenyl nucleus, i.e., that even a small substituent such as fluoro would markedly increase the interplanar angle.

The results of these studies are now well known and need only be briefly summarized. Using a range of substituents, X (F, Cl, Br, I, CH_3, NO_2), much larger decreases in both nematic and smectic thermal stabilities arose than when the same substituents were introduced into lateral positions in the same or similar molecules such that no steric effects were imposed. For example, a nonsterically affecting 3'-fluoro substituent diminishes the smectic and nematic thermal stabilities of 4'-n-octyloxybiphenyl-4-carboxylic acid by only 0.5° and 9°, respectively. A 2-fluoro substituent in the octyl ether of XLVIII decreases smectic thermal stability by 71° and nematic thermal stability by 49°. Similar results were obtained for compounds of type XLIX and L. This shows that if the unsubstituted biphenyl nucleus is not planar in the mesophases, the interplanar angle is significantly less than that for the 2-fluoro derivative. Thus, even small steric effects have large effects on nematic and particularly smectic thermal stabilities, and the original assumption was justified.

When the N–I temperatures for the Schiff's bases of one particular type (XLVIII, XLIX, or L) involving the full range of 2-substituents were plotted against an axis related to the broadening effect alone of the 2-substituent, an approximately straight line was obtained. However, the slope was much greater than that obtained for a range of similarly substituted mesogens in which *no* steric effect occurred. Some extra factor which increased in proportion to substituent size was decreasing the nematic thermal stability at a faster rate than when molecular broadening alone was involved. This was obviously the effect of the twist about the interannular bond increasing both the thickness of the biphenyl region and the intermolecular separation, and diminishing the anisotropy of molecular polarizability.

Similar plots of S–N or S–I transition temperatures gave a more complex picture, cf. p. 40). A scatter of points about a mean line was obtained and the conclusions reached were that dipole moments and possibly also stronger "crystal forces" operative in the more ordered smectic phase were imposing variable effects which did not arise for the simpler nematic phase. However, the slope of the average line was steeper than for the nematic phases, implying that the steric effect has a more serious disruptive effect on the lamellar smectic order.

Di-, tri-, and tetra-substituted derivatives related to the mono-substituted di-anils (L) were also prepared. Some were not sterically affected, and others were sterically crowded; in the latter case, extremely large decreases in N-I temperatures were obtained and no smectic phases were observed. As mentioned earlier, it was possible to demonstrate for the first time that the effect of introducing more than one substituent into a mesogen is approximately additive.

Since sterically affecting substituents cause large decreases in mesophase thermal stabilities, their presence is often disadvantageous, resulting in extinction of mesomorphic properties or at best formation of only monotropic phases. However, if sterically affecting substituents should diminish the thermal stabilities of the solid crystals *more* than the nematic thermal stabilities, such substituents could give mesogens which form their mesophases at low temperatures. This possibility was first recognized by Young, Aviram, and Cox [90] who examined stilbenes of type LI; whereas the planar [91], *trans*-stilbenes, with $R_3 = R_4 = R_5 = H$ and suitable substituents as R_1 and R_2, are high melting and often exhibit only monotropic mesophases, introduction of small globular substituents R_3, R_4, or R_5 might give nonplanar, low melting mesogens. Their most successful materials are listed in Table V.

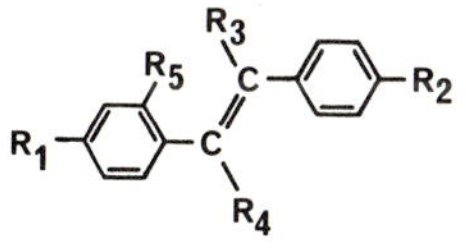

(LI)

TABLE V

Transition Temperatures for Compounds of Type LI

R_1	R_2	R_3	R_4	R_5	C–N (°C)	N–I (°C)
C_2H_5O	n-C_4H_9	Cl	H	H	29	58
C_2H_5O	n-C_8H_{17}	Cl	H	H	32	61
C_2H_5O	2-methylhexyl	Cl	H	H	22	35

A mixture of 60 mol % and 40 mol %, respectively, of the first two compounds in Table V had a nematic range of 8°–59°C. Consequently, steric effects can be turned to advantage, although with these α-chlorostilbenes their value is severely limited because they are unstable to light of <400 nm, which rapidly turns the nematic phases red.

Steric factors probably also account for the comparatively low melting points of α-cyanostilbenes of type LII. For example, with $R = n\text{-}C_6H_{13}$ and

$$RO-\underset{(LII)}{\bigcirc}-\overset{H}{\underset{\underset{N}{\overset{\|}{C}}}{\overset{|}{C}}}\overset{}{\underset{}{C}}-\bigcirc-OR'$$

$R' = CH_3$, the compound has C–N, 60.5°C and N–I, 70°C [8]. These compounds are of interest because of their strong negative dielectric anisotropy, and similar considerations have led to studies of α-cyano–Schiff's bases [92].

C. Branching of Terminal Alkyl Chains

Although it has long been recognized that branching of a terminal alkyl group decreases smectic and nematic thermal stabilities relative to those of the open chain compound, only recently have the effects been studied systematically.

Studies [47, 48] of methyl branching at the penultimate carbon of n-alkyl esters (LIII) revealed an effect similar, but smaller in magnitude, to that

$$X-\bigcirc-CH=N-\bigcirc-CH=CH-CO.O(CH_2)_nCH(CH_3)_2$$

$$(LIII)$$

observed for the analogous ω-phenylalkyl esters. The results for X = CN are given in Table VI and show an alternation of the N–I temperatures greater than that for n-alkyl series. When X was a group favoring smectic properties (e.g., $CH_3CO.O$), the smectic thermal stabilities did not alter greatly with n, whereas the nematic thermal stabilities for even n values were so low that they became "monotropic" (metastable) with respect to the smectic phases, and direct S_A–I transitions occurred (Fig. 18). Similar effects were obtained with X = Ph. The alternation in phase type (from S_A–I to S_A–N+N–I) which

TABLE VI

Transition Temperatures for
Compounds of Structure LIII[a]

n	C–N or I (°C)	N–I (°C)
0	142.5	(< 134.5)[b]
1	130	142.5
2	102.5	110
3	89	119.5

[a] $X = CN$.
[b] Temperature of solidification; no nematic phase detected.

occurs along the series (Fig. 18) and the rather high alternation of the N–I temperatures when $X = CN$ (Table VI) are analogous to the effects observed for ω-phenylalkyl esters (see p. 34), except that the n value corresponding to a lower nematic thermal stability or a direct S_A–I change is now even. This would be expected if the reason for the effects is the slight L-shape bestowed on the molecule by the protruding CH_3 group. Since this CH_3 group is on

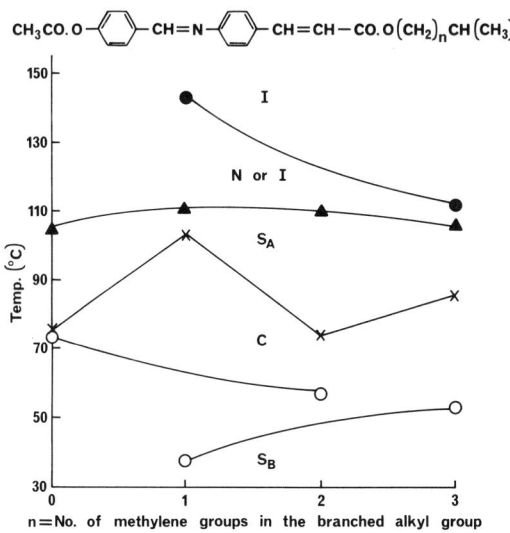

Fig. 18. Plot of liquid crystal transition temperatures against the number of methylene groups in the branched alkyl group of ω-isopropylalkyl 4-(p-acetoxybenzylideneamino)cinnamates (● = N–I; ▲ = S_A–I or N; ○ = S_B–S_A; × = C–S_A).

the penultimate carbon, whereas the larger Ph group of the ω-phenylalkyl esters is on the ultimate carbon, the effects should be opposite for even or odd n values.

Of more general interest were results obtained [47, 48] for alkyl 4-(p-substituted benzylideneamino)cinnamates in which a branching CH_3 was moved progressively to the penultimate carbon of the chain. Some typical results are given in Table VII.

TABLE VII

Transition Temperatures (°C) for Esters of the Type

$$X-\bigcirc-CH=N-\bigcirc-CH=CH-CO.O\,Alkyl$$

X	Alkyl	C–N or S_E	S_E–S_B	S_B–S_A	N or S_A–I
CN	$CH_2CH_2CH_2CH_2CH_3$	73.5	—	—	136.5
CN	$CH(CH_3)CH_2CH_2CH_2CH_3$	64.5	—	—	(<20)[a]
CN	$CH_2CH(CH_3)CH_2CH_2CH_3$	92.5	—	—	112
CN	$CH_2CH_2CH(CH_3)CH_2CH_3$	82	—	—	103
CN	$CH_2CH_2CH_2CH(CH_3)_2$	89	—	—	119.5
Ph	$CH_2CH_2CH_2CH_2CH_3$	92	101.5	168	204
Ph	$CH(CH_3)CH_2CH_2CH_2CH_3$	70	128	168	180
Ph	$CH_2CH(CH_3)CH_2CH_2CH_3$	68	113.5	157	190.5
Ph	$CH_2CH_2CH(CH_3)CH_2CH_3$	77.5	109	167.5	196
Ph	$CH_2CH_2CH_2CH(CH_3)_2$	93	108	168.5	199

[a] Temperature of solidification; no nematic phase detected.

For nematogenic n-alkyl esters (e.g., X = CN), 1-methylation greatly decreases nematic thermal stability, giving nonmesomorphic products or monotropic nematic phases. Movement of the CH_3 to the 2-, 3-, or 4-position has a smaller effect on nematic thermal stability, which possibly diminishes as the CH_3 is moved to the penultimate carbon. For smectogenic esters (e.g., X = Ph), 1-methylation has a much smaller effect on S_A–I temperatures, and as the CH_3 is moved to the end of the chain, the effect becomes smaller. S_B thermal stabilities are in general not greatly affected by branching and the more highly organized S_E phase has its thermal stability *enhanced particularly* by a 1-methyl group.

The large effect of a 1-methyl group on nematic thermal stability has been explained by a steric effect on the ester C=O group. Apparently, however, the branching CH_3, even in the 1-position, can be accommodated readily in a smectic packing and may even contribute to the smectic stability when the phase is highly organized. Again we have examples of smectic and nematic thermal stabilities being differently affected by changes in geometry.

To check whether these results were general, irrespective of the environment of the alkyl group, some branched ethers of structure LIV were prepared [48]. The results were similar, except that now a 1-methyl group also has a

AlkylO—⟨ ⟩—CH=N—⟨ ⟩—CH=CH—CO.OPh   X—⟨ ⟩—O—C(=O)—⟨ ⟩—Y

(LIV)   (LV)

large effect on the S_A–N transition temperature, but still less than that on the N–I temperature. A 1-methyl group in the ether function strongly interacts sterically with an ortho-hydrogen of the aromatic ring and a large effect on nematic thermal stability is expected. However, when a number of molecular models are placed side by side as in a S_A packing, no gaps exist to accommodate the protruding 1-methyl group as occurs for the 1-methylalkyl esters. Consequently a large effect on S_A thermal stability is observed.

These results show that we must be careful about generalizing on results for alkyl chain branching, and it is valuable that more recent studies of chain branching effects have been made to complement and supplement the results so far discussed. For example, esters of type LV where *either* X = n-alkoxy and Y = $CH_2.CH(CH_3)_2$ or —$CH(CH_3)CH_2CH_3$ *or* X = —$CH_2.CH(CH_3)_2$ or —$CH(CH_3)CH_2CH_3$, and Y = n-alkoxy have been compared [93] with the esters having n-alkyl groups. Apparently unaware of the work on branched *ethers*, the authors examined these compounds to establish the effect of branching in closer proximity to the aryl ring than occurs in the branched esters. None of the compounds containing the —$CH(CH_3)CH_2CH_3$ group was liquid crystalline, confirming the marked effect of a 1-methyl group on nematic thermal stabilities. Most of the esters containing a 2-methyl branch were liquid crystalline, but the N–I temperatures were always lower than those of the n-alkyl analogs by 10°–15°. It is difficult to extract information from the results about the effects on smectic thermal stability, but apparently a 1-methyl group must sharply diminish this as for the ethers, and a 2-methyl group may diminish smectic thermal stability rather more than nematic.

Branching of alkyl groups attached directly to the aromatic ring has also been studied [94] for the system LVI and comparisons have been made with

RO—⟨ ⟩—CH=N—⟨ ⟩—$(CH_2)_n CH_2 CH (CH_3) CH_2 CH_3$

(LVI)

the corresponding unbranched compounds; the consistent effect is a decrease in nematic (cholesteric) and/or smectic thermal stabilities. The decreases in transition temperatures vary somewhat with R, but are fairly constant in the majority of cases around 20°, irrespective of whether the smectic or nematic (cholesteric) phase is concerned or the value of n (from 0 to 2). Obviously (see LVI), this study did not include 1-methyl substitution, and the results do not inform us about the effects of moving the branching methyl group along a *given* chain.

In summary, branching of terminal alkyl chains consistently lowers the thermal stabilities of nematic and S_A phases, the effects of a 1-methyl group being particularly serious irrespective of whether the alkyl group is in an ester or an ether function or is directly attached to a ring. Smaller effects accompany 2-, 3-, etc., methylation and the relative effects on smectic and nematic thermal stabilities apparently depend on the environment of the alkyl group. Some studies of branched chain mesogens have been undertaken in the hope of obtaining lower melting mesogens for applications, and further data on branched chain compounds supporting the above conclusions can be found in Young et al. [*90*] and Goldmacher [*95*].

VIII. THE ROLE OF THE CENTRAL GROUP IN SIMPLE MESOGENS

For mesogens suitable for applications, a most important consideration is the temperature range of the mesophase. At present, concern is mainly with materials giving nematic or cholesteric mesophases, devices based on smectic mesophases being less important—but see Taylor and Kahn [*96*]. Our concern is therefore with nematogens or cholesterogens that melt at low temperatures giving mesophases stable in the room temperature range and persisting when heated above this range. Ideally, the mesogen should melt no higher than $-10°C$ and its mesophase should persist until at least $+60°C$. This desirable range cannot yet be met by any *single* mesogen. However, there are ways around this problem, and the aim is to produce nematogens or cholesterogens which melt as low as possible and form their mesophases over the widest possible range. To achieve this, mesogens of simple molecular structure must be employed. Since simple, *monomeric p*-disubstituted benzenes are not mesomorphic and aliphatic mesogens are unsuitable for a number of reasons, the least complex structure for a low melting mesogen is of type LVII, where

A—◯—X—◯—B

(LVII)

A and B are suitable terminal substituents, and X is the central group. For example, when A = CH$_3$O, B = (CH$_2$)$_3$CH$_3$, and X = —CH=N—, we have MBBA, a nematogen having C–N, 22°C and N–I, 47.5°C (see also Arora et al. [97]).

Many central, linking groups X have been examined with combinations of different terminal substituents. The N–I transition temperatures change with the nature of X in a fairly predictable manner determined by the effect of X on the anisotropy of the molecular polarizability and the stereochemistry. If X permits strong conjugative interactions between the rings, these tend to be coplanar and the anisotropy of molecular polarizability will be high provided that the molecule adopts a lathlike configuration. The groups X which favor thermally stable mesophases most are therefore those that involve π-electrons and preserve a lathlike molecular shape. For example, if X is —CH=CH—, the *trans*-1,4-di-*n*-alkoxystilbenes are mesogens, but the nonlinear *cis*-isomers are not. Saturated groups such as —CH$_2$—CH$_2$— are, however, too flexible. To obtain stable mesophases with saturated groups, we have to use a group X like *trans*-1,4-cyclohexylene or, better, 1,4-bicyclooctylene. Although the cyclohexylene ring is not rigid, it makes the molecule less flexible than when X is alkylene, and while not permitting conjugation between the rings, it enhances molecular length and axial polarizability. Naturally, both saturated central groups are less efficient in producing high N–I temperatures than a *p*-phenylene ring which is planar, rigid, and allows conjugation to occur.

Using literature values (e.g., [37, 41, 43, 90, 95, 98, 99]) of the N–I temperatures of a range of mesogens and a statistical analysis of data by Knaak, Rosenberg, and Servé [100], it is possible to construct an *average* order of efficiency for central groups X in promoting N–I transition temperatures. With minor exceptions, dependent upon the end groups A and B in LVII, the order is

X = —⟨phenylene⟩— > —⟨cyclohexylene⟩— > —CH=CH—> —N=N—> —CH=N—>
                                              (trans)   ↓   ↓
                                                            O   O

—C≡C—> —N=N—> —CH=N—> —CO.O—> none

Thus, thermally stable mesophases are produced by substituted *p*-terphenyls and *trans*-stilbenes, whereas mesophases of lower thermal stability are formed by substituted benzylideneanilines, phenyl benzoates, and biphenyls (X = none). Stereochemical considerations are certainly important, for the *trans*-stilbenes are planar, the azoxybenzenes are slightly twisted, and the benzylideneanilines are considerably twisted; it is not known whether the rings in biphenyl mesogens are coplanar in the mesophases (see p. 45). The relatively low position in the order for X = —C≡C— is surprising. In crystalline *trans*-stilbene [91] and diphenylacetylene [101] the molecules are planar, and one would associate a high axial polarizability with the —C≡C— linkage. It is possible, however, that when crystal forces are removed on melting, the rings become noncoplanar. This may be less serious than in Schiff's bases, since the cylinder of electron density of —C≡C— may still permit conjugative interactions to occur even when the rings are noncoplanar.

Of as great importance as the N–I transition temperatures, however, are the C–N temperatures of mesogens of type LVII required for applications.

TABLE VIII

Some Examples of Low Melting Mesogens of Type LVII

No.	A	X	B	C–S or N (°C)	S–N (°C)	N–I (°C)	Ref.
1	CH_3O	N=N	C_4H_9-n	32	—	48	[100]
2	CH_3O	N=N ↓ O	C_4H_9-n	42	—	77	[100]
3	CH_3O	CH=N	C_4H_9-n	20	—	47[a]	[100]
4	n-C_4H_9	N=N ↓ O	C_4H_9-n	14	—	28	[43]
5	n-C_5H_{11}	N=N ↓ O	C_5H_{11}-n	22	—	65	[43]
6	CH_3O	C≡C	C_7H_{15}-n	39	—	54	[99][b]
7	C_2H_5O	CH=CCl	C_4H_9-n	29	—	58	[90]
8	n-$C_{10}H_{21}O$	CO.O	C_3H_7-n	59.5	62.5	66	[8]
9	n-$C_6H_{13}O.CO.O$	CO.O	OC_6H_{13}-n	45	—	73.5	[8]
10	n-C_5H_{11}	None	CN	22.5	—	35	[16, 37, 40]
11	n-C_6H_{13}	None	CN	13.5	—	27	[16, 37, 40]
12	n-C_7H_{15}	None	CN	28.5	—	42	[16, 37, 40]
13	n-C_8H_{17}	None	CN	21	32.5	40	[16, 37, 40]
14	n-C_3H_7	CH=N	CN	46	—	62	[104]
15	n-C_4H_9O	CH=N	CN	65	—	108	[8]
16	n-$C_5H_{11}O$	CO.O	CN	85–87	—	(77)[c]	[105]
17	n-C_7H_{15}	CO.O	CN	40	—	54	[106][d]

[a] More usually quoted as C–N, 22°C; N–I, 47.5°C.
[b] For recent results on tolanes, see Dubois *et al.* [107].
[c] Monotropic transition in parentheses.
[d] Constants by G. W. Gray *et al.* (unpublished results).

Molecular Geometry

Melting temperatures of mesogens do not, however, conform to any regular pattern, and it is not possible to select any central group X and a combination of substituents A and B that will necessarily give a particularly low melting point. By and large, high N–I temperatures are associated with high C–N temperatures, but apart from the first three central groups in the above order, low melting mesogens can be found with any of the other central groups; examples are given in Table VIII. The nature of the groups A, B, and X is varied with the dielectric anisotropy (both in magnitude and sign) required of the mesogen for particular applications. A negative dielectric anisotropy can be achieved, for example, with A = alkoxy, B = alkyl, and X = —CH=N—. A strong positive dielectric anisotropy can be achieved by making one of the terminal groups A or B = —C≡N. Alternatively, mesogens may be required that have a dielectric anisotropy which changes sign at higher frequencies [102, 103]. There is a tendency to use n-butyl, n-pentyl, n-hexyl, n-heptyl, or n-octyl as one of the end groups, as it is noted empirically that low melting points are most common under these circumstances.

Compounds 3, 4, 5, 10, and 11 in Table VIII provide room temperature nematic materials, in certain cases because the formation of crystals on cooling involves supercooling. Compound 13 gives a room temperature smectic A [16, 60]. While these compounds are invaluable in providing single-component room temperature liquid crystals suitable for a variety of physical studies, none of the mesophases of the compounds in Table VIII approaches the required nematic range from $-10°$ to $+60°C$.

This can be overcome by using mixtures of low melting nematogens. It is important that *eutectic* compositions should be employed (i) to ensure the maximum depression of the C–N temperatures and (ii) to avoid segregation from the mixtures of higher melting components present in excess of the eutectic amounts. Eutectic compositions can be determined experimentally or obtained by calculation [108–110], knowing the C–N temperatures and enthalpies of fusion for all the components. Table IX gives some examples; most mixtures of this kind supercool many degrees below the C–N temperatures before crystallization occurs; in some cases it is difficult to achieve total crystallization so that C–N temperatures can be measured. The results in Table IX are given only as illustrations of the effects that can be achieved. The first two mixtures and the last in Table IX would have a strong positive dielectric anisotropy and be useful for twisted nematic displays (see also Titov [112]) and the third mixture would be of negative dielectric anisotropy as required for dynamic scattering. The α-chlorostilbene mixture is photochemically unstable, and mixtures of negative dielectric anisotropy in which the central group X is N=N(O), CH=N or CO.O are commonly employed for dynamic scattering. Mixtures with different A, X, and B combinations can also be used to obtain a dielectric anisotropy tailored to particular needs,

TABLE IX

Transition Temperatures for Mixtures of Mesogens of Type LVII

A	X	B	Composition	C–N (°C)	N–I (°C)	Ref.
n-C_4H_9	CO.O	CN	33.3^a	25	50	[106]
n-C_7H_{15}	CO.O	CN	66.7^a			
n-C_3H_7	CH=N	CN	33.3^a	−15	50	[111]
n-C_6H_{13}	CH=N	CN	66.7^a			
C_2H_5O	CH=CCl	n-C_4H_9	60 mol%	8	59	[90]
C_2H_5O	CH=CCl	n-C_8H_{17}	40 mol%			
n-C_5H_{11}	None	CN	55 mol%	−2	54.5	[16]
n-$C_5H_{11}O$	None	CN	15 mol%			
n-$C_7H_{15}O$	None	CN	13 mol%			
n-$C_8H_{17}O$	None	CN	17 mol%			

^{a} It is not specified in the reference whether the composition is expressed in mol % or wt %.

and in the case of benzoate esters, mixtures can be produced that change the sign of dielectric anisotropy with frequency [102, 103].

The question of chemical/photochemical stability is, of course, vital if display devices with good lifetime characteristics are required. The problems of chemical/photochemical stability in relation to the molecular structures of mesogens have been discussed at length in recent publications, as have the attendant problems of color and freedom from toxicity [8, 16, 36, 37, 40, 41]. Mesogens containing many of the central groups listed earlier can be faulted in some way, e.g., CH=N, colored and chemically unstable (particularly to hydrolysis); N=N(O) and N=N colored, and azo compounds are photochemically unstable; CH=CCl, photochemically unstable, etc. The only really satisfactory conditions are those in which the central group is absent and the two phenyl rings are directly linked, as in biphenyl. Thus we have the dilemma that the best properties in relation to chemical/photochemical stability and color are met by mesogens containing no central linking unit and having the lowest N–I temperatures. To obtain reasonably high N–I temperatures from simple 4,4′-disubstituted biphenyls, one terminal group must be CN which is high in the nematic terminal group efficiency order. Therefore biphenyls with simple substituents A and B are only suitable at present for producing nematic mesophases of positive dielectric anisotropy. For such purposes, they are very good indeed, except that the best nematic range attainable with mixtures of 4′-n-alkyl- and -alkoxy-4-cyanobiphenyls is not wide enough (see Table IX). This difficulty is overcome by adding to the mixtures a component having the central group X = *p*-phenylene, i.e., a 4″-*n*-alkyl-4-cyano-*p*-terphenyl [*16, 40, 41*]. Such compounds have a similar, high chemical/photochemical stability to the 4′-*n*-alkyl- and -alkoxy-4-cyanobiphenyls, but very much higher N–I temperatures, e.g., compound LVIII has C–N, 130°C; N–I, 239°C. Naturally, the C–N temperature is also

n–C₅H₁₁–⟨◯⟩–⟨◯⟩–⟨◯⟩–CN   A–⟨◯⟩–C(=O)–O–⟨◯⟩–C(=O)–O–⟨◯⟩–B

(LVIII)   (LIX)

raised by the central *p*-phenylene ring. However, the enthalpy of fusion of LVIII is low and eutectic compositions with various 4′-*n*-alkyl- and -alkoxy-4-cyanobiphenyls containing around 10% of LVIII have lower C–N and higher N–I temperatures than those that can be achieved with biphenyls alone. The desirable constants of C–N, −10°C; N–I, +60°C can be met by such mixtures, e.g., the mixture E8 (marketed by B.D.H. Chemicals, Ltd., Poole, Dorset, England) has C–N, ∼ −12°C; N–I, 70.5°C.

Problems in obtaining wide range mixtures of *negative* dielectric anisotropy which are colorless and stable are more difficult. If color is not a worry, azoxy compounds are satisfactory [*113*]. If colorless films are required, the best central group is probably CO.O, although response times may suffer. With esters [*114*], the range can also be widened by incorporating in the mixtures three ring mesogens such as LIX [*105*]; the same principle can be applied to cyano esters [*106, 115, 116, 116a*] and cyano Schiff's bases [*106, 117*] of positive dielectric anisotropy such as those in the first two mixtures in Table IX. The ester function, X = CO.O, is capable of being attacked by nucleophiles by a similar mechanism to that by which hydrolysis of the CH=N link occurs, but very much more slowly. Therefore, although less readily attacked by nucleophiles than are Schiff's bases, esters are not as stable under particular conditions as say the biphenyl or *p*-terphenyl mesogens. Nonetheless, it is still necessary to have some central group X present to produce suitable nematic phases of negative dielectric anisotropy, and so the problems of stability in relation to dynamic scattering materials remain with us to some extent.

Similar stability problems apply to chiral mesogens of negative dielectric anisotropy for cholesteric memory displays [*9*]. Cholesteryl esters are frequently used, even though their chemical/photochemical stabilities leave something to be desired. References [*118*] and [*119*] are also relevant.

The problem is easily solved for chiral mesogens of positive dielectric anisotropy for cholesteric–nematic phase change displays. Chemically and photochemically stable cholesterogens can be obtained by using branched alkyl groups containing an asymmetric center in the 4'-*n*-alkyl- and -alkoxy-4-cyanobiphenyls [*40*, *120*] or the corresponding *p*-terphenyls [*16*]. For example, compound LX has C–S$_2$, 120°C; S$_2$–S$_A$, 137°C; S$_A$–Ch, 163°C;

$$(-)-CH_3CH_2CH(CH_3)CH_2CH_2CH_2-\bigcirc-\bigcirc-\bigcirc-CN$$

(LX)

Ch–I, 186°C (S$_2$ has not yet been classified), and a very important, recent addition to the list of such materials is the low melting, high twisting power cholesterogen 4'-(2"-methylbutyl)-4-cyanobiphenyl [*120*]. Incorporation of such a compound in eutectic composition with mixtures of alkyl- and alkoxy-cyano-biphenyls and -*p*-terphenyls gives colorless, stable, wide range, long pitch cholesterics which function well in phase change displays (see also White and Taylor [*121*]). Such stable chiral mesogens, e.g., the readily prepared compound 4'-(2"-methylbutyloxy)-4-cyanobiphenyl, can also be used at low concentration (ca. 1%) in nematic cyanobiphenyl mixtures to prevent areas of reversed twist in twisted nematic displays.

IX. FEATURES OF MOLECULAR STRUCTURE FAVORING FORMATION OF SMECTIC OR NEMATIC/CHOLESTERIC PHASES

The question frequently arises as to whether a proposed mesogen will exhibit smectic or nematic properties. While this basic question relating to knowledge of molecular structure and geometry and their effects on liquid crystal properties can usually be answered with some degree of certainty, recent experimental work shows that quite trivial changes in molecular architecture can have far-reaching effects on the balance between nematic and smectic tendencies [*6*, *18*, *48*]. Therefore, one must be cautious in diagnosing the probable behavior of a new mesogen. Even greater caution would be wise in predicting the type of smectic phase(s) likely to be exhibited; we have much to learn about this problem [*60*, *89*, *122*, *123*].

In this section, the broad guidelines relating to smectic vs. nematic tendencies are covered and the reader's attention is drawn to some of the subtleties.

Assuming that the molecules of our mesogen are lath-shaped, then, before smectic properties can be exhibited, the molecules must form a layer crystal lattice. If the crystal lattice has an overall interdigitated arrangement of molecules, a smectic mesophase will not be formed on heating the crystal.

Molecular Geometry

The factors which relate molecular structure to crystal lattice type are uncertain, and so our diagnosis of whether a compound will be smectic or nematic begins (in the absence of X-ray data about the crystal lattice) with an assumption regarding the crystal lattice. Therefore, when we predict that a compound will be smectic, we mean that, provided a layer crystal lattice exists in the solid, then, because the molecules possess particular features, smectic properties will probably arise when the crystal melts.

However, even if a layer crystal lattice exists, smectic properties need not be exhibited. If the melting point of the solid is high, a parallel arrangement of the lathlike molecules may not persist and the isotropic liquid may be formed direct. Smectic properties need not even arise if the parallel molecular orientation persists on melting a layer crystal lattice, for if the lathlike molecules translate in the direction of their long axes, an overall interdigitated arrangement will result and the phase will be nematic. In addition to the requirement that a layer lattice should exist must be added the requirements that the molecular interactions should be (i) not too strong, (ii) sufficiently anisotropic in order that the molecules remain in a layer arrangement in which the layers can move over one another (strong lateral and weak terminal interactions). In fact, our qualitative judgment as to the occurrence of smectic or nematic properties is based on an assessment of the extent to which this anisotropy of molecular interactions is affected by particular features of structure. It becomes obvious why uncertainties exist in predicting such a basic matter as the phase type likely to be exhibited.

Important factors which affect the anisotropy of molecular interactions are the terminal substituents. If these enhance the axial polarizability, i.e., are high in the nematic terminal group efficiency order, then smectic properties are suppressed. Therefore terminal nitro, cyano, and methoxy groups strongly promote nematic properties. On the other hand, terminal functions which contribute strongly to the resultant dipole acting across the molecular long axis promote smectic properties. Thus molecules with terminal —CO.Oalkyl or —CH=CH—CO.Oalkyl groups are predominantly smectic, as are compounds with terminal —CO.NH$_2$ or —O.CF$_3$ groups. Ionized functions, e.g., —COO$^-$M$^+$ and —NH$_3^+$X$^-$, usually give purely smectic behavior.

It is also relevant to remember that replacement of a terminal ring hydrogen in a mesogen by any substituent which does not broaden the molecule or cause it to deviate from linearity too greatly will enhance the N–I temperature, whereas replacement of a terminal ring hydrogen by substituents such as cyano, nitro, or methoxy may depress the smectic tendencies of the system. In some cases, the system can apparently adapt to such a situation by forming an interdigitated lamellar smectic A, e.g., in the longer chain 4'-*n*-alkyl- and -alkoxy-4-cyanobiphenyls [59, 60]. Terminal ring substituents such as —Ph, —NH.CO.CH$_3$, and O.CO.CH$_3$, though strongly promoting both smectic and nematic properties, enhance the smectic characteristics to the greater extent.

The effects of increasing the length of a terminal *n*-alkyl chain on smectic and nematic tendencies are well established [3, 6]. Whether the *n*-alkyl group is directly linked to a ring or is present as a terminal ether or ester group, smectic properties are enhanced by increasing its length. The exact points in homologous series at which smectic properties first appear depend to some extent on the melting points of the mesogens; this particular property of a series cannot therefore be related definitely to molecular structure. However, it is noted that in series of *n*-alkyl esters in which the other terminal group is nitro or cyano, despite the —CO.O alkyl group, smectic properties may not appear until long chains, e.g., C_8 or C_9, are used [47]. This is because cyano and nitro groups are high in the *nematic* terminal group efficiency order. On the other hand, if the other terminal group is say —Ph or —O.CO.CH$_3$, smectic properties will probably occur for the entire series from the methyl ester upward [47, 48]. As described earlier, once smectic properties have begun in the series, further increases in the length of the *n*-alkyl chain often increase the S–N temperatures rapidly at first and then more gradually, until a point is reached in the series at which purely smectic compounds are formed.

At this stage, the reader is reminded of the effects of alkyl chain branching on the balance between smectic and nematic properties of mesogens. The conclusion was that a 1-methyl branch has a marked effect on nematic and smectic thermal stabilities; in 1-methylalkyl esters, the effect on nematic stability is much greater, however, than that on smectic (S_A) thermal stability. Smaller effects accompany 2-, 3-, etc., methylation. A 1-, 2-, or 3-methyl group in an alkyl ester chain often has little effect on smectic B and E phases, and may even enhance their thermal stabilities.

The effect of methyl branching at the penultimate carbon of a chain, i.e., —$(CH_2)_n CH(CH_3)_2$, is particularly important. Quite large alternations in nematic thermal stabilities and much smaller changes in smectic (S_A) thermal stabilities occur as *n* changes from even to odd. As a result, as *n* increases, we get an alternation in phase type from smectic and nematic (*n*, odd) to purely smectic (*n*, even) as the series is ascended. The same effect occurs with ω-phenylalkyl groups, but the purely smectic properties are observed when *n* is odd. Thus we have the subtle situation that nematic properties can be replaced by purely smectic properties merely by adding or subtracting a methylene group in a chain such as —$(CH_2)_n CH(CH_3)_2$ or —$(CH_2)_n Ph$.

Care must also be taken over the location of a particular function in a molecule. For example, a terminal —CO.OR has been stated to be a smectic-favoring group. However if the function —CO.O— links up two rings as in the benzoate esters [*105, 114*], i.e., if it is part of the central core of the molecule, the compounds do not have particularly marked smectic tendencies.

$$\text{n-C}_6\text{H}_{13}\text{O}-\underset{}{\bigcirc}-\overset{\overset{\displaystyle O}{\parallel}}{\text{C}}\underset{\text{O}-\underset{}{\bigcirc}-\text{OC}_8\text{H}_{17}\text{-n}}{}$$

(LXI)

For example, compound LXI [*105*] is nematic from its melting point of 54°C until 89°C, despite the presence of two long chains. It is probable that conjugation of the ester function with both rings reduces the contribution to the dipole moment acting across the long molecular axis, and diminishes the smectic characteristics in relation to those of *alkyl* esters of aromatic acids. Such conjugative interactions will enhance the anisotropy of the polarizability and favor nematic phase formation.

Difficulties in assessing the contribution which a particular dipolar group makes to the magnitude and direction of the dipole moment of the molecule as a whole, and therefore the influence which that group has on the smectogenic or nematogenic character of the compound, have been discussed at length elsewhere [*3, 5, 6*].

Finally, the effects [*6*] of introducing one or more than one lateral substituent into the core part of the mesogen molecule in relation to smectic vs. nematic tendencies are summarized.

If the substituent broadens the molecule markedly, the tendency to give a smectic phase will be decreased more than the tendency to give a nematic phase if the substituent is (i) weakly dipolar like CH_3 or (ii) moderately dipolar but large, like Br or I. Such effects are greater when the mesogen molecule is quite short, and vice versa.

If the substituent does not broaden the molecule, e.g., as in the 5-substituted 6-*n*-alkoxy-2-naphthoic acids, liquid crystal thermal stabilities can be enhanced. When the substituent is dipolar, smectic thermal stability is increased more than nematic.

These effects are associable with the size and dipolar properties of the substituent. The size factor is the all important one, however, when the substituent exerts a steric effect, e.g., as in 2-substituted biphenyl derivatives. The twisted molecules are much less diposed to form liquid crystals, and smectic thermal stability is decreased much more than nematic.

The structural and geometric factors which influence the nematic thermal stabilities of mesogens are, of course, exactly those that influence the cholesteric thermal stabilities of optically active mesogens. (The N–I temperature of a racemic modification is the same as the Ch–I temperature of each of the optically active isomers [*5, 124*], the cholesteric terminal group efficiency order is the same as the nematic order [*5*], and the Ch–I temperatures for ω-phenylalkyl esters of sterols alternate markedly [*74, 75*].) Therefore the guidelines laid down for the smectic vs. nematic tendencies of mesogens also apply to the smectic vs. cholesteric tendencies of chiral mesogens (cholesterogens)—see also Vora in [*58*].

X. MOLECULAR STRUCTURAL EFFECTS ON THE LIQUID CRYSTAL PROPERTIES OF STEROL DERIVATIVES

Derivatives of sterols (particularly cholesterol) are simply good mesogens which, being optically active, are cholesterogens. However, mesogens derived from sterols are of particular interest since they provide accessible materials for applications in surface thermography, and their role in biological functions and membrane systems is a challenging area for investigation [*125*]. It seems worthwhile therefore to devote a section of this review to considerations of what is known about how the smectic and cholesteric tendencies of such derivatives vary with the nature of the sterol system.

This subject was recently briefly reviewed by Gray [*6*], but since then, Elser [*13*], in a longer review on cholesteric liquid crystals in this volume, has detailed the structures of known mesogenic sterols and recorded whether they give enantiotropic or monotropic smectic or cholesteric phases, and whether the cholesteric phases reflect colored light. As Elser emphasizes, however, it is not always easy to reach definite conclusions from literature data as to whether a particular steryl system is conducive to mesophase formation or not. If only the steryl alkanoates have been made and found to be noncholesterogens, this does not mean that the corresponding benzoates or cinnamates might not exhibit cholesteric phases. In this section therefore, the general conclusions [*6*] relating the structures of sterols and the liquid crystal behavior of their derivatives are reinspected, as far as possible, in the light of the information presented by Elser, to find whether it is necessary to modify earlier conclusions [*6*] or possibly to extend them.

At one time almost all cholesteric derivatives of sterols were derived from cholesterol (cholest-5,6-en-3β-ol)-structure (LXII; X = OH). The suitability of this particular system for mesophase formation is reflected in the cholesteric

(LXII)

properties of the cholesteryl *n*-alkanoates, *O*-alkyl carbonates, and ω-phenylalkanoates, and the α,ω-polymethylene-bis-cholesteryl carbonates [*71, 126*]. The group X in LXII can be a simple substituent such as SH, Cl, Br, I, O.CO.Cl, O.CO.N$_3$, CO$_2$H, CO$_2$(CH$_2$)$_n$CH$_3$, or CH(CO$_2$CH$_3$)$_2$ and cholesteric phases (enantiotropic or monotropic) are found [*13*]. In all cases, the substituent X is a 3β-substituent, and derivatives of thiocholesterol (X = SH; cholest-5,6-en-3β-thiol) are also cholesteric, e.g., *S*-cholesteryl alkanethioates, alkyl thiocarbonates, and ω-phenylalkanethioates. The effects of replacing *O* by *S* in these systems on the smectic and cholesteric thermal stabilities have been discussed [*77, 127, 128, 129*]. Where homologous series are involved, smectic properties become more pronounced relative to the cholesteric properties as the series are ascended. There is, however, a consistent difference compared with homologous series of nonoptically active mesogens in that cholesteric properties are never completely replaced by smectic properties even for the longest chain homologs studied. A cholesteric phase, though sometimes of very narrow range, always persists.

The influence of modifying the cholesterol skeleton was examined by Wiegand [*130*] who established that cholestanol (LXIII with X = OH;

(LXIII)

5α-cholestan-3β-ol) also gives cholesteric derivatives. The system does not have a 5,6-double bond. The molecular polarizability will therefore be lower and where comparisons can be made, cholestanyl derivatives have lower Ch–I temperatures than the cholesteryl analogs [*3, 13*]. The 5α-cholestan-3β-*n*-alkanoates, alkyl carbonates, *S*-alkyl thiocarbonates, and ω-phenylalkanoates are series which exhibit smectic and cholesteric or only cholesteric properties. The nature of the 3β-substituent, X in LXIII cannot be varied greatly while retaining cholesteric properties [*13*]. Compared with the list of 3β-substituents which may be used in LXII, X may apparently be only Cl or O.CO.Cl, and it is not certain that the phases are cholesteric.

Wiegand [*130*] also showed that derivatives of epicholesterol (a 3α-sterol) and coprostanol (a sterol in which rings A and B—see LXIII—are cis) did not give liquid crystals. He concluded that mesophases are formed by sterol derivatives only when the substituent is in the 3β-position, and when rings A and B are quasi-planar. More recently, Knapp and Nicholas [*131*] pointed out that sterols like epicholesterol and coprostanol are "spatially kinked" and that this will disfavor mesophase formation. The required quasi-planar nature arises in 5α-cholestan-3β-ol where rings A and B are trans. In cholesterol, rings A and B cannot be said to be trans but, as models show, the rings are quasi-planar. Wiegand concluded that with this quasi-planar arrangement, the 3β-substituent extends the molecular long axis and favors mesophase formation. Of many sterol derivatives listed by Elser [*13*], almost all involve 3β-substituents, the only "exceptions" being derivatives of sterols in which the 3-carbon is sp^2 hybridized. The 3-substituent therefore lies in the plane of the olefinic system and again extends the molecular axis. The fact that the acetate and benzoate of cholesta-3,4;5,6-dien-3-ol (LXIV) may be cholesteric is not therefore surprising. However, three other sterol derivatives with double bonds at the 3-carbon give normal melting.

The general conclusion [*130*] that a quasi-planar arrangement of rings A and B and a 3β-substituent that extends the long axis are prerequisites for phase formation in steryl derivatives seems therefore to remain valid.

(LXIV)

The effects of moving the 5,6-double bond of cholesterol have been examined [*6*]. The conclusion reached from Wiegand's work was that the double bond could be in the 7,8-, 8,9-, or 8,14-positions. Assuming the phases observed were cholesteric, the clearing temperatures of the benzoates were about the same as for cholesteryl benzoate in the first two cases, but over 30° lower in the third case. However, Elser [*13*] points out that *cholesteric* properties have been confirmed only for derivatives of 5α-cholest-8,14-en-3β-ol; in the other two cases, only clearing temperatures are reported. In fact the phases of the 5α-cholest-7,8-en-3β-yl *n-alkanoates* are smectic. It was also concluded that location of the double bond at the 14,15-position gave no mesophases.

These conclusions are still valid, but from Elser's data additional information can now be listed. Normal melting behavior occurs for derivatives with the double bond in the 1,2-, 3-4, 4,5-, or 6,7-position, but cholesteric properties are found for 5α-cholest-2,3-en-3β-yl 4'-*n*-alkoxybenzoates, and 4α,14α-dimethyl-5α-cholest-9,11-en-3β-yl acetate. Therefore in addition to cholesterol (5,6-double bond), cholesteric properties have been *confirmed* when the double bond is in the 2,3-, 9,11-, or 8,14-position.

Molecular Geometry

From Wiegand's results [130] on cholestadienol derivatives, double bonds at 5,6;7,8 or 6,7;8,9 give mesomorphic properties, definitely cholesteric in the former case, and higher clearing temperatures arise. When in the 7,8;14,15-positions, no mesophases were found (and a large decrease in clearing temperature can be inferred from the low melting point). From Elser's review [13], we can now add that derivatives of cholestadienols with double bonds at 3,4;5,6 and 7,8;9,11 may be mesomorphic (but not definitely cholesteric), whereas with double bonds at the 1,2;3,4-, 4,5;6,7-, 4,5;7,8-, 5,6;8,9-, 5,6;16,17-, or 8,9;14,15- positions, normal melting occurs.

Knapp and Nicholas [131] pointed out that the orientation of the 17β-side chain relative to the rest of the molecule is different when a double bond is at the 14,15-position and this probably explains the absence of mesophases in cholestenol and cholestadienol derivatives with such a double bond. The absence of mesophases for cholesta-14,15;24,25-dien-3β-yl benzoate can probably be attributed to the 14,15-double bond rather than the 24,25-double bond [6]. Supporting evidence [131, 132] is that cycloartanyl palmitate (LXV) and the corresponding ester with a 24,25-double bond are cholesteric.

$C_{15}H_{31}$·CO·O

(LXV)

Subsequently, Elser [13] pointed out that two double bonds (one in a ring and one in the 17β-chain) in the 7,8;22,23-positions give nonmesomorphic behavior. A cloudy melt is obtained with a cholestatrienol derivative (cholesta-6,7;8,14;9,11-trien-3β-yl octadecanoate).

Based on the earlier review [6], it was stated that (i) cholesteric thermal stability was sensitive to the position of a single olefinic bond if this is moved *out* of ring B, and (ii) introduction of a second olefinic bond does not always give the expected increase in cholesteric thermal stability.

With reference to point (i) we must now note that 5α-cholest-6,7-en-3β-yl benzoate is quite low melting (128°–129°) relative to cholesteryl benzoate and yet it gives no mesophase [133]. Therefore, the system is apparently sensitive to the location of the double bond *even in ring B* when the 6,7-position is involved. The reported cholesteric properties of 5α-cholest-2,3-en-3β-yl 4'-n-alkoxybenzoates [58] may mean that a 2,3-double bond encourages mesophase formation, but transition temperatures are not available. The same situation applies *vis à vis* the mesomorphic properties reported for 4α,14α-dimethyl-5α-cholest-9,11-en-3β-yl acetate since a comparison with the analogous 5,6-en compound is not possible. Point (ii) still stands in the light of the reported clearing temperatures for 5α-cholesta-7,8;9,11-dien-3β-yl benzoate [*134*] and cholesta-3,4;5,6-dien-3-yl acetate [*135*]; in the latter case the clearing temperature is only 10°–15° higher than that of cholesteryl acetate.

Finally we might consider the nature of the 17β-side chain. In the earlier survey [*6*], the following general observations were made mainly in relation to double bonds.

(i) As shown by the mesomorphic properties of 5α-cholesta-8,9;24,25-dien-3β-yl benzoate and cycloartenyl palmitate (LXVI) (cholesteric), a 24,25-double bond does not eliminate mesophases, but thermal stabilities are lowered—see also p. 65.

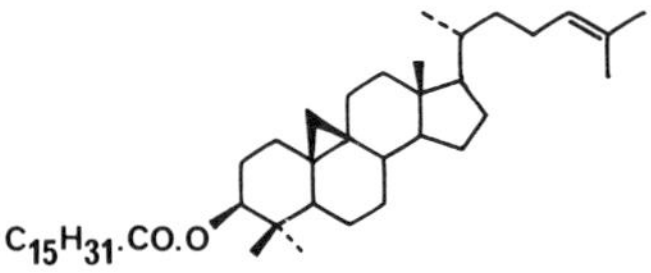

$C_{15}H_{31}.CO.O$

(LXVI)

(ii) 24-Methylenecycloartanyl esters and cycloeucalenyl esters are smectic or nonmesogenic, whereas the 24-methyl analogs are cholesteric [*131*].

(iii) Stigmasteryl esters which differ from cholesteryl esters in having a 22,23-double bond and a 24-ethyl are smectic, not cholesteric [*131*].

(iv) β-Sitosteryl esters which differ from cholesteryl esters in having a 24-ethyl give only monotropic smectic phases [*136, 137*].

(v) 24α-Methylcholesta-5,6;7,8;22,23-trien-3β-yl alkanoates are purely smectic [*138*].

In addition we can now note that cholesta-7,8;22,23-dien-3β-yl [*139*] and cholesta-5,6;24,25-dien-3β-yl [*140*] esters give no mesophases, whereas cholesteric properties have been observed with esters of cholesta-5,6;20,21-dien-3β-ol [*141*], cholesta-5,6;22,23-dien-3β-ol [*140*], and cholesta-5,6;25,26-dien-3β-ol [*141*]. The position of a double bond in the 17β-side chain therefore gives widely different results.

Elser's review [*13*] also draws attention to modifications in both nature and length of the 17β-substituent which can be tolerated without loss of cholesteric properties. The results are summarized below for selected 3β-acyloxy derivatives with the same cyclic system as cholesterol, but having the 17β-substituent changed from $CH_3CH(CH_2)_3CH(CH_3)_2$ to:

$CH_3CH \cdot C_nH_{2n+1}$	$n = 6$ and 7—mesophases of unspecified type
	$n = 0$ to 5 —cholesteric mesophases
$CH_3CH \cdot CO \cdot C_nH_{2n+1}$	$n = 2$ to 4 —cholesteric mesophases
	$n = 1$ —no mesophases
$CH_3CH(CH_2)_nCO_2CH_3$	$n = 0$ to 2 —cholesteric mesophases
$CH_3C{=}O$	cholesteric mesophase

No mesophases were observed, however, for esters of androst-5,6-en-3β-ol (LXVII)—17β-substituent = H, androst-5,6-en-3β-ol-17-one or 17β-carbomethoxyandrost-5,6-en-3β-ol.

(LXVII)

We can conclude that the 17β-substituent [$CH_3CH(CH_2)_3CH(CH_3)_2$] of cholesterol and related sterols is not a critical feature for preservation of cholesteric properties and its nature can be changed quite radically, cf. the 3β-substituent in the cholest-5,6-en-3β-yl system. Changes to the fused cyclic system of cholesterol have more pronounced effects; although a fused cyclopropane ring can be added to the sterol skeleton (LXVI) (see also Atallah and Nicholas [142]), the quasi-planar nature of rings A and B must not be affected or cholesteric properties are lost. The fused ring system is, of course, the core part of a steryl mesogen, and the 3β- and 17β-substituents are the terminal substituents. As noted in earlier sections, changes in terminal substituents often, though not always, give gentle, regular trends in mesomorphic properties, whereas alteration of core parts often leads to radical changes in behavior. In this general sense, steryl mesogens behave normally with change in structure, but the complex stereochemistry of the system makes any detailed interpretation of the effects very difficult.

REFERENCES

1. G. W. Gray and P. A. Winsor, in "Liquid Crystals and Plastic Crystals" (G. W. Gray and P. A. Winsor, eds.), Vol. 1, p. x. Ellis Horwood, Chichester, England, 1974.
2. P. A. Winsor, in "Liquid Crystals and Plastic Crystals" (G. W. Gray and P. A. Winsor, eds.), Vol. 1, Chapter 5. Ellis Horwood, Chichester, England, 1974.

3. G. W. Gray, "Molecular Structure and the Properties of Liquid Crystals." Academic Press, New York, 1962.
4. G. W. Gray, *Mol. Cryst. Liquid Cryst.* **1**, 333 (1966).
5. G. W. Gray, *Mol. Cryst. Liquid Cryst.* **7**, 127 (1969).
6. G. W. Gray, in "Liquid Crystals and Plastic Crystals" (G. W. Gray and P. A. Winsor, eds.), Vol. 1, Chapter 4.1. Ellis Horwood, Chichester, England, 1974.
7. M. Davies, *Chem. Brit.* **10**, No. 6, 223 (1974).
8. G. W. Gray, *Mol. Cryst. Liquid Cryst.* **21**, 161 (1973).
9. Eastman Kodak Co., Eastman Liquid Crystal Products, Kodak Publ. JJ-14. Rochester, New York (1975).
10. J. P. Van Meter, *Eastman Org. Chem. Bull.* **45**, No. 1, 1 (1973).
11. J. P. Schroeder, in "Liquid Crystals and Plastic Crystals" (G. W. Gray and P. A. Winsor, eds.), Vol. 1, Chapter 7.3. Ellis Horwood, Chichester, England, 1974.
12. G. W. Gray, in "Liquid Crystals and Plastic Crystals" (G. W. Gray and P. A. Winsor, eds.), Vol. 1, Chapter 7.1. Ellis Horwood, Chichester, England, 1974.
13. W. Elser, This volume, p. 73.
14. D. Demus, G. Kunicke, J. Neelson, and H. Sackmann, *Z. Naturforsch. A* **23**, 84 (1968.)
15. G. W. Gray, B. Jones, and F. Marson, *J. Chem. Soc.* 393 (1957).
16. G. W. Gray, *Proc. Int. Conf. Liquid Cryst., 5th, 1974*; *J. Phys.* (*Paris*) **36**, C1, 337 (1975).
17. D. Demus, M. Klapperstück, R. Rurainski, and D. Marzotko, *Z. Phys. Chem.* (*Leipzig*), **246**, 385 (1971).
18. D. Coates and G. W. Gray, *Proc. Int. Conf. Liquid Cryst., 5th, 1974*; *J. Phys.* (*Paris*) **36**, C1, 365 (1975).
19. J. Canceill, C. Gros, J. Billard, and J. Jacques, *Raman Int. Liquid Cryst. Conf., 1973*; *Pramana, Suppl.* **1**, 397 (1975).
20. F. Reinitzer, *Monatsh. Chem.* **9**, 421 (1888).
21. W. Kast, in "Landolt-Börnstein," 6th ed., Vol. II, Part 2a, p. 266 *et seq.* Springer-Verlag, Berlin and New York, 1960.
22. D. Demus, H. Demus, and H. Zaschke, "Flüssige-Kristallen in Tabellen." V.E.B. Deutscher Verlag für Grundstoffindustrie, Leipzig, 1974; see also D. Demus, *Z. Chem.* **15**, 1 (1975).
23. L. Verbit, private communication, 1974.
24. J. D. Bernal and D. Crowfoot, *Trans. Faraday Soc.* **29**, 1032 (1933).
25. A.-M. Levelut and M. Lambert, *C. R. Acad. Sci., Ser. B* **272**, 1018 (1971); J. Doucet, M. Lambert, A.-M. Levelut, L. Liebert, and L. Strzelecki, *Proc. Int. Conf. Liquid Cryst. 5th, 1974*; *J. Phys.* (*Paris*) **36**, C1, 13 (1975).
26. J. Doucet, A.-M. Levelut, and M. Lambert, *Mol. Cryst. Liquid Cryst.* **24**, 317 (1973).
27. H. Sackmann and D. Demus, *Fortschr. Chem. Forsch.* **12**, 349 (1969); *Mol. Cryst. Liquid Cryst.* **21**, 239 (1973); This Series. To be published.
28. A. De Vries, *Proc. Int. Conf. Liquid Cryst., 5th, 1974*; *J. Phys.* (*Paris*) **36**, C1, 1 (1975). *Mol. Cryst. Liquid Cryst.* **24**, 337 (1973).
29. W. Maier and A. Saupe, *Z. Naturforsch. A* **13**, 564 (1958); *A* **14**, 882 (1959); *A* **15**, 287 (1960).
30. G. W. Smith, *Int. Sci. Technol.* **61**, 72 (1967); *General Motors Corp., Res. Publ.* GMR-1545 (1974); This Series **1**, 189 (1975).
31. P. A. Winsor, in "Liquid Crystals and Plastic Crystals" (G. W. Gray and P. A. Winsor, eds.), Vol. 1, Chapter 2.2 and refs. cited therein. Ellis Horwood, Chichester, England, 1974.
32. G. W. Gray and P. A. Winsor, eds., in "Liquid Crystals and Plastic Crystals," Vol. 1, Chapter 1, p. 12. Ellis Horwood, Chichester, England, 1974; *Mol. Cryst. Liquid Cryst.* **26**, 305 (1974).

33. J. D. Brooks and G. H. Taylor, "Chemistry and Physics of Carbon" (P. L. Walker, Jr., ed.), Vol 4, p. 243. Dekker, New York, 1968; L. S. Singer, R. T. Lewis, A. T. Lauria, and S. L. Strong, *Proc. Int. Conf. Liquid Cryst., 5th, 1974*, to be published.
34. T. Clark, D. E. Johnston, H. Mackle, M. A. McKervey, and J. J. Rooney, *J. Chem. Soc. Chem. Commun.* 1042 (1972); T. Clark, H. Mackle, M. A. McKervey, and J. J. Rooney, *ibid.* 7 (1973).
35. H. Kelker and R. Hatz, *Ber. Phys. Chem.* **78**, No. 9, 819 (1974).
36. G. W. Gray, K. J. Harrison, and J. A. Nash, *Electron. Lett.* **9**, 130 (1973).
37. G. W. Gray, K. J. Harrison, J. A. Nash, J. Constant, D. S. Hulme, J. Kirton, and E. P. Raynes, *in* "Liquid Crystals and Ordered Fluids" (J. F. Johnson, and R. S. Porter, eds.), Vol. 2, p. 617. Plenum, New York, 1974.
38. P. Culling, G. W. Gray, and D. Lewis, *J. Chem. Soc.* 2699 (1960).
39. H. Schubert and H. Dehne, *Z. Chem.* **12**, 241 (1972).
40. G. W. Gray, K. J. Harrison, and J. A. Nash, *Raman Int. Liquid Cryst. Conf., 1973*; *Pramana, Suppl.* **1**, 381 (1975).
41. G. W. Gray, K. J. Harrison, and J. A. Nash, *J. Chem. Soc. Chem. Commun.* 431 (1974).
42. M. J. S. Dewar and R. S. Goldberg, *J. Org. Chem.* **35**, 2711 (1970).
43. J. van der Veen, W. H. de Jeu, A. H. Grobben, and J. Boven, *Mol. Cryst. Liquid Cryst.* **17**, 291 (1972); W. H. de Jeu and J. van der Veen, *Philips Res. Rep.* **27**, 172 (1972); J. van der Veen, W. H. de Jeu, M. W. M. Wanninkhof, and C. A. M. Tienhoven, *J. Phys. Chem.* **77**, 2153 (1973); W. H. de Jeu and Th. W. Lathouwers, *Z. Naturforsch. A* **29**, 905 (1974).
44. M. J. S. Dewar and R. S. Goldberg, *J. Amer. Chem. Soc.* **92**, 1582 (1970).
45. M. J. S. Dewar, A. C. Griffin, and R. M. Riddle, "Liquid Crystals and Ordered Fluids" (J. F. Johnson and R. S. Porter, eds.), Vol. 2, p. 733. Plenum, New York, 1974; M. J. S. Dewar and R. M. Riddle, *J. Amer. Chem. Soc.* **97**, 6658 (1975); M. J. S. Dewar and A. C. Grffin, *ibid.* **97**, 6662 (1975).
46. H. Schubert, R. Dehne, and V. Uhlig, *Z. Chem.* **12**, 219 (1972).
47. G. W. Gray and K. J. Harrison, *Mol. Cryst. Liquid Cryst.* **13**, 37 (1971).
48. G. W. Gray and K. J. Harrison, *Symp. Faraday Soc.* **5**, 54 (1971).
49. H. Schubert and H. Zaschke, *J. Prakt. Chem.* **312**, 494 (1970).
50. H. Schubert, *Wiss. Z. Univ. Halle*, XIX, '70 M, H.5, S.1.
51. J. A. Nash and G. W. Gray, *Mol. Cryst. Liquid Cryst.* **25**, 299 (1974).
52. R. A. Champa, *Mol. Cryst. Liquid Cryst.* **19**, App. 2, 246 (1973).
53. R. A. Champa, *Mol. Cryst. Liquid Cryst.* **16**, 175 (1972); **19**, 233 (1973); *in* "Liquid Crystals and Ordered Fluids" (J. F. Johnson and R. S. Porter, eds.), Vol. 2, p. 507. Plenum, New York, 1974.
54. C. S. Oh, *Mol. Cryst. Liquid Cryst.* **19**, 95 (1972).
55. G. W. Smith, *General Motors Corp. Res. Publ.* GMR-1354 (1973); G. W. Smith, Z. G. Gardlund, and R. J. Curtis, *Mol. Cryst. Liquid Cryst.* **19**, 327 (1973); G. W. Smith and Z. G. Gardlund, *J. Chem. Phys.* **59**, 3214 (1973).
56. W. R. Young, I. Haller, and L. Williams, *in* "Liquid Crystals and Ordered Fluids" (J. F. Johnson and R. S. Porter, eds.), p. 383. Plenum, New York, 1970.
57. J. S. Dave and R. A. Vora, *in* "Liquid Crystals and Plastic Crystals" (G. W. Gray and P. A. Winsor, eds.), Vol. 1, Chaptre 4.2. Ellis Horwood, Chichester, England, 1974.
57a. J. van der Veen, *J. Phys. (Paris)* **36**, C1, 375 (1975).
58. L. Verbit, private communication (1974) to W. Elser; see also R. A. Vora, Thesis, M.S. University of Baroda, Baroda, India, 1974.
59. J. E. Lydon and C. J. Coakley, *Proc. Int. Conf. Liquid Cryst., 5th, 1974*; *J. Phys (Paris)* **36**, C1, 45 (1975).

60. G. W. Gray and J. E. Lydon, *Nature (London)* **252**, 221 (1974).
61. W. H. de Jeu, J. van der Veen, and W. J. A. Goossens, *Solid State Commun.* **12**, 405 (1973).
62. H. Stenschke, *Solid State Commun.* **10**, 653 (1972).
63. J. I. Kaplan, *J. Chem. Phys.* **57**, 3015 (1972).
64. G. W. Gray, J. B. Hartley, A. Ibbotson, and F. Marson, *J. Chem. Soc.* 4359 (1955).
65. H. Kelker and B. Scheurle, *J. Phys. (Paris)* **30**, C4, No. 11–12, 104 (1969).
66. J. A. Castellano, J. E. Goldmacher, L. A. Barton, and J. S. Kane, *J. Org. Chem.* **33**, 3501 (1968).
67. G. W. Gray and A. Mosley, *J. Chem. Soc., Perkin Trans. II*, 97 (1976).
68. J. L. W. Pohlmann and W. Elser, *Mol. Cryst. Liquid Cryst.* **8**, 427 (1969).
69. W. Elser and R. D. Ennulat, *J. Phys. Chem.* **74**, 1545 (1970).
70. D. Coates, K. J. Harrison, and G. W. Gray, *Mol. Cryst. Liquid Cryst.* **22**, 99 (1973).
71. W. Elser, J. L. W. Pohlmann, and P. R. Boyd, *Mol. Cryst. Liquid Cryst.* **20**, 77 (1973).
72. D. Coates and G. W. Gray, *Phys. Lett. A* **45**, No. 2, 115 (1973).
73. H. J. Dietrich and E. L. Steiger, *Mol. Cryst. Liquid Cryst.* **16**, 263 (1972).
74. R. D. Ennulat, *Mol. Cryst. Liquid Cryst.* **8**, 247 (1969).
75. W. Elser, J. L. W. Pohlmann, and P. R. Boyd, *Mol. Cryst. Liquid Cryst.* **15**, 175 (1971).
76. J. L. W. Pohlmann, W. Elser, and P. R. Boyd, *Mol. Cryst. Liquid Cryst.* **26**, 59 (1974).
77. W. Elser, J. L. W. Pohlmann, and P. R. Boyd, *Mol. Cryst. Liquid Cryst.* **27**, 325 (1974).
78. G. W. Gray and B. Jones, *J. Chem. Soc.* 683 (1954); 236 (1955).
79. I. Teucher, C. M. Paleos, and M. M. Labes, *Mol. Cryst. Liquid Cryst.* **11**, 187 (1970); H. Hirata, S. N. Waxman, I. Teucher, and M. M. Labes, *ibid.* **20**, 343 (1973).
80. H. B. Bürgi and J. D. Dunitz, *Helv. Chim. Acta* **53**, 1747 (1970).
81. J. van der Veen and A. H. Grobben, *Mol. Cryst. Liquid Cryst.* **15**, 239 (1971).
82. J. van der Veen and T. C. J. M. Hegge, *Angew. Chem.* **86**, 378 (1974).
83. Z. G. Gardlund, R. J. Curtis, and G. W. Smith, *in* "Liquid Crystals and Ordered Fluids" (J. F. Johnson and R. S. Porter, eds.), Vol. 2, p. 541. Plenum, New York, 1974.
84. S. L. Arora, J. L. Fergason, and T. R. Taylor, *J. Org. Chem.* **35**, 4055 (1970).
85. J. P. Van Meter and B. H. Klanderman, *Mol. Cryst. Liquid Cryst.* **22**, 285 (1973).
86. D. J. Byron, G. W. Gray, and B. M. Worrall, *J. Chem. Soc.* 3706 (1965).
87. W. R. Young, I. Haller, and D. C. Green, *IBM Res. Rep.* RC 3827 (1972); *J. Org. Chem.* **37**, 3707 (1972); W. R. Young and D. C. Green, *IBM Res. Rep.* RC 4121 (1972); *Mol Cryst. Liquid Cryst.* **26**, 7 (1974).
88. T. Gunjima, Y. Nakagawa, and Y. Masuda, *Proc. Int. Conf. Liquid Cryst., 5th, 1974*, to be published.
89. A. Biering, D. Demus, G. W. Gray, and H. Sackmann, *Mol. Cryst. Liquid Cryst.* **28**, 275 (1974).
90. W. R. Young, A. Aviram, and R. J. Cox, *Angew. Chem.* **83**, 410 (1971); *IBM Res. Rep.* RC 3559 (1971).
91. J. M. Robertson and I. Woodward, *Proc. Roy. Soc. Ser. A* **162**, 568 (1937).
92. J. van der Veen and W. H. de Jeu, *Mol. Cryst. Liquid Cryst.* **27**, 251 (1974); J. van der Veen and T. C. J. M. Hegge, *Proc. Int. Conf. Liquid Cryst., 5th, 1974*, to be published.
93. M. E. Neubert, L. T. Carlino, R. D'Sidocky, and D. L. Fishel, *in* "Liquid Crystals and Ordered Fluids" (J. F. Johnson and R. S. Porter, eds.), Vol. 2, p. 293. Plenum, New York, 1974.
94. Y. Y. Hsu and D. Dolphin, *in* "Liquid Crystals and Ordered Fluids" (J. F. Johnson and R. S. Porter, eds.), Vol. 2, p. 461. Plenum, New York, 1974.
95. J. E. Goldmacher, German Patent, 2,026,280 (1970).
96. G. N. Taylor and F. J. Kahn, *J. Appl. Phys.* **45**, 4330 (1974).

97. S. L. Arora, J. L. Fergason, and T. R. Taylor, *Proc. Int. Conf. Liquid Cryst.*, *5th*, *1974*, to be published.
98. L. Verbit and R. L. Tuggey, in "Liquid Crystals and Ordered Fluids" (J. F. Johnson and R. S. Porter, eds.), Vol. 2, p. 307. Plenum, New York, 1974.
99. S. Malthète, M. Leclercq, J. Gabart, J. Billard, and J. Jacques, *C. R. Acad. Sci. Ser. C* **273**, 265 (1971).
100. L. E. Knaak, H. M. Rosenberg, and M. P. Servé, *Mol. Cryst. Liquid Cryst.* **17**, 171 (1972).
101. J. M. Robertson and I. Woodward, *Proc. Roy. Soc. Ser. A* **164**, 436 (1938).
102. G. Baur, A. Stieb, and G. Meier, in "Liquid Crystals and Ordered Fluids" (J. F. Johnson and R. S. Porter, eds.), Vol. 2, p. 645. Plenum, New York, 1974.
103. H. K. Bucher, R. T. Klingbiel, and J. P. Van Meter, *Appl. Phys. Lett.* **25**, No. 4, 186 (1974).
104. R. E. Spillett, private communication (1974).
105. J. P. Van Meter and B. H. Klanderman, *Mol. Cryst. Liquid Cryst.* **22**, 271 (1973); J. P. Van Meter and A. K. Seidel, *J. Org. Chem.* **40**, 2998 (1975).
106. A. Boller, H. Scherrer, M. Schadt, and P. Wild, *Proc. IEEE* **60**, 1002 (1972).
107. J. C. Dubois, A. Zann, and A. Couttet, *Mol. Cryst. Liquid Cryst.* **27**, 187 (1974).
108. D. S. Hulme, E. P. Raynes, and K. J. Harrison, *J. Chem. Soc. Chem. Commun.* 98 (1974).
109. J. Billard, *Bull. Soc. Fr. Minéral Cristallogr.* **95**, 206 (1972); M. Domon, Thèse, Thome I and II, Université des Sciences et des Techniques de Lille, Lille, France, 1973; see also E. C.-H. Hsu, J. L. Haberfield, J. F. Johnson, and E. M. Barrall II, *Mol. Cryst. Liquid Cryst.* **27**, 269 (1974) and E. C.-H. Hsu and J. F. Johnson, *ibid.* **27**, 95 (1974).
110. E. C.-H. Hsu and J. F. Johnson, *Mol. Cryst. Liquid Cryst.* **27**, 95 (1974).
111. Hofmann-La Roche, Basle, Switzerland, Information Leaflet (1974).
112. V. V. Titov, E. I. Kovshev, A. I. Pavluchenko, V. T. Lazareva, and M. F. Grebenkin, *Proc. Int. Conf. Liquid Cryst.*, *5th*, *1974*; *J. Phys. (Paris)*, **36**, C1, 387 (1975).
113. "Licristal (Liquid Crystals)", E. Merck, Darmstadt, Germany, current information leaflets.
114. R. Steinsträsser, *Z. Naturforsch. B* **27**, 529 (1972).
115. J. P. Van Meter and A. K. Seidel, *Proc. Int. Conf. Liquid Cryst.*, *5th*, *1974*, to be published.
116. J. P. Van Meter, R. T. Klingbiel, and D. J. Genova, *Proc. Int. Conf. Liquid Cryst.*, *5th*, *1974*; *Solid State Commun.* **16**, 315 (1975).
116a. D. Coates and G. W. Gray, *Mol. Cryst. Liquid Cryst.* **31**, 275 (1975).
117. J. Billard, J. C. Dubois, and A. Zann, *Proc. Int. Conf. Liquid Cryst.*, *5th*, *1974*; *J. Phys. (Paris)* **36**, C1, 355 (1975).
118. J. A. Castellano, C. S. Oh, and M. T. McCaffrey, *Mol. Cryst. Liquid Cryst.* **27**, 417 (1974).
119. H. W. Gibson, *Mol. Cryst. Liquid Cryst.* **27**, 43 (1974).
120. G. W. Gray, K. J. Harrison, and J. A. Nash, *Electron. Lett.* **9**, 616 (1973); G. W. Gray and D. G. McDonnell, *ibid.* **11**, 556 (1975).
121. D. L. White and A. N. Taylor, *J. Appl. Phys.* **45**, 4718 (1974).
122. J. W. Goodby and G. W. Gray, *J. Phys. (Paris)*, in press (1976).
123. S. Diele, *Phys. Status Solidi A* **25**, K183 (1974).
124. M. Leclercq, J. Billard, and J. Jacques, *Mol. Cryst. Liquid Cryst.* **8**, 367 (1969).
125. D. Chapman, in "Liquid Crystals and Plastic Crystals" (G. W. Gray and P. A. Winsor, eds.), Vol. 1, Chapter 6.1. Ellis Horwood, Chichester, England, 1974.
126. W. Elser, J. L. W. Pohlmann, and P. R. Boyd, *Mol. Cryst. Liquid Cryst.* **20**, 87 (1973).

127. W. Elser, *Mol. Cryst. Liquid Cryst.* **8**, 219 (1969).
128. W. Elser, J. L. W. Pohlmann, and P. R. Boyd, *Mol. Cryst. Liquid Cryst.* **11**, 279 (1970).
129. W. Elser, R. D. Ennulat, and J. L. W. Pohlmann, *Mol. Cryst. Liquid Cryst.* **27**, 375 (1974).
130. Ch. Wiegand, *Z. Naturforsch. B* **4**, 249 (1949).
131. F. F. Knapp and H. Nicholas, *J. Org. Chem.* **34**, 3328 (1969).
132. F. F. Knapp and H. Nicholas, *J. Org. Chem.* **33**, 3995 (1968).
133. O. Wintersteiner and M. Moore, *J. Amer. Chem. Soc.* **72**, 1923 (1950).
134. C. F. Hammer and R. Stevenson, *Steroids* **5**, 637 (1965).
135. H. Reich and A. Lardon, *Helv. Chim. Acta* **29**, 671 (1946).
136. J. L. W. Pohlmann, *Mol. Cryst. Liquid Cryst.* **8**, 417 (1969).
137. A. Kuksis and J. M. R. Beveridge, *J. Org. Chem.* **25**, 1209 (1960).
138. F. F. Knapp and H. Nicholas, *Mol. Cryst. Liquid Cryst.* **10**, 173 (1970).
139. K. Sakai and K. Tsuda, *Chem. Pharm. Bull.* **11**, 529 (1963).
140. W. Bergmann and J. P. Dusza, *J. Org. Chem.* **23**, 1245 (1958).
141. W. Elser, J. L. W. Pohlmann, and P. R. Boyd, *Mol. Cryst. Liquid Cryst.* **13**, 255 (1971).
142. A. M. Atallah and H. J. Nicholas, *Mol. Cryst. Liquid Cryst.* **24**, 213 (1973).

SELECTIVE REFLECTION OF CHOLESTERIC LIQUID CRYSTALS

W. Elser and R. D. Ennulat

U. S. Army Electronics Command
Night Vision Laboratory
Fort Belvoir, Virginia

    I. Introduction 73
   II. Physics of Cholesteric Mesophases 74
       A. The Cholesteric Mesophase and Its Textures 74
       B. Selective Reflection 80
       C. Theories of Cholesteric Mesophases 93
 III. Chemistry of Cholesteric Liquid Crystals 97
       A. Various 3β-Substituted Cholest-5-enes and 5α-Cholestanes . . 99
       B. Constitution and Mesomorphic Behavior 110
       C. Nonsteroidal Cholesteric Liquid Crystals 121
       D. Stability 123
       E. Summary and Conclusion 124
 IV. Temperature Dependence of Selective Reflection 125
       A. Single Compounds 125
       B. Mixed Liquid Crystals 132
  V. Application 137
       A. Methods of Temperature Sensing 137
       B. Liquid Crystal Devices 145
       C. Nondestructive Testing 152
       D. Medical Application 154
       References 161

I. INTRODUCTION

Since the celebrated review by Friedel [1] in 1922, three major overviews have been published, which formed the basis of research on thermotropic liquid crystals for much of the last decade—the review paper by Brown and Shaw in 1957 [2] and the books by Gray in 1962 [3] and by Gray and Winsor in 1974 [3a]. The subsequent increase of research activity culminated in the

"International Liquid Crystal Conferences" of 1965 [4], 1968 [5], 1970 [6], and 1972 [7] and also in the conferences on "Ordered Fluids and Liquid Crystals" in 1965 [8], 1970 [9], and 1974 [10]. Recently de Gennes [11] reviewed in detail most of the theories on liquid crystals. In addition a number of survey papers were published, which were aimed at various groups of readers ranging from scientists interested in general information [12] to engineers primarily concerned with applications [13].

It is our assignment to address those aspects of the physics and chemistry of cholesteric liquid crystals that are relevant with respect to temperature dependence and temperature sensitivity of the selective reflection of visible light. In addition we will discuss some of the significant applications of temperature-sensitive cholesteric liquid crystals in engineering and medicine.*

II. PHYSICS OF CHOLESTERIC MESOPHASES

The brilliant color play, now known as "selective reflection" or "dispersive reflection," is one of the most significant characteristics of cholesteric mesophases. Actually, the observation of this unusual interference effect of visible light in a liquidlike substance may have been primarily responsible for the discovery of the mesomorphic state and for the coining of the more descriptive term "liquid crystal" by Lehmann [14]. But, more importantly, this distinct phenomenon, together with associated optical effects, reveals the basis of molecular arrangement in cholesteric mesophases.

The absence of the color band, however, does not necessarily preclude the existence of a cholesteric mesophase. For example, the selective reflection may not be observed because it occurs in the infrared or because the sample may not be in the state of alignment that exhibits selective reflection. Therefore, features other than selective reflection have to be considered to identify properly cholesteric mesophases in new materials and to induce that manifestation or texture of this phase which exhibits selective reflection.

A. The Cholesteric Mesophase and Its Textures

1. Mesophases and Molecular Arrangement

Mesomorphic states or mesophases are characterized according to the structure of the molecular arrangement. Consequently, definite phase transitions are observed whenever a substance changes to the mesomorphic state or from one mesophase to the other. This property allows the detection of mesophases by differential scanning calorimetry [15]. But the determination

* The abstract literature has been surveyed until the end of 1974.

of the structure of a mesophase is a more difficult matter. For example, X-ray diffraction patterns are hard to interpret because these structures have a much lower symmetry than three-dimensional translation lattices. Furthermore, the samples cannot always be maintained in sufficiently uniform alignment along a preferred direction because of the liquid nature of the substance and the complex interaction between the substance and its boundaries. The most practical approach to phase identification is the microscopic analysis of those optical properties which are unambiguously related to the structure of the molecular arrangement. These characteristic optical properties are usually observed only in certain textures.

2. Types of Mesophases

So far only two classes of thermotropic mesophases can be definitely distinguished by characteristic textures [*11*]: the smectic mesophases in which the molecules are arranged in layers and the nematic mesophases in which the molecules are in parallel alignment. The former is observed in three or possibly seven distinct variations, while the latter is either a conventional nematic phase (from hereon called nematic mesophase) or a twisted nematic phase (from hereon called cholesteric mesophase), in which the nematic structure is twisted with a handedness characteristic of the particular compound. This twist commonly occurs if the molecules are chiral (i.e., helical molecules and molecules having asymmetry centers). In mixtures at least one component has to be chiral; it can even be the component that does not by itself form a nematic mesophase. The complexity of the molecular interaction responsible for the creation of the twisted structure is implied by the fact that the handedness of the molecular optical activity or helix alone does not determine the handedness of the helical structure [*16*].

3. Relation between Mesophases and Other Phases [*11*]

A transition between cholesteric and nematic mesophases has not been accomplished yet by simply changing the temperature of a pure compound. But external fields are capable of causing such transitions. Furthermore, mixtures of *levo-* and *dextro*-cholesteric phases exhibit nematic behavior at the temperature at which the mixture is racemic. However, above and below this temperature the mixture exhibits cholesteric behavior of opposite handedness. These observations indicate that the transition between cholesteric and nematic phases may be induced only by agencies of the proper asymmetry and not by an isotropic change in temperature.

The temperature interval of the cholesteric mesophase is limited at the high temperature by a transition to the isotropic liquid and at the low temperature by a transition to either the solid state or the smectic mesophase. The twist of the helix decreases rapidly as the temperature approaches the point of transition to the smectic state. Apparently the structure untwists before it undergoes this phase transition. It is in this pretransitional region where the cholesteric mesophase exhibits selective reflection of extremely high temperature dependence.

4. TEXTURES OF CHOLESTERIC MESOPHASES

Thin films of cholesteric mesophases exhibit unexplored textures, the focal conic texture, and the planar texture.

a. *Unexplored Textures.* These cannot be properly defined because their features are ambiguous. Most is known about the texture described by Gray [3] as a "homeotropic" condition and by others as the "blue phase" [17]. Apparently the term homeotropic condition relates to the structureless grayish appearance under the microscope, while the term blue phase relates to the dull purple appearance of the same texture observed under certain conditions. For example, the cover slip on top of a thin cholesteric film apparently collects the obliquely scattered radiation and guides it to the rim where it exits as a bluish circle. We observed that appearance and disappearance of this bluish ring coincides with the clearing point. For many compounds this effect is the best optical indicator of this transition to the isotropic phase. Below the clearing point the focal conic texture may either grow directly into the blue phase* or may be formed by the coalescing birefringent bâtonnets [1, 3]. Furthermore, displacing the cover slide can convert this blue texture directly into the planar texture. The results of scattering experiments conducted on the blue textures of cholesteryl tetradecanoate indicate scattering centers ranging in size from 500 to 800 Å [17]. However, more evidence is needed to determine whether this blue texture represents an unstable, pretransitional region just below the clearing point or whether it is a stable texture with a definable static structure. Most other effects occurring in the temperature region of the blue texture are very inconsistent. For example, we observed a second color band of reflected light just below the clearing point [18], which, depending on the compound, changes colors sluggishly with temperature or not at all. In some compounds the color regions look like ink blotches, in others those regions are uniform and shaped like angular platelets. These domains are usually very sensitive to mechanical disturbances but very insensitive to temperature changes. These effects are the least understood appearances of the cholesteric liquid crystal.

b. *Focal Conic Textures.* These are well-established structures [1, 3, 11] which are distinguished by their birefringence. Although the smectic and cholesteric focal conics incorporate the different molecular arrangements

---

* We think that the term "blue phase" is misleading because these texture changes are not reversible with temperature and because differential scanning calorimetry did not reveal any thermodynamic phase transition within the cholesteric mesophase for over 200 compounds investigated by our group. Therefore, we prefer the name blue texture.

characteristic of the respective mesophases, the appearance of both types of textures has much in common. They can be distinguished under the polarization microscope only because of the difference of the position of the optical axis [1, 15]—that is, precisely by that property which is directly associated with the molecular arrangement. The application of a shear stress or of certain electrical fields [11] can convert the focal conic texture into the planar texture whenever surface properties or other obstacles do not prevent this orientation. It will be shown later that the focal conic texture is essentially a planar texture whose helical axis is predominantly parallel to the surface.

c. *Planar Texture.* This exhibits the building principle of cholesteric mesophases in the purest form. It consists of molecules aligned parallel to each other and to the substrate plane. Because of the chirality of the molecules the direction of the preferred molecular alignment rotates about the surface normal as one progresses from the bottom to the top of the sample layer. Thus, the preferred direction of the molecules outlines a helix of a handedness which is characteristic for a given compound. Notice that this helical structure does not have physically defined layers.

The optical properties of the planar texture can be summarized as follows:

1. It exhibits selective reflection of light.

2. Its optical activity is orders of magnitude larger than that of the constituent molecules.

3. The optical activity for wavelengths above and below the selective reflection band has opposite signs.

4. *Levo* structures reflect left-handed circularly polarized light without reversal of handedness and transmit only right-handed circularly polarized light. *Dextro* structures exhibit the reverse effect.

Theoretical and experimental studies have proved that these unusual optical effects have one common cause: the helical structure of the planar texture.

Although certain unexplored textures may still exist, the most stable and prevailing textures of the cholesteric mesophase have the same helical structure. It is therefore possible to clearly identify cholesteric mesophases even if selective reflection is not apparent.

5. Helix-Forming Tendency of Molecules

Although mesomorphic compounds differ widely in chemical constitution, they have certain features in common: the molecules have to be elongated or rod shaped to be suitable for structures of predominantly orientational order; the molecules have to interact weakly enough to avoid crystallization and yet strongly enough to establish the order characteristic of the mesomorphic state. Both the lack of knowledge about the details of the molecular arrangement and the complexity of the interaction forces made it impossible to derive the mesomorphic behavior of a compound from first principles.

For smectic and nematic mesophases Gray [3] succeeded in correlating changes of molecular features, introduced by systematic, chemical substitutions, with variations of the mesomorphic properties. He found that the amount of lateral attraction relative to that of the terminal attraction between neighboring molecules is significant. If the former predominates, the molecules prefer layer structures and thus the smectic mesophase. If the latter predominates, the molecules most likely form nematic phases. However, a certain amount of lateral attraction is still required to enforce short-range molecular parallelism in an environment of disordering thermal agitation. For specific classes of compounds Gray was able to increase or decrease transition temperatures in the melt by the proper choice of substituents.

Cholesteric mesophases are generally influenced by molecular modifications affecting nematic mesophases. But it is not possible to achieve useful correlations between molecular properties and cholesteric behavior without also considering changes of the helix-forming tendency of the molecules. Two empirical concepts have been developed to assess this important feature: the helical twisting power [19] and the effective rotary power of molecules [20].

The helical twisting power (HTP) of a solute [19] is derived from the temperature shift ΔT of the nematic temperature T_n of a compensated mixture of cholesteric liquid crystals (that is, a mixture of left- and right-handed cholesteric liquid crystals which is racemic at T_n) caused by adding a solute of concentration c.

The extrapolation

$$1/Z \equiv m \, \Delta T_n(c)\big|_{c=1} \qquad (1)$$

defines the HTP of the solute (m is a constant characteristic of the solvent system). This limit process is meaningful if the relation between ΔT_n and c is linear. Under this condition HTP is equal to the inverse of the pitch Z of the pure solute at the temperature $T_n + \Delta T_n$.

The effective rotary power [20] is derived from the pitch p of the mixture

$$1/p \equiv \Sigma \alpha_i \theta_i$$

where α_i and θ_i are weight percentage and effective rotary power of component i. Depending on the intent of the investigation, one usually selects a left- or a right-handed cholesteric reference material (a compound or a mixture [21]) of desirable pitch and temperature and measures the change of the pitch caused by the addition of the test material as a function of concentration. If crystallization limits the range of concentration or if the additive is not cholesteric itself, an effective rotary power is defined by linear extrapolation of the inverse pitch to the 100% point of the additive. In the case of linear concentration dependence, both the helical twisting power and the effective rotary power are approximately equal for cholesteric compounds [22].

The linear superposition principle of the effective rotary power expressed in the last equation is approximately valid for mixtures of similar cholesteric constituents which do not change their molecular features due to mutual interaction. Using Goossens' theory (see Section II,C,2) it can be shown [23, 24] that the dissimilarity of the constituents alone is sufficient to account for the nonlinear concentration dependence of the effective rotary power and its relative minimum. But the cholesteric matrix can also modify the helical twisting power of a solute. For example, it was observed that cholesteryl iodide [21] and cholesteryl 2-(2'-butoxyethoxy)ethyl carbonate [25] can adopt the handedness of the cholesteric solvent helix. Since these compounds have a very small effective rotary power, the induction effects must be rather small.

Addition of nematogenic compounds to a cholesteric material generally increases the pitch at a given temperature with concentration [24]. For small concentrations of the additive, the twisting ability of the cholesteric constituent does not appear to be influenced by induction effects [26].

The helical twisting power of certain solute molecules depends on the molecular configuration. For example, the branched-chain cholesteryl 2-methylpentanoate (monotropic smectic) exhibits in a compensated cholesteric material different helical twisting powers for the *l*- and *d*-enantiomers [27]. A similar effect was observed on azobenzene dissolved in a cholesteric matrix. Photo-induced cis-trans and trans-cis isomerization resulted in a corresponding variation of the pitch [28, 29]. This effect can be used to modulate the color of selectively reflected light.

Several investigators [30–33] have found that the cholesteric matrix can impose a helical arrangement on achiral solute molecules. As a consequence, induced circular dichroism occurs in achiral compounds. However, no correlation was found between this circular dichroism and the solute and the pitch of the solvent [31].

Friedel [1] found that the addition of nonmesomorphic compounds with asymmetric carbon atoms to a nematic compound can result in a cholesteric mixture. Recent investigations showed that the same effect can be achieved by dissolving chiralic compounds which do not have asymmetric carbon atoms [34]. The handedness of the resulting cholesteric mesophase is determined by the chirality of the solute.

These examples demonstrate the usefulness of the empirical measures describing the molecular ability to form helical superstructures. But these results may also have significant theoretical implications because according to Goossens' theory the twist between the directors of adjacent molecular layers is proportional to the ratio of the asymmetric (or chiral) and the anisotropic contributions to the molecular interaction.

B. Selective Reflection

The intensity of selectively reflected, monochromatic light depends on the angles formed by the incident and observed light beams with the helical axis, on pressure and temperature, on purity and composition, and on alignment of the sample.

1. Experimental Considerations

The extrinsic variables are restricted primarily by experimental consideration. For example, the convergence of the incident light, the area of illumination, and the acceptance angle of the observed light have to be made small enough to ease the interpretation of the experimental results (e.g., angular resolution) and yet large enough to measure the light intensity at an acceptable signal-to-noise ratio. The experiments indicate that a useful compromise can be achieved. As expected from thermodynamic considerations [35], high pressure increases the temperature of the transition from the cholesteric to the smectic phase A and thus the temperature of the selective reflection. The latter was observed on mixtures of cholesteric compounds [35a]. For materials exhibiting a strong temperature dependence of selective reflection, special precautions are necessary to achieve a sufficiently fine control of temperature and temperature uniformity.

More serious are the effects of the intrinsic parameters such as chemical stability, impurity content of the sample, and the alignment of the planar texture. We observed [36] that an unidentified impurity, apparently formed by oxidation during the test, shifted the color band of cholesteryl nonanoate at a rate of about 0.1° per day to lower temperatures. (See also Kahn [37].) Furthermore, the intensity peak obtained for a given wavelength occurred at slightly different temperatures for heating and cooling. Removal of the impurity eliminated this hysteresis effect and moved the color band back to its original temperature range. We believe that the impurity depressed the phase transition, thus shifting the pretransitional color band and increasing the tendency toward undercooling. It should also be considered that dilute gases in contact with the sample may shift the color band significantly in a reversible or irreversible manner [38] or even suppress the color band permanently [39]. Furthermore, the composition of mixtures exhibiting cholesteric mesophases has a significant influence on temperature level and interval of the visible selective reflection.

The sample should consist of a uniformly aligned planar texture characterized by one helical axis normal to the surface. In addition, the sample should be thick enough to exhibit a selective reflectivity close to unity. In spite of careful sample preparation, one usually obtains a mosaic of uniform planar texture domains ranging in lateral dimensions from 10 to 50 μm. In poorly aligned samples the individual domains exhibit different colors [40]

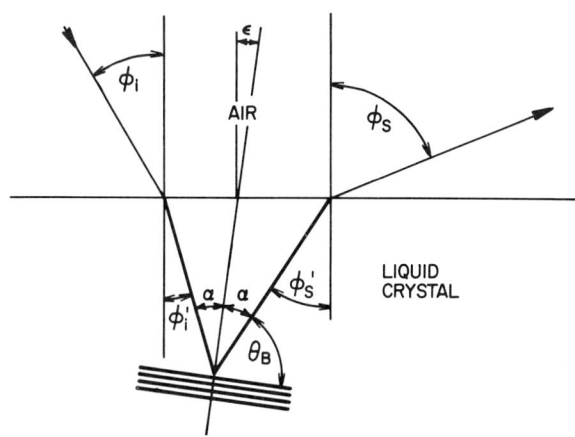

Fig. 1. Geometry for selective reflection in planar texture. Stack of parallel lines indicates planes of Bragg scattering sites. [Modified from Adams et al., J. Chem. Phys. **50**, 2458 (1969).]

because of the angular dependence of the selective reflection. However, even extremely temperature-sensitive samples can be aligned to such a degree that for normal illumination all individual domains selectively reflect monochromatic light within a narrow cone normal to the surface [36]. Under this condition we can estimate the range of the angular deviation between the helical axes and the common surface normal. Considering that single domains selectively reflect at specular angles [41], we obtain from Fig. 1 the relation

$$\varepsilon = \frac{1}{2}\left\{\sin^{-1}\left(\frac{\sin\phi_s}{n}\right) - \sin^{-1}\left(\frac{\sin\phi_i}{n}\right)\right\} \quad (2)$$

where ϕ_i and ϕ_s are the angles of incidence and observation and n is the average index of refraction of the liquid crystal. Obviously, the largest deviation ε occurs if the light ray incident at the largest possible angle travels in the domain parallel to the helical axis, that is, if

$$\phi_{s_{max}} = -\phi_{i_{max}}$$

Using the data reported in our work [36] (i.e., $\phi_{i_{max}} = 7°$, $n \simeq 1.5$), we conclude that the helical axes of the domains in our sample were aligned at least to within 4.7° of the surface normal. Adams, Haas, and Wysocki [42] conclude from their measurement of the angular dependence of the selectively reflected light intensity that the helical axes are close to the normal of the surface. As we will see, this small misalignment already washes out those features of the spectral selective reflectance which are characteristic of single domains.

A few investigators have succeeded in preparing large single-domain samples suitable for the experimental test of the theory of selective reflection. They avoided many problems by using materials exhibiting an extremely low temperature dependence of the pitch. Berreman and Scheffer [43] prepared their samples by mixing a nematic and an optically active material, while Dreher *et al.* [44] used room temperature mixtures of cholesteryl derivatives. The latter group of investigators sandwiched a 21-μm thick liquid crystal between two parallel glass plates. Since parallelism of the surface contacting the liquid crystals resulted in the best alignment, they used thick plates to avoid warping and polished the glass surfaces to a flatness of a quarter wavelength over an area of several square centimeters. Apparently, uniform sample thickness diminished the occurrence of pitch distortions observed in Cano wedges [45]. The best samples were obtained by cleaning the surfaces with an organic solvent (and not with chromic acid) and by rubbing them in one direction. As pointed out by Chatelain for nematic phases [46], this procedure may

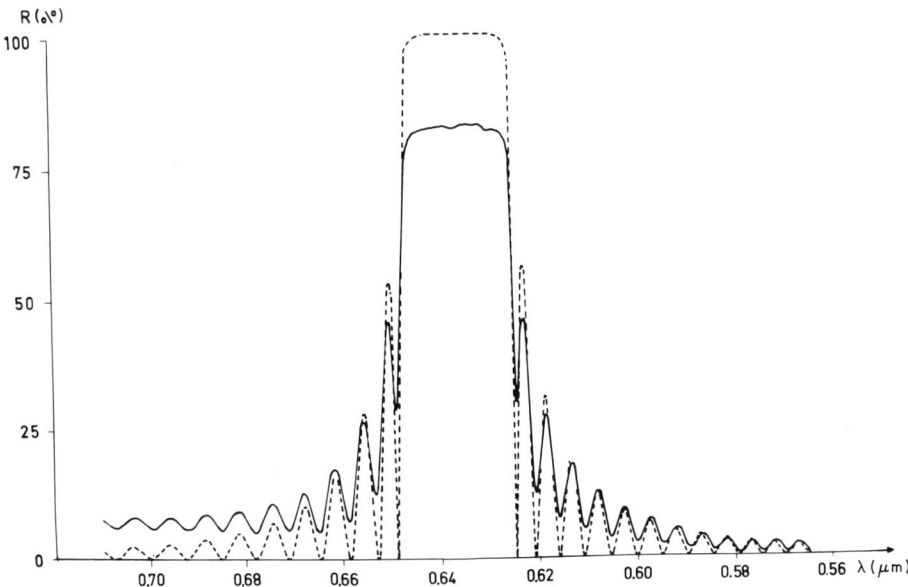

Fig. 2. Reflection spectrum of a mixture of cholesteryl acetate, chloride, and nonanoate (weight ratio 6:15:20) at 24° at normal incidence. Dashed curve: computed spectrum with pitch $p = 0.4273$ μm and thickness of layer of 21.0 μm. Solid curve: experimental spectrum. Intensity in arbitrary units. [Reproduced from Dreher *et al.*, *Mol. Cryst. Liquid Cryst.* **13**, 17 (1971), with permission.]

produce a thin film of fatty acid impurities, in which the molecules are parallel to the substrate in the direction of the rubbing. It is interesting to note that the aligning action of this dirt film was affected neither by moderate heating nor by slight wiping with a cloth wetted by a solvent. The liquid crystal was heated above the clearing point, applied to one of the glass substrates, and subsequently covered by the other one. (It was found that insertion of the liquid crystal by capillary action into the gap between the glass plates did not render good samples.) The residual focal conic texture was eliminated by mechanical shearing of the liquid crystal. This procedure did not affect focal conic bands (i.e., oily streaks) caused by dust particles or by surface imperfections of the substrates. But these effects were avoided by careful preparation techniques. Reasonable agreement between experiment and theory was achieved only with samples which had been stored for at least one day. Both the long equilibration time implied by the latter and the strong influence of the preparation procedure on the sample quality indicate that the preparation of single-domain samples is still an art rather than a science.

2. Typical Experimental Results

Figure 2 shows the spectral selective reflectance obtained with a single-domain sample. Notice the good agreement with the theoretical curve with respect to the nearly rectangular shape of the main maximum and the positions of the secondary maxima and minima. As the comparison with Fig. 3

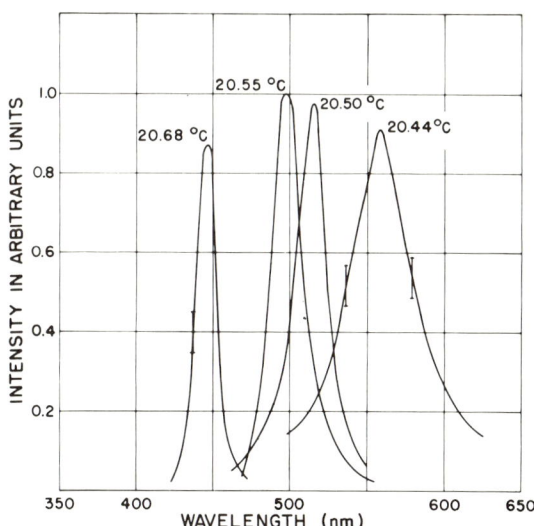

Fig. 3. Intensity of selectively reflected light as a function of wavelength. Material: cholesteryl oleyl carbonate. [Reproduced from Ennulat, *Mol. Cryst. Liquid Cryst.* **13**, 337 (1971), with permission.]

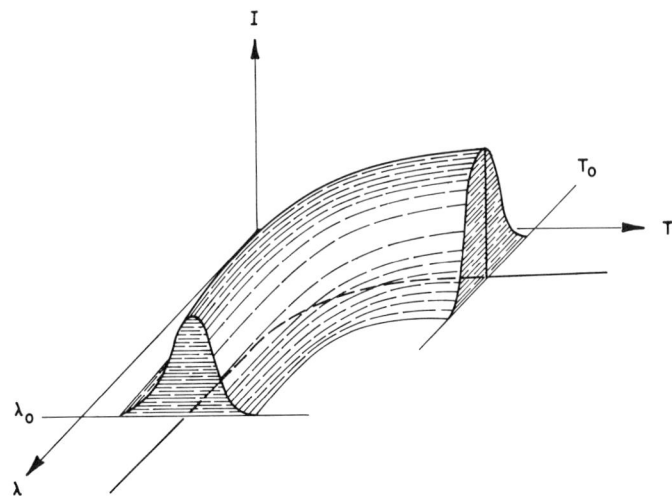

Fig. 4. Intensity of light selectively reflected by the planar texture as a function of wavelength and temperature. [Reproduced from Ennulat, *Mol. Cryst. Liquid Cryst.* **13**, 337 (1971), with permission.]

shows, polydomain samples do not exhibit any of these characteristic features. Apparently the misalignment of the domains by only a few degrees is sufficient to wash out the side maxima and to distort the main reflection band into a bell-shaped curve. Since the sample quality could not be assessed by microscopic inspection, Dreher [41] considered the square shape of the main maximum and the existence of the side maxima as the essential criteria of a well-aligned sample.

To date, the temperature dependence of the selective reflection has been determined only for polydomain samples. Figure 4 depicts the schematic dependence of reflected intensity versus temperature and wavelength for light traveling approximately parallel to the sample normal. Notice that the intensity dependence on temperature at constant wavelength also forms a bell-shaped curve and that the projection of the intensity maximum onto the wavelength–temperature plane results in a hyperbolic-like curve. Figures 3, 5, and 6 represent results of measurements [36]. The samples were illuminated by a 14° cone of monochromatic light of 12 Å bandwidth which was perpendicular to the sample surface. The intensity of the selectively reflected light was measured over the same solid angle. As already discussed, we estimated a maximum angle of misalignment of 4.7°. The illuminated region of the sample contained several thousand domains, which all selectively reflected light within the solid angle of observation. Since the selective reflectance depends on the thermal history of the sample, we always made our measure-

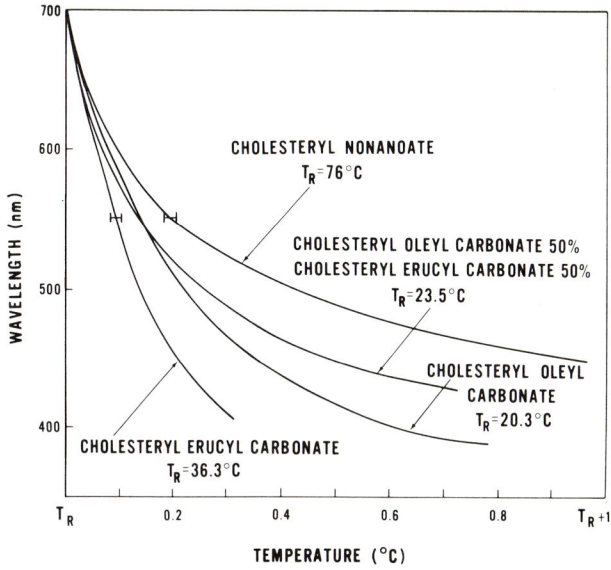

Fig. 5. Wavelength of maximum selective reflection as a function of temperature. [Reproduced from Ennulat, *Mol. Cryst. Liquid Cryst.* **13**, 337 (1971), with permission.]

ments after slowly cooling the sample from the isotropic state at the same cooling rates. Thus we imposed on our samples the same thermal history. We believe that this is the most practical procedure, considering that it takes at least one day to establish thermal equilibrium in a single-domain sample.

Figure 5 shows the dependence of the wavelength at peak intensity (from now on called peak wavelength) on the temperature. A single-domain sample should exhibit the same results because—as we will see—the angular dependence of the selective reflection is negligible under the measurement conditions stated above. Figure 3 shows the spectral response associated with various temperatures. Although the shapes of these curves deviate significantly from those obtained from single-domain samples, their halfwidths are approximately equal to the halfwidth of the reflection band of a single domain [*47*]. Again, this agrees with the fact that the polydomain samples were very well aligned. Figure 6 indicates the high temperature dependence of the selectively reflected monochromatic light. For example, at the wavelength of 0.650 μm the halfwidth of the intensity curve amounts to only 0.025°.

Fergason [*48, 49*] measured and calculated the angular dependence of the peak wavelength (see also Magne [*50*] and Böttcher [*230*]). He assumed that a stack of fictitious planes separated by the pitch of the helix furnishes Bragg reflection sites that are imbedded in a medium having the average index of refraction n of the liquid crystal. As shown in Fig. 1, a simple geometry

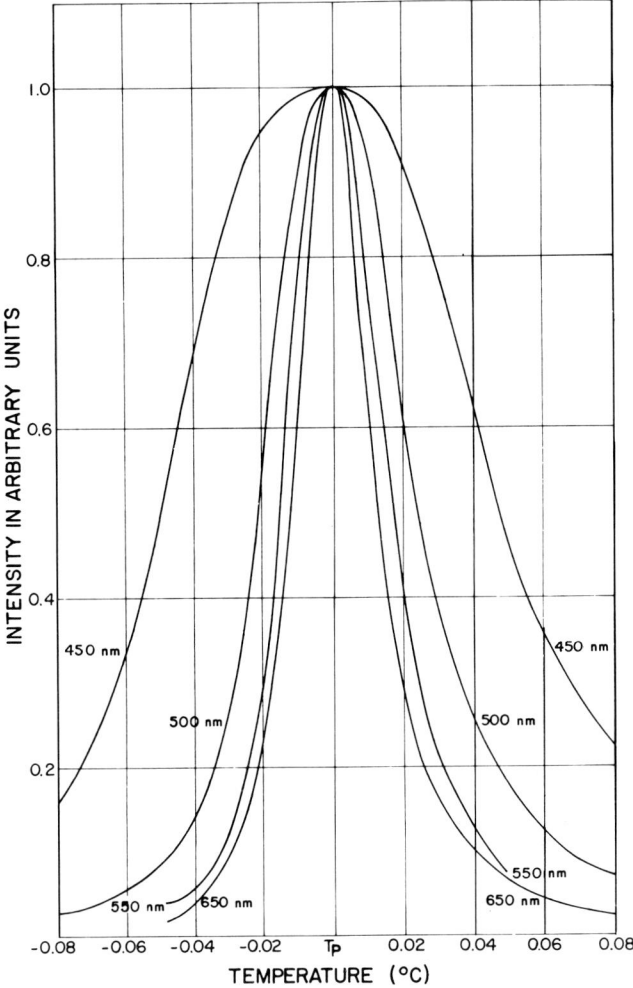

Fig. 6. Normalized intensity of selectively reflected light as a function of temperature. [Reproduced from Ennulat, *Mol. Cryst. Liquid Cryst.* **13**, 337 (1971), with permission.]

consideration based on that model gives the relation

$$\lambda_p = \lambda_n \cos \frac{1}{2} \left\{ \sin^{-1}\left(\frac{\sin \phi_i}{n}\right) + \sin^{-1}\left(\frac{\sin \phi_s}{n}\right) \right\} \tag{3}$$

where λ_n is the peak wavelength obtained from light traveling along the helical axis and ϕ_i and ϕ_s are angles of incidence and observation, respectively. This formula agrees well with the measurements except for very large angles of

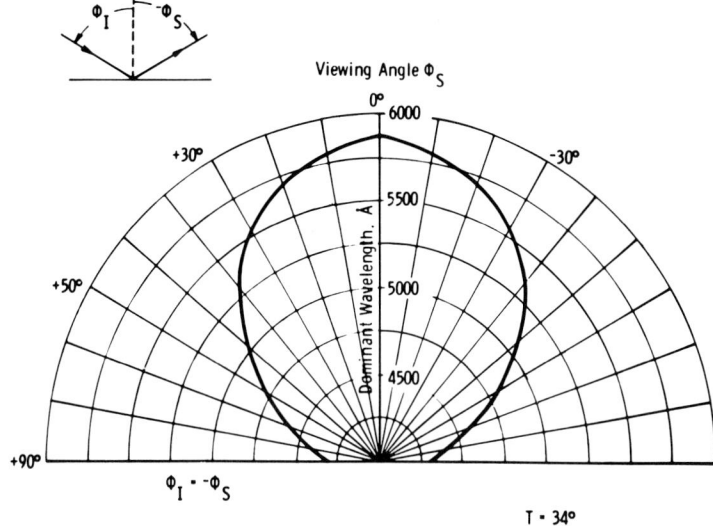

Fig. 7. Wavelength of maximum scattering as a function of viewing angle. [Reproduced from Fergason, *Mol. Cryst. Liquid Cryst.* **1**, 293 (1966), with permission.]

incidence [41]. The experimental results in Fig. 7 indicate that for increasing angles of incidence the peak wavelength is shifted toward the blue.

Equation (3) allows the determination of the wavelength shift caused by the maximum misalignment ε of the domains in the sample. Using the data reported in our work [26] ($\phi_i = -\phi_s = 7°$), we obtained a shift of

$$\lambda_p - \lambda_n = -23 \text{ Å}$$

which is about twice as large as the spectral bandwidth of the light used.

Adams, Haas, and Wysocki [42] have observed selectively diffracted light transmission in samples deposited on a partially reflective substrate. Figure 8 depicts this effect. They also observed that both this selective transmission and the selective reflection exhibit the same angular dependence of the peak wavelength. Much more fundamental is their finding that focal conic textures display selective reflection when a mirror substrate is used. As shown in Fig. 9, this effect can be explained by assuming that the focal conic texture consists of a polydomain planar texture whose helical axes lie predominantly in the sample plane. According to Fig. 10 the selective transmission observed under certain conditions can be explained by the same structure. The latter is also compatible with the fact that birefringent patterns are observed with the microscope when transmitting or absorbing substrates are used. These effects support Friedel's hypothesis of the molecular arrangement in cholesteric focal conics [51]. He concluded from his observations that the layers required

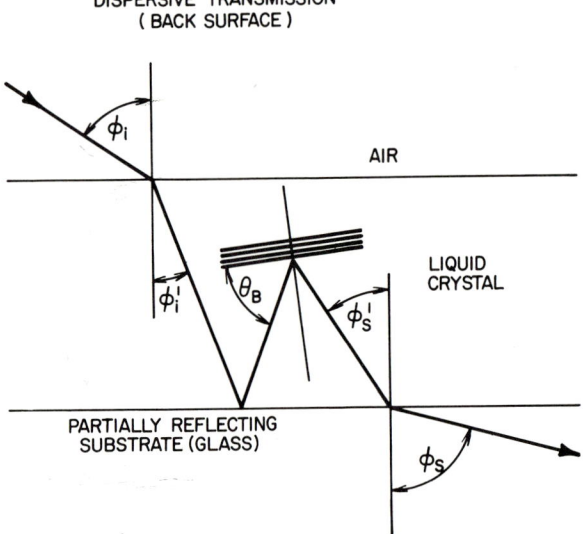

Fig. 8. Geometry for dispersive transmission in planar texture. [Reproduced from Adams et al., *J. Chem. Phys.* **50**, 2458 (1969), with permission.]

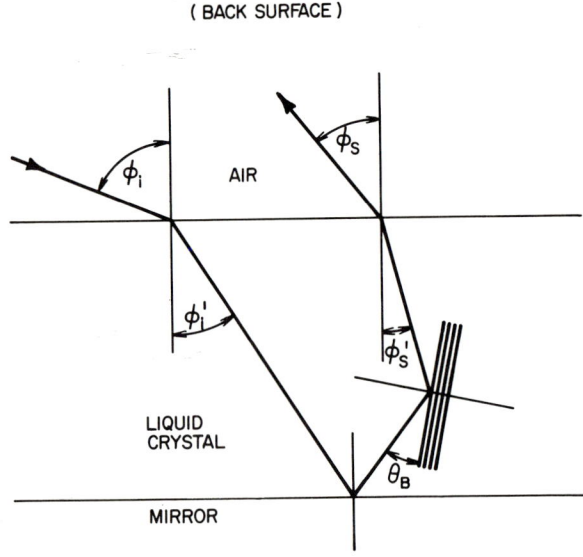

Fig. 9. Geometry for selective reflection in focal conic texture. [Reproduced from Adams et al., *J. Chem. Phys.* **50**, 2458 (1969), with permission.]

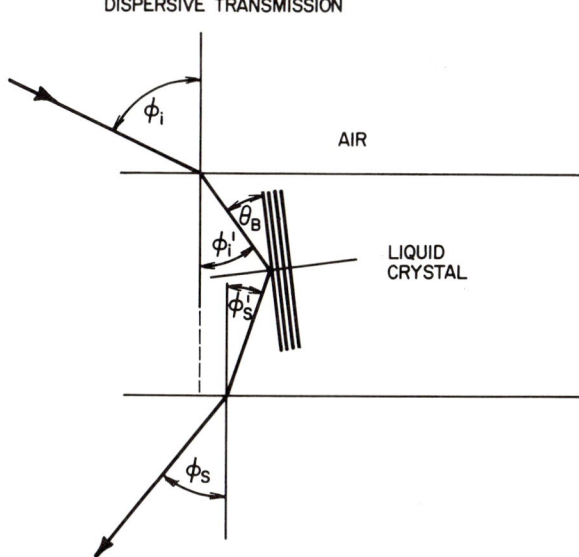

Fig. 10. Geometry for selective transmission in focal conic texture. [Reproduced from Adams et al., *J. Chem. Phys.* **50**, 2458 (1969), with permission.]

to explain the focal conic properties should consist of planar texture elements which have their helical axes parallel to the local normal of the Dupin cyclides comprising the focal conic structure. In such a molecular arrangement the helical axes of a large number of domains would make a small angle with the sample plane. The results prove that the focal conic textures are embodiments of essentially the same molecular building principle, namely, the helical arrangement of the molecules.

3. Phenomenological Theories

The propagation of light along the helical axis of a single planar texture domain was first treated by Oseen [52] as part of a more comprehensive theory of liquid crystals, and later by De Vries [53]. Both investigators used Friedel's molecular model of the planar texture described above and assumed molecules of uniaxial symmetry. De Vries pursued the theory in more detail. He found that the optical features characteristic of the planar texture can be calculated from the birefringence, the average index of refraction, and the pitch of the liquid crystal. Furthermore, he derived the fundamental relation

$$\lambda_p = np \tag{4}$$

linking the pitch p of the structure and the average index of refraction n with the peak wavelength. (Oseen [52] had already obtained the related expression

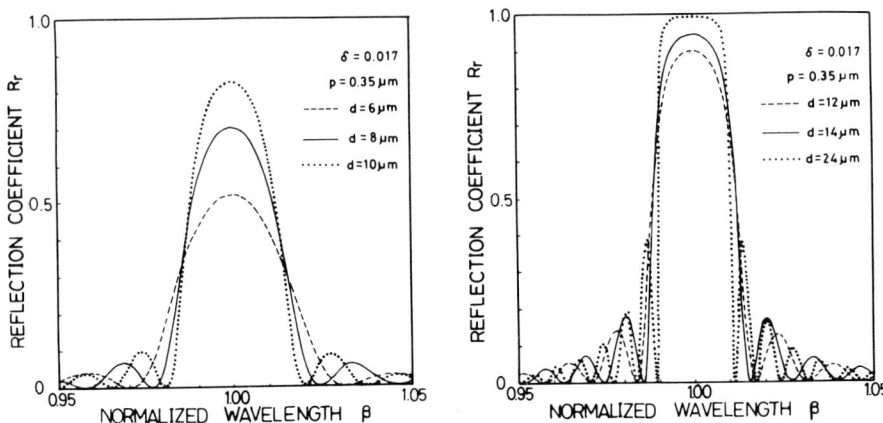

Fig. 11. Reflection coefficient for right-handed circularly polarized wave as a function of normalized wavelength. $\beta = \lambda/p$; λ = wavelength; p = pitch; d = thickness; δ = anisotropy constant. [Reproduced from Aihara and Inaba, *Opt. Commun.* **3**, 77 (1971), with permission.]

$pn_1 > \lambda > pn_2$ describing the wavelength region of selective reflection; n_1 and n_2 are the principal indices of refraction.) In recent years attempts were made to simplify and improve this theory [54–58]. Several investigators [59, 44] calculated and experimentally confirmed the type of the spectral selective reflectance profile shown in Fig. 2 for light traveling along the helical axis.

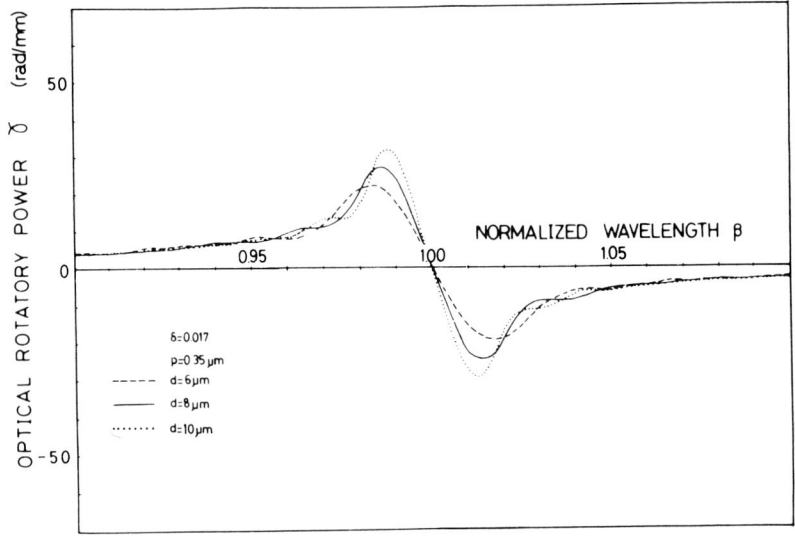

Fig. 12. Optical rotary dispersion of transmitted light. (Parameters as defined in Fig. 11.) [Reproduced from Aihara and Inaba, *Opt. Commun.* **3**, 77 (1971), with permission.]

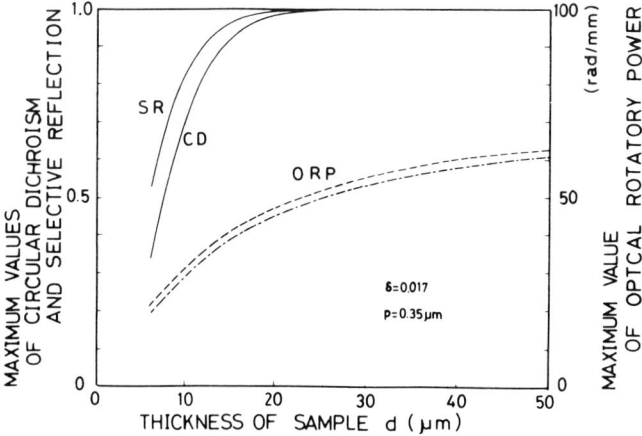

Fig. 13. Variation of the maximum values of selective reflection (SR) coefficient, circular dichroism (CD), and optical rotatory power (ORP) with the sample thickness. [Reproduced from Aihara and Inaba, *Opt. Commun.* **3**, 77 (1971), with permission.]

Others [60–62] determined theoretically the influence of sample thickness and dielectric anisotropy on selective reflectivity (Figs. 11, 13, and 14), optical rotary dispersion (Figs. 12, 13, and 14), and circular dichroism (Figs. 12, 13, and 14). The theory, extended to include molecular absorption [63, 64], achieved good qualitative agreement with experimental results. The treatment of the case where the helical axis is perpendicular to both the electric field vector and the direction of propagation [65] yielded higher-order diffraction maxima which appear to be in good agreement with experiments [66].

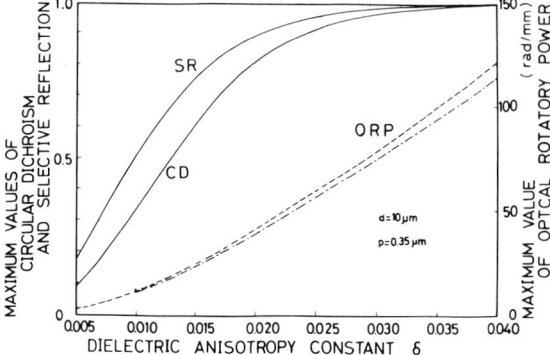

Fig. 14. Variation of the maximum values of selective reflection (SR) coefficient, circular dichroism (CD), and optical rotatory power (ORP) with the dielectric anisotropy constant. [Reproduced from Aihara and Inaba, *Opt. Commun.* **3**, 77 (1971), with permission.]

The problem of selective reflection of light incident at oblique angles to the helical axis was solved by numerical techniques in different ways. Taupin [67] obtained approximate solutions for the case of a semi-infinite liquid crystal by truncating an infinite set of simultaneous equations. Berreman and Scheffer [43] used a 4×4 matrix technique to solve the problem for finite sample thickness. Their numerical results agree qualitatively with their carefully measured data. However, they feel that the discrepancy between calculated and measured intensities may be significant. Kats [68] derived approximate solutions by using the theory of parametric resonance in the reflective region and a perturbation theory for wavelengths sufficiently removed from the selective reflection region.

Böttcher [69] extended the original theory of De Vries for arbitrary angles of incidence. He derived an eigenvalue equation, which can also be obtained by simply replacing the pitch p in De Vries' eigenvalue equation by the pitch in the direction of incidence, i.e., by the pitch divided by the cosine of the angle of incidence. However, he attained only qualitative agreement between measured and calculated reflection and transmission coefficients.

Belyakov and Dmitrienko [70] obtained approximate analytical solution of Maxwell's equations by applying the dynamic theory of diffraction. The latter was generalized by the introduction of an anisotropic permittivity that is characteristic of cholesteric liquid crystals [68]. The authors report good agreement of the theory with the wavelength dependence of the optical rotatory power measured by Chistyakov [71] and with the results published by Berreman and Scheffer [43].

Dreher and Meier [72] developed a general theory of wave propagation in helical structures. Assuming negligible absorption, they obtain results bearing a striking resemblance to the wave propagation in solid crystals. The solutions are Bloch-type waves because they are plane waves with an amplitude function that is periodic with the structure. Furthermore, their wavelength region consists of forbidden bands separated by allowed regions. Waves with allowed wavelengths propagate in the liquid crystal, while those with forbidden regions are totally reflected. Figure 15 shows these forbidden wavelength regions as a function of the directional quantity $(n_g \sin \psi)/n$, where ψ is the angle of incidence, n_g the index of refraction of the semi-infinite space bordering the liquid crystal film on either side, and n the average index of refraction of the liquid crystal. (Contrary to our nomenclature, λ_p in Fig. 15 is the normalized wavelength.) Shown is the main reflection band ranging from λ_I to λ_{II} on the wavelength axis (i.e., the short- and long-wavelength limit of shaded region in Fig. 15) and ending on the ordinate at m_I and m_{II}. For large angles of incidence the main reflection band splits into two branches. Only a finite number of the selective reflection bands of higher order are shown. The two almost-horizontal curves starting at m_I and m_{II} describe ordinary total reflection

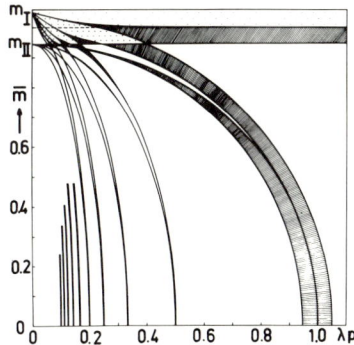

Fig. 15. Chart of stability. Forbidden bands are shaded. Computed for $\varepsilon = 2.27$; $\varepsilon_1 - \varepsilon_2 = 0.48$. [Reproduced from Dreher and Meier, *Phys. Rev. A* **8**, 1616 (1973), with permission.]

effects. This diagram is analogous to the Brillouin zone representation of solids. The following outstanding features derived from the figure are in good qualitative agreement with the experiment.

1. Only one selective reflection band exists for light of normal incidence, while higher orders occur for obliquely incident light.
2. The first-order reflection band splits into two branches for large angles of incidence.
3. The wavelength region of selective reflection shifts toward shorter wavelengths with increasing angle of incidence.
4. The angle of incidence is equal to the angle of selective reflection within the plane of incidence.

Figure 16 represents the part of a measured wavelength diagram.

C. Theories of Cholesteric Mesophases

1. CONTINUUM THEORIES

In his attempt to explain the mesomorphic behavior by the molecular interactions between pairs of molecules, Oseen [52] found that purely electrostatic and magnetic forces are not sufficient. He concluded that other yet-unknown forces must be responsible for the existence of mesomorphic states. Nevertheless, he was able to derive an expression for the free energy by assuming that these unknown forces between molecules decrease rapidly enough with increasing distance and that the density of the material remains constant. Frank [73] refounded this theory by proposing at the outset that torque stresses are proportional to curvature strains. He showed that such a relationship is compatible with the fluid nature and the orientational order of mesophases. Except for two additional terms, he obtained the same free energy

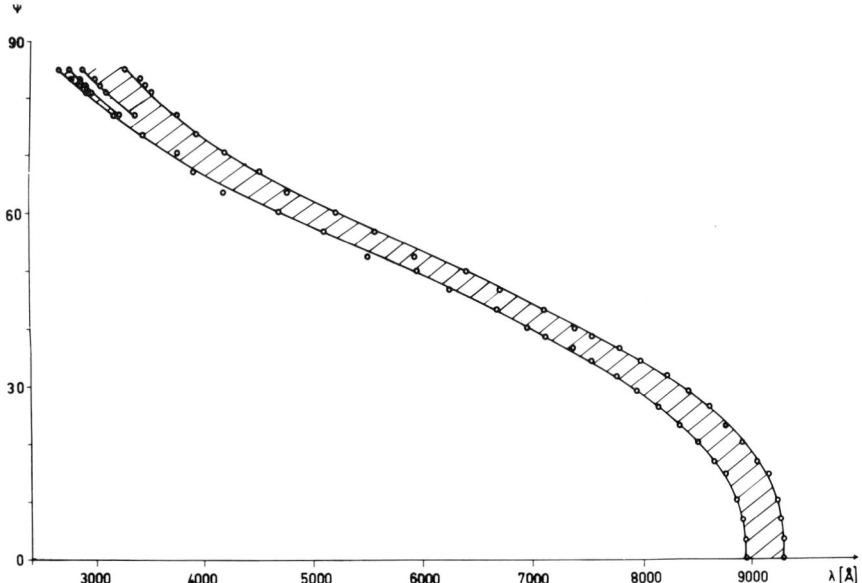

Fig. 16. Experimental stability chart of a mixture of cholesteryl oleate, cholesteryl nonanoate, and cholesteryl chloride (11:40:29 by weight) at room temperature. [Courtesy Dr. G. Meier.]

expression as Oseen. Several investigators derived important relationships and explained singularities occurring in textures by minimizing the free energy for various applied fields and mechanical stresses and by exploiting the existence of certain symmetry conditions. Recent results of this work have contributed significantly to the understanding of liquid crystals. De Gennes reviewed and summarized these theoretical and experimental developments from a fundamental point of view [11].

2. Mean Field Theories

Maier and Saupe [74] overcame Oseen's difficulties for nematic mesophases by considering dispersion forces for the molecular interaction and by developing a mean field theory. They obtained an internal field that depends on the molecular volume, the anisotropy of the molecular polarizability, and the degree of order, which explains essential features of nematic mesophases. Goossens [75] extended this theory to cholesteric mesophases by also incorporating dipole–quadrupole interactions. It appears that dipole–dipole interactions are necessary to achieve the molecular parallelism in mesophases and that, in addition, dipole–quadrupole interactions are essential for causing the twist in cholesteric mesophases. Goossens was able to estimate the order of magnitude of the twist. In addition, he showed that in cholesteric mesophases composed of nematogenic and optically active materials the molecular optical activity alone does not determine the handedness of the pitch. This conclusion was confirmed by experimental results [16].

Alben [76] studied the unwinding of the helix occurring at the transition from the cholesteric to the smectic phase by coupling the director field describing the local, preferred direction in the cholesteric state with that of the incipient smectic structure. Applying a mean field theory, he concluded that both the twist coefficient K_{22}, defined in Frank's theory [73], and the pitch of the helix increase with decreasing temperature while the ratio of these qualities is approximately constant. He suggested testing this theory by measuring the temperature dependence of the externally applied field E required to completely unwind the helix. He showed that the relation $E \simeq 1/[p(T)]^{1/2}$ would be compatible with the proposed mechanism. This theory would explain why the peak wavelength of the selective reflection and its temperature coefficient increase as the temperature approaches the transition to the smectic mesophase. However, it is not possible yet to predict the temperature dependence of the pitch.

Similar pretransitional effects could occur at the transition from the cholesteric mesophase to the isotropic liquid. We believe that this may explain the shift of the peak wavelength observed for cholesteryl 2-(2 -ethoxyethoxy)ethyl carbonate from 2.2 μm at 15° to 3.6 μm at 30° [77], a temperature which is only 1° below the clearing point [78]. It is questionable that the twist unwinds completely at this phase transition, since substantial optical activity proportional to $(T-T_0)^{-1/2}$ was observed in an interval of up to 6° above the clearing point with another compound [79]. These results can be explained with the aid of the Landau–De Gennes theory of short-range interaction in this pretransitional region. Thus it must be concluded that the short-range chiral orientational order persists into the isotropic phase.

3. Anharmonic Force Theory

Keating [80] explains the existence of the helical structure by assuming anharmonic interaction forces. In his model the molecules conduct thermally excited rational oscillations which tend to twist the directors of neighboring molecular sheets with respect to each other. Since the restoring forces are anharmonic, the time average of these twisting motions does not vanish. Keating uses complicated, mostly qualitative arguments to show that this residual twist leads to the helical arrangement of molecules typical for cholesteric mesophases. He obtained an approximate solution of the equation of the undamped anharmonic oscillator and determined the time average of the twisting motion. After several other operations he arrives at the following expression for the temperature dependence of the peak wavelength in the vicinity of the cholesteric–smectic phase transition

$$\lambda_p = A/T[1+B/(T-T_0)]^2 \qquad (5)$$

where A, B, and T_0 are parameters. This equation was derived under the condition that the harmonic restoring forces in the vicinity of a phase transition are proportional to $(T-T_0)^{-1}$. The predicted temperature dependence of the peak wavelength agrees well with the experimental results. We were able to match our experimental data within the measurement uncertainty of ± 0.2 m°C (m°C = 0.001 °C) [47].

Böttcher [81] modified this model and explained the generation of the pitch as follows. He assumes that the first layer of the liquid crystal molecules is absorbed parallel to the substrate and oriented along a certain direction. The layer generates an internal field that acts on the quasi-free molecules of the next layer as a restoring force of the thermally excited oscillations. Because of the asymmetry of the molecules these oscillations are nonlinear and the time average of the oscillation angle in the plane is finite. The latter results in an angular twist of the preferred molecular direction of the second layer with respect to that of the first layer. The second layer in turn generates a field for the third layer. As a consequence, the latter experiences a twist in the same direction. Because of this next-neighbor interaction the whole liquid crystal assumes the helical structure. It should be pointed out that these layers are smeared out by the thermally excited agitation of the molecules. This model renders a temperature-sensitive pitch of the helical structure because the time average of the rotational molecular oscillations depends on the temperature. Böttcher uses the equation of the damped anharmonic oscillation to derive an approximate solution for the rotational oscillations and obtains for the peak wavelength

$$\lambda_p = (C_1/T)(1+C_2\eta)$$

where C_1 and C_2 are constants and η is the viscosity. Using the viscosity data measured in the vicinity of the phase transition at low shear rates [82], he obtains a good fit to the approximation [83]

$$\eta = \eta_0 \exp[C/(T-T_0)]$$

where C and T_0 are constants. Under these conditions his formula for λ_p agrees well with the available experimental results [50, 84].

A mixture of cholesteryl nonanoate and cholesteryl chloride exhibits the exponential relation [28]

$$p = E + F\exp[-G(T-T_0)]$$

where E, F, and G are parameters. This behavior may be explained by using Keating's theory to relate the angular twist between neighboring molecular layers with the average number of smectic clusters occurring in the pretransitional region. According to Frenkel's theory [87] on heterophase fluctuations, the number of clusters increases exponentially as the temperature approaches the cholesteric–smectic phase transition.

Pindak et al. [85] derived the relation

$$\lambda_p = np = n(p_n + [D/(T-T_0)^v])$$

with

$$p_n \equiv p_0[1 + \alpha(T-T_0)]$$

and

$$0.5 < v < 1$$

This formula is based on the suggestion by De Gennes [86] that the twist elastic constant follows the relation

$$K_{22} = K_0 + [C/(T-T_0)^v]$$

and on Alben's theoretical prediction [76] that

$$K_{22}/p = \text{const}$$

Both assumptions apply only to the pretransitional region. A good curve fit to the experimental data was obtained only when the parameters p_0 and α were determined independently.

It is very difficult to obtain a consistent, unambiguous curve fit, because of the nonlinearity of the pitch–temperature relationship and because of the many free parameters of the theoretically predicted functions. This may explain why several investigators [37, 47, 76] measured the relationship

$$\lambda_p = \lambda_0 + [A/(T-T_0)^2] \tag{6}$$

More accurate measurements, as well as independent methods for the determination of at least some of the free parameters, are necessary to confirm conclusively any of these theories.

III. CHEMISTRY OF CHOLESTERIC LIQUID CRYSTALS

Cholesteryl oleyl carbonate [88] is one of the most popular cholesteric liquid crystals used in temperature sensors because it has a high temperature coefficient of the selective reflectance at about room temperature. Moreover, the addition of certain components displaying the cholesteric mesophase at higher temperatures results in mixed liquid crystals, in which the narrow, visible, selective reflection region is shifted toward higher temperatures. Since this shift depends on the concentration, it is possible to obtain a series of temperature-indicating liquid crystals covering a wide range of operating temperatures. The major drawback of these materials is the gradual degradation of the selective reflection due to autoxidation of the cholesteryl oleyl carbonate. Even the amount of UV radiation present in daylight or in the light of ordinary fluorescent lamps is sufficient to speed up this decomposition process.

The synthesis of more stable compounds of desirable temperature sensitivity is a difficult task because theories cannot provide sufficient information relating molecular features to mesomorphic properties. The alternative is an empirical approach based on systematic structural modifications in compounds of known mesomorphic behavior. By studying the influence of these modifications on the mesomorphic properties it is hoped that a better cholesteric liquid crystal can be obtained. But such substitutions cannot readily be made because only a limited number of suitable steroids of high purity are available. Considering the present inability to predict the mesomorphic properties of new compounds, it appears to be premature to embark on an expensive and time-consuming synthetic program for new steroids. Instead, commercially available sterols are being used as precursors in the effort to obtain temperature-sensitive cholesteric liquid crystals.

The empirical approach is feasible only if the impurities of the investigated compounds do not significantly influence the mesomorphic properties. Consequently, these materials have to be synthesized from extremely pure starting materials. In some cases, especially mild reaction conditions have to be developed to avoid side products which are difficult to remove. Such measures can reduce the impurity level below the threshold of chromatographic testing methods. Plausible arguments and past experience have led to the hypothesis that a homologous series of pure materials should exhibit a systematic relationship between chain length and transition temperatures in the melt [3]. The reverse then implies that the compounds of a series are of sufficient purity if such a definite relationship exists. Using this criterion we were able to recognize the influence of unknown impurities which were not initially detected by analytical tests.

Simple derivatives of cholesterol, such as the alkanoates, are suitable starting points for the empirical investigation because these materials are available in sufficient purity and their mesomorphic properties are well known. One part of the investigation concentrates on changes in the 3β-side chain, while the other part deals with more complex variations in the sterane moiety. The comparison of the mesomorphic features observed within a homologous series, as well as the comparison of homologs belonging to different homologous series, should indicate the changes of the mesomorphic behavior caused by the variation of molecular features.

In the following we will review this work and, in addition, discuss nonsteroidal cholesteric liquid crystals and the chemical stability of the important cholesteric liquid crystals.

A. Various 3β-Substituted Cholest-5-enes and 5α-Cholestanes*

Cholesterol (**1a**) is not mesomorphic, but minor variations of the substituents at the 3β-position group have led to cholesteric mesophases. For example, the replacement of oxygen by sulfur results in the monotropic cholesteric cholest-5-en-3β-thiol (thiocholesterol, **1b**) [*89*], which selectively reflects in the red on quenching of the isotropic melt [*90*]. Furthermore, with the exception of 3β-fluorocholest-5-ene (**1c**), the cholesteryl halides (**1d–1f**) exhibit cholesteric mesophases [*89*]. Table I shows that their clearing temperatures increase with decreasing electronegativity of the halide. The effective rotary power of these compounds follows the relation [*89*]

$$1/\lambda_p = 1/np = 7.5d - 15.6$$

where d is the distance in Å between the centers of the C-3 and the halide atoms. In the case of cholesterol, d is equal to the sum of the C–O bond length and the projection of the O–H bond onto the C–O direction. The latter contribution allows for the fact that the C–O and O–H bonds are at an angle and that the O–H bond rotates freely about the C–O axis. It is surprising that the application of this geometric definition of d to 3β-groups of longer chain length is sufficient to ensure compliance with the above relation for cholest-5-ene, cholest-5-en-3β-thiol, and the methyl ether, ethyl ether, formate, acetate, and chloroformate of cholesterol [*89*]. If the value of d is smaller than 2.1 Å, the effective rotary power is negative and the compounds form right-handed helices. This occurs in cholest-5-ene, cholesterol, cholest-5-en-3β-thiol, and the cholesteryl halides with the exception of 3β-iodocholest-5-ene, which has a very weak tendency to form a left-handed helix.

Cholesteryl chloroformate (**1g**) is monotropic cholesteric and displays the color band upon quenching of the isotropic melt with ice water [*90*]. Replacing the oxygen by sulfur increases the cholesteric stability because cholesteryl chlorothiolcarbonate (**1h**) is enantiotropic cholesteric and selectively reflects the visible spectrum above its melting point [*90*]. (The missing derivative, cholesteryl chlorothioncarbonate, could not be obtained because cholesterol and thiophosgene form only dicholesteryl thioncarbonate [*91*].) A similar effect is observed with the substitution leading from cholesteryl chloroformate to the azidoformate (**1i**) [*92*].

---

* The nomenclature used corresponds to that of *Chemical Abstracts*.

TABLE I: Mesomorphic Properties of Various 3β-Substituted Cholest-5-enes and 5α-Cholestanes[a]

1	R	Ch–I	mp	1	R	Ch–I	mp	2	R	Ch–I	mp
a	OH	—	150	g	O=C–Cl (O-)	86	117.5	a	OH	—	143
b	SH	53	98	h	S–C(=O)–Cl	133.5	128.6	b	Cl	—	114.0
c	F	—	94	i	O–C(=O)–N$_3$	104	91	c	O–C(=O)–Cl	58.3	99.6
d	Cl	64	96	j	COOH	258.6	224.5				
e	Br	69	100.5	k	H$_2$C COOCH$_3$	—	78				
f	I	95	106.5	l	HC(COOCH$_3$)$_2$	105	89				

[a] Abbreviations used: mp, melting point (°C); Ch–I, cholesteric–isotropic transition (°C).

In general, the attachment of larger substituents to the 3β-position is a more complicated case because steric factors, increases in the lateral attractions between molecules due to higher polarizability, and other properties of the substituent simultaneously influence the mesomorphic behavior. For example, cholest-5-ene-3β-carboxylic acid (Marker's acid, **1j**) has an enantiotropic cholesteric mesophase which may be dimerized, as the high transition entropy of the clearing point indicates [*93*]. Several of its alkyl esters exhibit cholesteric colors [*38, 12f*] and transition temperatures in the melt which are lower than those of the corresponding cholesteryl alkanoates [*94*]. But no mesophases have been reported for cholest-5-en-3β-yl acetic acid (**1k**) and its methyl ester, while dimethyl cholest-5-en-3β-yl malonate (**1l**), a branched ester, is mesomorphic [*95*].

Hydrogenation of the Δ^5 double bond of cholesterol results in nonmesomorphic 5α-cholestan-3β-ol (**2a**). Mesomorphism is suspected in 3β-chloro-5α-cholestane (**2b**) [*96*] and a monotropic cholesteric mesophase without a color band has been observed in the chloroformate (**2c**) [*97*]. These results and, more clearly, the observations on related homologous series which will be discussed later, indicate that the removal of the Δ^5 double bond of cholesterol seems to depress the temperature region of the cholesteric mesophase.

The effective rotary power of 5α-cholestan-3β-ol and of its chloride, formate, and acetate, follows the relation [*89*]

$$1/\lambda_p = 7.5d - 16.2$$

Notice that parallel shifting of this plot by -0.6 along the d axis results in the function pertaining to the corresponding cholesteryl derivatives. However, the right-handed 5α-cholestan-3β-yl derivatives have a smaller and the left-handed derivatives a larger pitch than the respective cholesteryl derivatives [*89*]. Apparently the additional asymmetric center at C-5 is responsible for the latter and for the parallel shift of the effective rotary power.

Cholesteryl cinnamates, 5α-cholestan-6-one-3β-yl benzoates and a number of substituted benzoates and alkoxybenzoates of cholesterol and 5α-cholestan-3β-ol have been recently investigated [*98*], but complete results have not been published yet. Several papers discuss the influence of aromatic substitution in benzoates [*99*] and cinnamates [*99, 100*] of cholesterol on the mesomorphic and thermodynamic properties. Cholesteryl ferrocenylacetate and 5-ferrocenylpentanoate [*101*], as well as 2,2-dimethylpropionate (pivalate) and 1-adamantanecarboxylate [*102*], are reported to be not mesomorphic. Apparently, bulky 3β-substitutes prevent the formation of mesophases.

Dicholesteryl dicarboxylates exhibit cholesteric mesophases above 150°. If the isotropic melts are cooled rapidly to room temperature, they form vitreous glasses with a frozen-in mesophase [*103*]. A similar behavior was found in cholesteryl hydrogen phthalate [104, 105], where a glass transition temperature was determined by heat capacity measurements. The addition of a compound such as cholesteryl benzoate converts these "glassy" liquid crystals [104] into "cholesteric solids" [105]. These have a solid consistency and yet selectively reflect the solar spectrum like ordinary cholesteric mesophases without change for years [105, 106].

HOMOLOGOUS SERIES

During the past seven years the number of investigated homologous series increased from 1 to about 17. In most of the mesomorphic compounds the transition temperatures in the melt are reproducible within about 1°, partly because of the high purity attained by modern preparation methods and partly because of better test equipment. Although temperature-programmable miscroscope stages increased the accuracy of the optical determination of the transition temperatures, differential temperature analysis (DTA) and differential scanning calorimetry (DSC) are used more and more to certify thermodynamically the phase transitions and, in addition, to measure the transition enthalpies. For our purpose it is expedient to characterize the mesomorphic behavior primarily in terms of types of mesophases, transition temperatures in the melt, and existence of color bands in planar textures. Compounds that exhibit a narrow spectrum of selectively reflected light are tabulated in Table VIII in Section IV,A.

The known homologous series are compiled in Table II. In the series I–VII and XIV–XVII no impurities could be detected by thin-layer chromatography. This implies a purity of at least 99 mol %. The purity of the other series is unknown, but it is evidently sufficient. As Figs. 17–21 indicate, these series exhibit a dependence of transition temperatures on chain length which is as clearly defined as that of the compounds of known purity. In the following we will briefly comment on each one of the homologous series.

a. *Esters. Cholesteryl alkanoates* (I): This is the most investigated cholesteric homologous series. Recent data generated by different investigators [107–111] are in good agreement because most of the homologs are commercially available with a reasonable purity. Except for the first two members of the series, the chain length dependence of the transition temperatures is characteristic of homologous series of many sterol derivatives: the clearing points decrease, while the smectic–cholesteric transition temperatures first rise and then fall with increasing chain length (see Fig. 17). The occurrence of the unusually low clearing point of the formate may be due to the coupling between the formyl proton and the 3α-proton [112]. Notice that only the octanoate and the higher members have a smectic mesophase. It is unlikely that smectic mesophases of shorter chain length were overlooked, because the transition enthalpy decreases with decreasing chain length to as low as 110

TABLE II

HOMOLOGOUS SERIES OF DERIVATIVES OF CHOLESTEROL AND 5α-CHOLESTAN-3β-OL

No.	1 R	No.	1 R	No.	2 R
I	$CH_3-(CH_2)_n-\overset{O}{\underset{\|\|}{C}}-O$	VIII	$CH_3-(CH_2)_n-O-(CH_2)_2-\overset{O}{\underset{\|\|}{C}}-O$	XIV	$CH_3-(CH_2)_n-\overset{O}{\underset{\|\|}{C}}-O$
II	$CH_3-(CH_2)_n-\overset{O}{\underset{\|\|}{C}}-S$	IX	$CH_3-(CH_2)_n-O-\underset{\text{(phenyl)}}{\bigcirc}-\overset{O}{\underset{\|\|}{C}}-O$	XV	$\underset{\text{(phenyl)}}{\bigcirc}-(CH_2)_n-\overset{O}{\underset{\|\|}{C}}-O$
III	$\underset{\text{(phenyl)}}{\bigcirc}-(CH_2)_n-\overset{O}{\underset{\|\|}{C}}-O$	X	$CH_3-(CH_2)_n-O-\underset{\text{(phenyl)}}{\bigcirc}-CH=CH-\overset{O}{\underset{\|\|}{C}}-O$	XVI	$CH_3-(CH_2)_n-O-\overset{O}{\underset{\|\|}{C}}-O$
IV	$\underset{\text{(phenyl)}}{\bigcirc}-(CH_2)_n-\overset{O}{\underset{\|\|}{C}}-S$	XI	$CH_3-(CH_2)_n-O-\underset{\text{(naphthyl)}}{\bigcirc\bigcirc}-\overset{O}{\underset{\|\|}{C}}-O$	XVII	$CH_3-(CH_2)_n-S-\overset{O}{\underset{\|\|}{C}}-O$
V	$CH_3-(CH_2)_n-O-\overset{O}{\underset{\|\|}{C}}-O$	XII	$CH_3-(CH_2)_n-O-\underset{\text{(naphthyl)}}{\bigcirc\bigcirc}-CH=N-\underset{\text{(phenyl)}}{\bigcirc}-\overset{O}{\underset{\|\|}{C}}-O$		
VI	$CH_3-(CH_2)_n-S-\overset{O}{\underset{\|\|}{C}}-O$	XIII	$CH_3-(CH_2)_n-O-\underset{\text{(phenyl)}}{\bigcirc}-CH=N-\underset{\text{(phenyl)}}{\bigcirc}-\overset{O}{\underset{\|\|}{C}}-O$		
VII	$CH_3-(CH_2)_n-O-\overset{O}{\underset{\|\|}{C}}-S$				

cal/mole for the octanoate [107]. The helical twisting power increases monotonically with chain length and exhibits an odd-even effect up to the hexanoate [19]. Only the nonanoate and the decanoate exhibit a color band within a narrow temperature range. Apparently selective reflection in the visible is not observed for the lower homologs because of the interfering freezing and for the higher homologs because of the large pitch which restricts selective reflection to the infrared region.

S-Cholesteryl alkanethioates (II): A general increase of both the smectic–cholesteric and the cholesteric–isotropic transition temperatures is observed in the alkanoates of cholest-5-en-3β-thiol relative to their cholesteryl analogs I (Fig. 17). Smectic mesophases and the color band occur at shorter acyl chain length [107, 113, 114].

5α-*Cholestan-3β-yl alkanoates* (XIV) are formally obtained by hydrogenation of the Δ^5 double bond of their cholesteryl analogs I. The transition temperatures are drastically lowered (Fig. 17). The formate and the hepta-

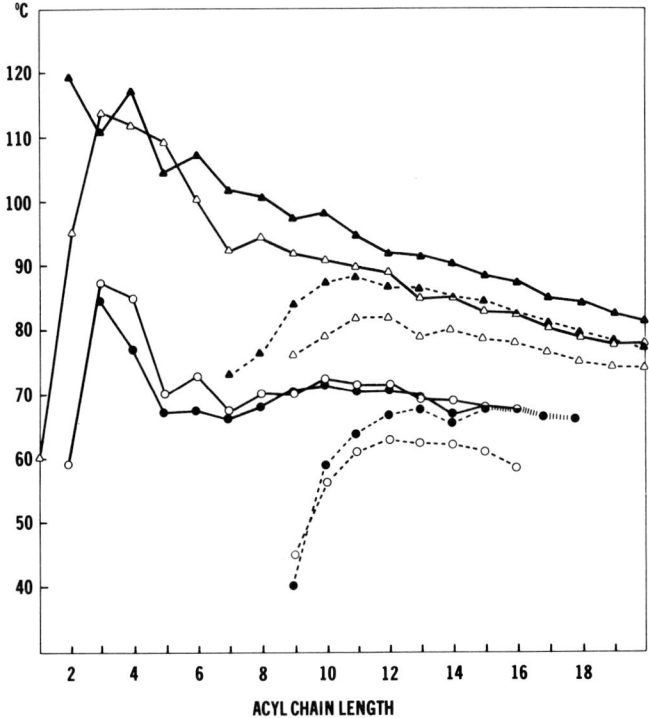

Fig. 17. Transition temperatures of alkanoates. Solid lines: cholesteric-isotropic phase transitions; broken lines: smectic-cholesteric phase transitions. △, Cholesteryl alkanoates, I; ▲, S-cholesteryl alkanethioates, II; ○, 5α-cholestan-3β-yl alkanoates, XIV; ●, 5α-cholest-8(14)-en-3β-yl alkanoates, 75.

decanoate through the eicosanoate are not mesomorphic and the acetate has an unusually low clearing point [*115*]. Only the undecanoate and the dodecanoate display a color band with a narrow temperature interval.

5α-Cholest-8(14)-*en*-3β-*yl alkanoates* (**75**): The transition temperatures in the melt are not much different from those of the corresponding 5α-cholestan-3β-yl alkanoates XIV (Fig. 17). However, the $\Delta^{8(14)}$ double bond does result in several significant differences. The smectic–cholesteric transition temperatures converge more strongly with the cholesteric–isotropic transition temperatures, and only monotropic smectic mesophases are observed in the hexadecanoate through the octadecanoate. In addition, all the cholesteric members of this series are strongly right-handed. These phenomena have not been reported for any other homologous series. Several homologs of medium acyl chain length also display colors but their temperature intervals have not been recorded [*116, 117*].

Cholesteryl ω-phenylalkanoates (III): A terminal phenyl group in cholesteryl alkanoates I causes dramatic changes of the cholesteric–isotropic transition temperatures of the resulting ω-phenylalkanoates. The values of the clearing points alternate with increasing acyl chain length and fall on two hyperbolic curves: an upper curve for odd and a lower inverted curve for even acyl chain length. The two branches approach each other (Fig. 18). Yet the smectic–cholesteric transition temperatures do not have such an odd-even effect [*107, 118, 119*]. Monotropic smectic mesophases were observed in the 8-phenyloctanoate, the 10-phenyldecanoate and successive homologs. Except for the phenylacetate, all of the first 16 homologs are cholesteric and most exhibit cholesteric colors. The 10-phenyldecanoate and its higher homologs selectively reflect the visible spectrum.

S-Cholesteryl ω-phenylalkanethioates (IV): A significant accentuation of the odd-even effect of the cholesteric–isotropic and the introduction of a less-pronounced one for the smectic–cholesteric transition temperatures is observed in this homologous series, which is obtained by replacing the ester group of III by the thiol ester group. All transition temperatures are in phase (Fig. 18). Several homologs reflect selectively the visible spectrum over very narrow temperature intervals [*120, 121*]. The 14-phenyltetradecanethioate is the most temperature-sensitive cholesteric liquid crystal reported to date [*47*]. It selectively reflects the spectral range of 0.4–0.7 μm within 0.07°.

5α-Cholestan-3β-*yl ω-phenylalkanoates* (XV): Derivatives of 5α-cholestan-3β-ol have generally lower transition temperatures in the melt than their cholesteryl analogs. The same is noticed in this series. The odd-even effect of the clearing points is shifted to lower temperatures and the occurrence of the first smectic member to higher acyl chain length (Fig. 18). All members are cholesteric except the first three even homologs [*120, 122*]. Starting with the 12-phenyldodecanoate, the higher even homologs display the visible spectrum within relatively narrow temperature intervals.

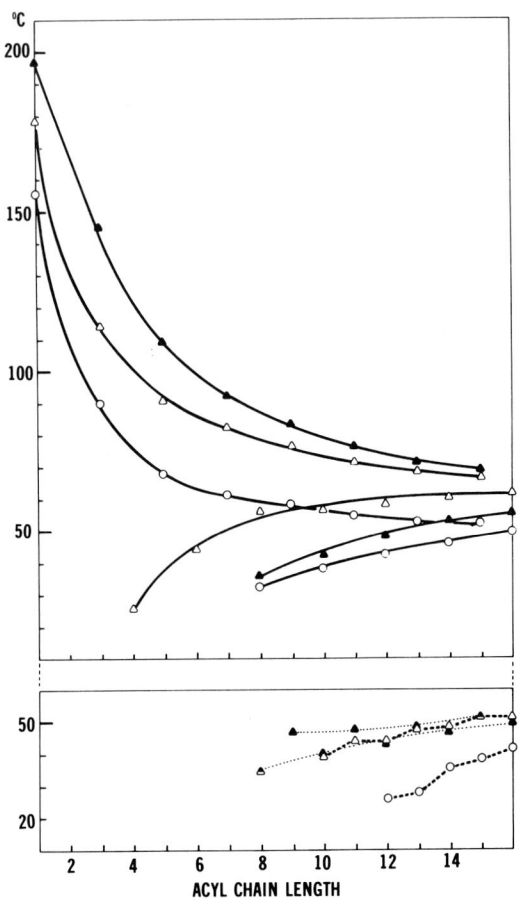

Fig. 18. Transition temperatures of ω-phenylalkanoates. Transitions as defined in Fig. 17. △, Cholesteryl ω-phenylalkanoates, III; ▲, S-cholesteryl ω-phenylalkanethioates, IV; ○, 5α-cholestan-3β-yl ω-phenylalkanoates, XV.

b. *Carbonates. Cholesteryl alkyl carbonates* (V): The mesomorphic properties of this series, obtained by replacing the ester group of I by the carbonate group, are quite different from those of the cholesteryl alkanoates (Fig. 19). The cholesteric–isotropic transition temperatures are decreased by at least 10° and the smectic–cholesteric transition temperatures by about 20°–30° [*123, 18*]. Consequently the cholesteric mesophases of the cholesteryl alkyl carbonates occupy a considerably larger temperature interval than the corresponding cholesteryl alkanoates. Contrary to the latter series, the first two members of the alkyl carbonates fit the cholesteric–isotropic transition temperature curve very well. Below the clearing point most of the carbonates form the platelet structure discussed in Section II, A, 4, a. The helical twisting power increases with increasing chain length and reaches a constant value for the butyl carbonate and higher members [22]. Since the peak wavelength associated with this maximum twisting power amounts to about 0.29 μm, one should expect a color band in the vicinity of the cholesteric–smectic phase transitions. This in fact is observed with the heptyl through nonadecyl carbonates. Cholesteryl butyl through octyl carbonate do not exhibit a color band, because these compounds do not have a smectic mesophase and thus no pretransitional region in which the helix can unwind.

Cholesteryl S-alkyl thiocarbonates (VI): A thioalkyl group replacing the alkoxy group of V modifies only slightly the cholesteric–isotropic transition temperatures of the resulting S-alkyl thiocarbonates. However, the smectic–cholestic transition temperatures of the S-pentadecyl thiocarbonate and its lower homologs are significantly increased (Fig. 19). In addition, the lowest homolog with a smectic mesophase is shifted from the eighth member to the fifth member of this series [*124, 125*]. Only a few homologs exhibit a color band.

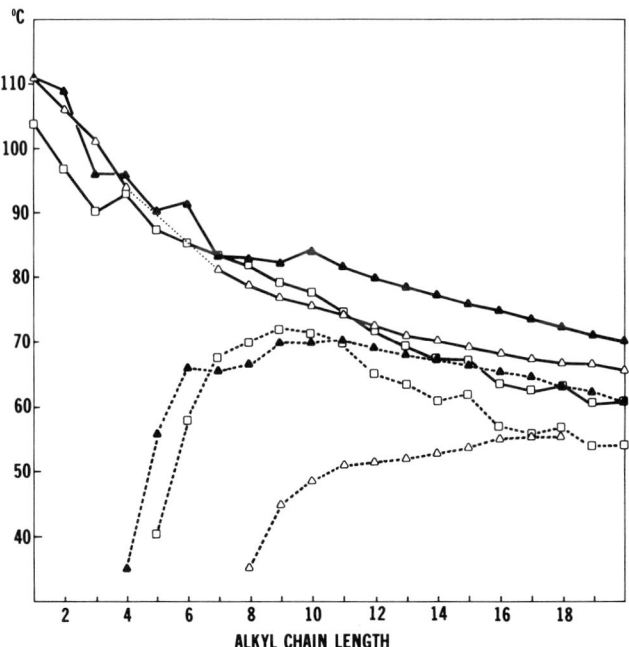

Fig. 19. Transition temperatures of carbonates. Transitions as defined in Fig. 17. △, Cholesteryl alkyl carbonates, V; □, cholesteryl S-alkyl thiocarbonates, VI; ▲, S-cholesteryl alkyl thiocarbonates, VII.

S-Cholesteryl alkyl thiocarbonates (VII): In this series, which is obtained by replacing the 3β-oxygen of V by sulfur, the smectic–cholesteric transition temperatures are increased much more than the clearing points. The minimum alkyl chain length for smectic homologs is shortened by four methylene groups (Fig. 19). But more importantly, the butyl thiocarbonate and its higher homologs selectively reflect the visible spectrum within about 0.1°. Some of these compounds are among the most temperature-sensitive cholesteric materials [*124, 90*].

5α-Cholestan-3β-yl alkyl carbonates (XVI): The observation that derivatives of 5α-cholestan-3β-ol have lower transition points in the melt than their cholesteryl analogs is supported by the decreased clearing point temperatures (about 20°) of this series. However, the smectic mesophase is suppressed completely (Fig. 20). Only a few members exhibit a color band [*113, 126*].

5α-Cholestan-3β-yl S-alkyl thiocarbonates (XVII): Figure 20 illustrates the difference between the carbonate group of XVI and the thiocarbonate group of XVII. The cholesteric–isotropic transition temperatures are hardly affected, but a smectic mesophase is introduced in *S*-nonyl through *S*-eicosyl thiocarbonate. Only four homologs selectively reflect the visible spectrum [*124, 127*].

c. *Alkoxy Esters. Cholesteryl β-alkoxypropionates* (VIII): A β-ether function in the original acyl chain of I results in monotropic cholesteric mesophases with cholesteric–isotropic transition temperatures about 10° lower than those of the corresponding cholesteryl alkanoates [*128*]. The compounds selectively reflect visible light and do not exhibit smectic mesophases.

Cholesteryl p-alkoxybenzoates (IX): The replacement of the two methylene groups of the previous series VIII by a phenylene group increases substantially

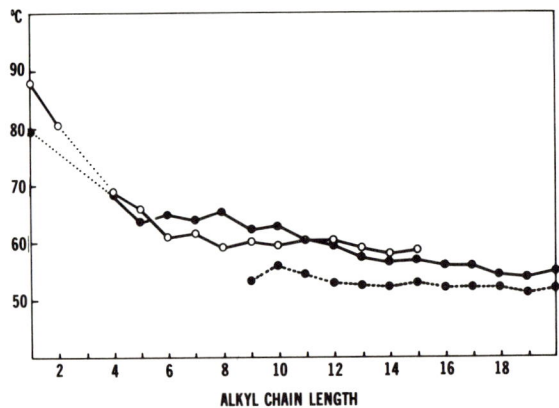

Fig. 20. Transition temperatures of carbonates. Transitions as defined in Fig. 17. ○, 5α-Cholestan-3β-yl alkyl carbonates, XVI; ●, 5α-cholestan-3β-yl *S*-alkyl thiocarbonates, XVII.

the lateral polarizability of the 3β-side chain and thus the lateral attraction between molecules. This is reflected by the observed increase of about 170° in the cholesteric–isotropic transition temperatures and the occurrence of smectic mesophases (Fig. 21). The visible spectrum is exhibited close to the smectic–cholesteric transitions, but the temperature intervals of the respective color bands have not been reported [*129*].

Cholesteryl p-alkoxycinnamates (X): The influence of a *trans*-cinnamoyl group on the mesomorphic behavior is demonstrated by this series (Fig. 21). The cholesteric–isotropic transition temperatures are increased by about an additional 20°. The homologs are enantiotropic cholesteric [*130*] and display

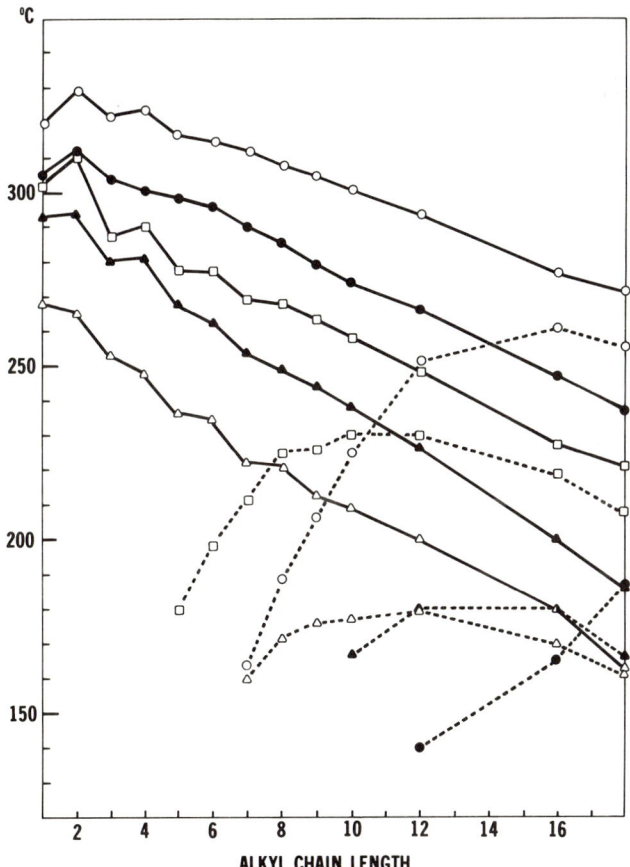

Fig. 21. Transition temperatures of alkoxy esters. Transitions as defined in Fig. 17. △, Cholesteryl *p*-alkoxy benzoates, IX; ▲, cholesteryl *p*-alkoxycinnamates, X; □, cholesteryl 6-alkoxy-2-naphthoates, XI; ●, cholesteryl *p*-{[(4-alkoxy-1-naphthyl)methylene]amino} benzoates, XII; ○, cholesteryl *p*-[(*p*-alkoxybenzylidene)amino]benzoates, XIII.

colors. The visible spectrum is exhibited only by those homologs that are both smectic and cholesteric.

Cholesteryl 6-alkoxy-2-naphthoates (XI): The transition temperatures in the melt of the compounds in this series are about 50° higher than those of the benzene analogs (IX) (Fig. 21). This may be explained by the fact that the naphthalene ring system in XI has a greater polarizability than the benzene ring in IX. All reported naphthoates (XI) are enantiotropic cholesteric and display colors close to the cholesteric–smectic transition [*131*].

Cholesteryl p-{[(4-alkoxy-1-naphthyl)methylene]amino}benzoates (XII): Although the cholesteric–isotropic transition temperatures are higher than those of the last series, the smectic–cholesteric transition temperatures are drastically reduced (Fig. 21). Furthermore, only the highest homologs have a smectic mesophase [*132*].

Cholesteryl p-[(p-alkoxybenzylidene)amino]benzoates (XIII): These compounds exhibit the highest cholesteric–isotropic and smectic–cholesteric transition temperatures reported to date. The 7-heptyloxy derivative has the remarkable cholesteric range of over 150°. The visible spectrum appears close to the smectic–cholesteric transition temperatures [*133*].

B. Constitution and Mesomorphic Behavior

The influence of the sterol moiety on the mesomorphic behavior can be determined only if one can separate the effects of the 3β- and the 17β-side chains from those caused by variations of the sterane structure. At present this distinction is difficult to achieve since in most cases only a few derivatives have been reported for a given modification of the sterol. Nevertheless, the comparison of these derivatives with the analogs of cholesterol and 5α-cholestan-3β-ol should surface at least some drastic effects, assuming that impurity influences are negligible. Since the latter may not always be the case and because some of the investigators did not properly identify the mesophases, we decided to present primarily the published data.

1. SUBSTITUTED 17β-SIDE CHAIN

Both alkyl and alkylidene substituents in the 17β-side chain of sterols increase the lateral polarizability of the molecule. It is therefore not surprising that this type of substitution increases the smectic stability. The results are summarized in Table III. Notice that the alkanoates tend to be more smectic than the respective benzoates and cinnamates.

In addition to 17β-alkyl substitution, only a few compounds with modifications in the sterane moiety have been investigated, such as introducing a Δ^7 double bond into **6**. The alkanoates of (22E, 24R)-ergosta-5,7,22-trien-3β-ol (ergosterol, **15**) are smectic [*152*], while the benzoate and several substituted

TABLE III
Substituted 17β-Side Chain

No.	R	Compound	Derivatives	Mesophase[a]	Reference
3		(24R)-Ergost-5-en-3β-ol (campesterol)	Alkanoates Substituted benzoates	S, Ch S, Ch	134 98
4		(24S)-Ergost-5-en-3β-ol (22,23-dihydrobrassicasterol)	p-Nitrobenzoate	Ch, c	135
5		(22E,24S)-Ergosta-5,22-dien-3β-ol (crinosterol)	Benzoate	Ch, c	136, 137
6		(22E,24R)-Ergosta-5,22-dien-3β-ol (brassicasterol)	Benzoate	NR	138
7		24-Methylcholesta-5,24(28)-dien-3β-ol (chalinasterol)	Benzoate	M	139
8		Cholest-5-en-3β-ol-22-one (22-ketocholesterol)	Acetate Benzoate	NR NR	140 141

(continued)

TABLE III—*continued*

No.	R	Compound	Derivatives	Mesophase[a]	Reference
9		Cholest-5-en-3β-ol-24-one (24-ketocholesterol)	Acetate	Ch, c	*139, 142*
10		(24S)-Stigmasta-5,22-dien-3β-ol (stigmasterol)	Alkanoates	S	*143, 144*
			Alkyl carbonates	S	*145*
			N-(p-Methoxyphenyl)-carbamate	Ch	*146*
			Benzoate, substituted benzoates	Ch	*98*
			Prop-2-yn-1-yl carbonate	Ch, c	*147*
			3-Methoxy-4-acyloxy-cinnamates	Ch, c	*148*
11		(24R)-Stigmasta-5,22-dien-3β-ol (poriferasterol)	Benzoate	M	*149*
12		(24R)-Stigmast-5-en-3β-ol (β-sitosterol)	Alkanoates	S	*150*
			N-(p-Methoxyphenyl)-carbamate	Ch	*146*
			3-Methoxy-4-acyloxy-cinnamates	Ch, c	*148*
13		(24S)-Stigmast-5-en-3β-ol (clionasterol)	Benzoate	Ch, c	*149*
14		(24E)-Stigmasta-5,24(28)-dien-3β-ol (fucosterol)	Alkanoates	S	*19, 151*

[a] Abbreviations used: S = smectic mesophase; Ch = cholesteric mesophase; c = colors; M = mesophase, not identified; — = not observed; NR = not reported.

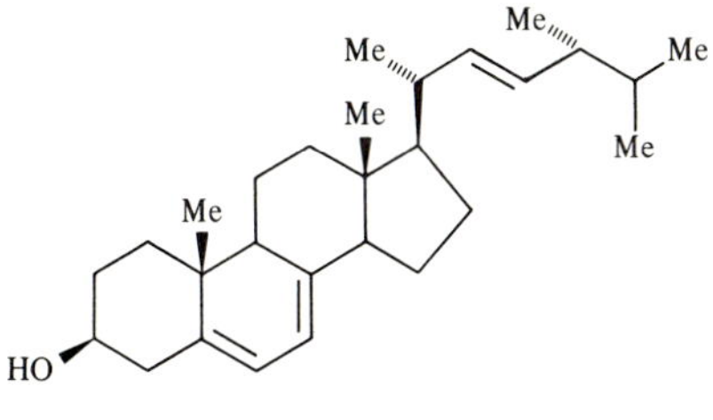

15 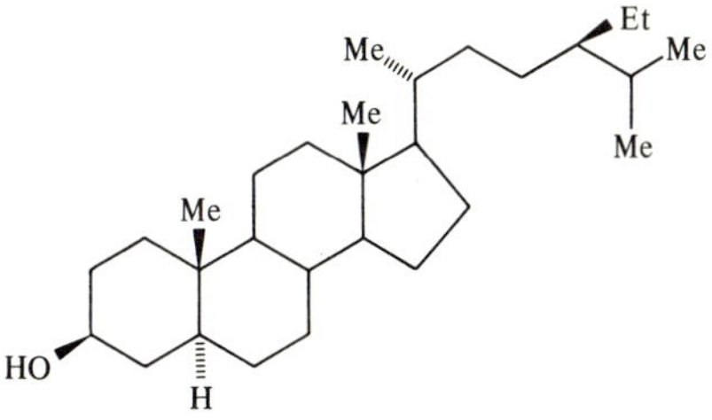   16

benzoates are cholesteric [98]. On complete hydrogenation of **10**, one obtains (24S)-5α-stigmastan-3β-ol (**16**), which has a smectic octadecanoate [151].

While alkanoates of **10** and **12** are only smectic, two cholesteric derivatives have recently been reported: N-(p-methoxyphenyl)carbamates [146] and 3-methoxy-4-acyloxycinnamates [148]. It would be quite interesting to prepare these two derivatives of the compounds listed in Table III to investigate the influence of the substituted 17β-side chain on the stability of the cholesteric mesophase.

2. Minor Changes at the 17β-Side Chain

Minor changes, which result in different polarizabilities, permanent dipole moments, or changes in the breadth of the molecule, can greatly influence the mesomorphic behavior, as exemplified by the compounds discussed in the following.

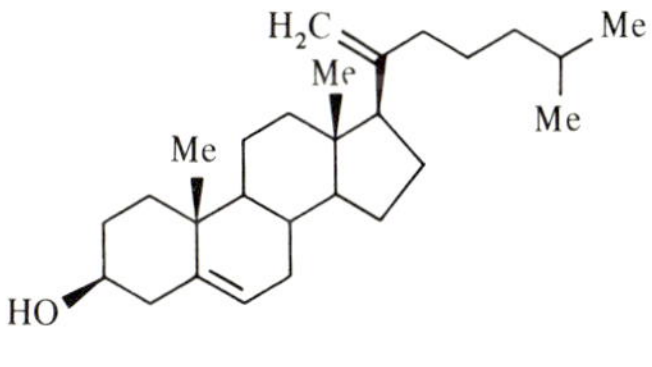

17 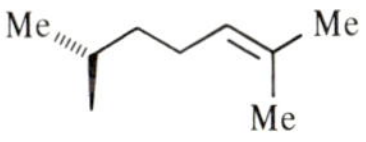   18 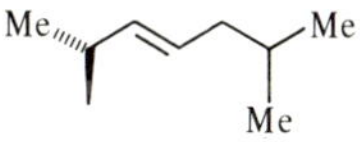   19
       20 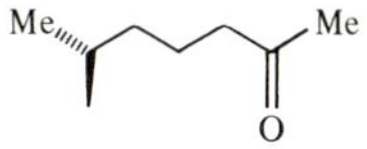   21 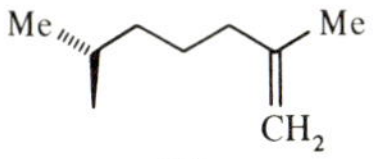
22 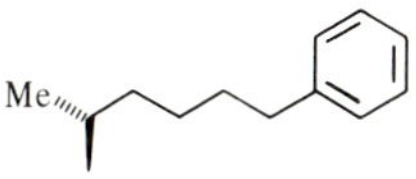   23 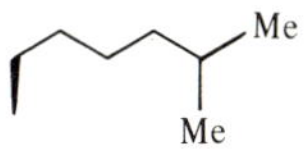   24 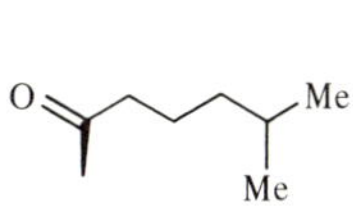
       25 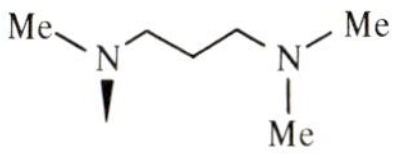   26

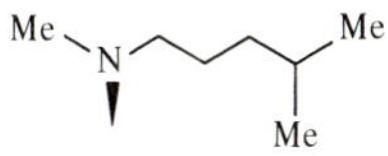

The acetate of cholesta-5,20-dien-3β-ol (**17**) is monotropic cholesteric and monotropic smectic and exhibits the visible spectrum of selectively reflected light [*134*]. No mesophases have as yet been reported for derivatives of cholesta-5,24-dien-3β-ol (desmosterol, **18**), but the benzoate of cholesta-5,22-dien-3β-ol (**19**) is enantiotropic cholesteric and displays colors [*153*]. The acetate of cholesta-5,25-dien-3β-ol (**20**) is monotropic cholesteric but does not display colors [*134*].

The *p*-nitrobenzoate of 27-norcholest-5-en-3β-ol-25-one (**21**) is enantiotropic cholesteric and displays colors [*154*] and the acetate is monotropic cholesteric [*134*]. However, the acetate of 21-norcholest-5-en-3β-ol-20-one (**22**) is not mesomorphic [*134*].

The benzoate of 17β-(1-methyl-5-phenylpentyl)androst-5-en-3β-ol (**23**) is enantiotropic mesomorphic [*155*]. The acetate of 21-norcholest-5-en-3β-ol (**24**), and the acetates and octadecanoates of 20-azacholest-5-en-3β-ol (**25**) and of 20,25-diazacholest-5-en-3β-ol (**26**) are not mesomorphic [*134*].

3. ELONGATED AND ABBREVIATED 17β-SIDE CHAIN

An intriguing question is the influence of the length of the 17β-side chain of sterols on the mesomorphic properties. The compounds listed in Table IV demonstrate that the 17β-side chain can be varied considerably in length without a loss of the mesomorphic character. No mesomorphic derivatives of androst-5-en-3β-ol (**39**) [*156, 159*] and of androst-5-en-3β-ol-17-one (**40**) have been reported [*159, 146*]. The octadecanoate of pregn-5-en-3β-ol-20-one (**41**) is not mesomorphic [*159*], but substituted benzoates exhibiting a cholesteric mesophase have been reported [*98*]. The *p*-methoxycinnamate of 5α-androstan-3β-ol (**42**) is monotropic cholesteric [*158*].

The *p*-methoxycinnamates of 17β-methyl-5α-androstan-3β-ol (**43**), 5α-pregnan-3β-ol (**44**), 17β-propyl-5α-androstan-3β-ol (**45**), and 17,17β-dimethyl-*D*-homo-18-nor-5α-androsta-13,15,17-trien-3β-ol (**46**) are enantiotropic cholesteric [*158*].

The octadecanoate of methyl 3β-hydroxychol-5-en-24-oate (**47**) is monotropic cholesteric and exhibits cholesteric colors [*159*] and several substituted benzoates are also cholesteric [*98*].

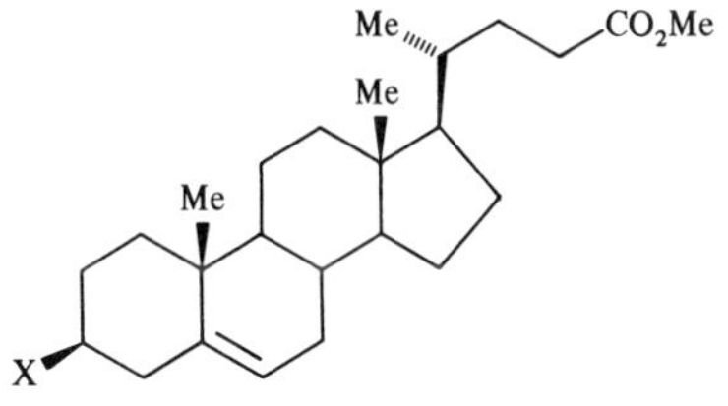

47 X = OH
48 X = Cl

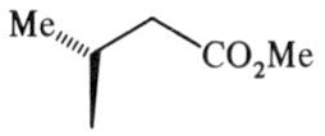

49

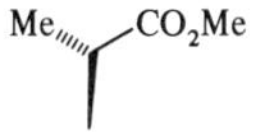

50

51

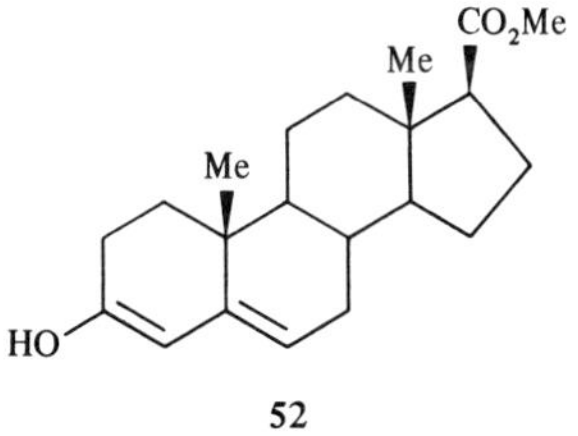

52

Methyl 3β-chlorochol-5-en-24-oate (**48**) is also cholesteric and displays colors [*160*]. The acetate of methyl 3β-hydroxy-24-norchol-5-en-23-oate (**49**) and the octadecanoate of 20β-carbomethoxypregn-5-en-3β-ol (**50**) are monotropic cholesteric [*159*]. While no mesophase could be detected in the octadecanoate of 17β-carbomethoxyandrost-5-en-3β-ol (**51**) [*159*], the acetate of 17β-carbomethoxyandrosta-3,5-dien-3-ol (**52**) has a clearing point [*161*].

TABLE IV

Elongated and Abbreviated 17β-Side Chain

No.	R	Compound	Derivatives	Mesophase[a]	References
27	$(CH_2)_6CH_3$	17β-(1-Methyloctyl)androst-5-en-3β-ol	Acetate, benzoate	M	155
28	$(CH_2)_5CH_3$	17β-(1-Methylheptyl)androst-5-en-3β-ol	Acetate, benzoate	M	155
29	$(CH_2)_4CH_3$	27-Norcholest-5-en-3β-ol	Acetate	Ch	134
30	$(CH_2)_3CH_3$	24-Methylchol-5-en-3β-ol	Acetate	Ch, c	156, 157
31	$(CH_2)_2CH_3$	Chol-5-en-3β-ol	Acetate	Ch	156, 157
32	CH_2CH_3	24-Norchol-5-en-3β-ol	Acetate	Ch	156, 157
33	CH_3	20β-Methylpregn-5-en-3β-ol	Acetate	—	156
34	H	Pregn-5-en-3β-ol	Nonanoate	S, Ch	156
			Octadecanoate	Ch	156
			Benzoate, cinnamate	Ch	158
35	$CO(CH_2)_3CH_3$	27-Norcholest-5-en-3β-ol-22-one	Acetate	Ch	156
36	$CO(CH_2)_2CH_3$	24-Methylchol-5-en-3β-ol-22-one	Acetate	Ch	156
37	$COCH_2CH_3$	Chol-5-en-3β-ol-22-one	Acetate	Ch	156
38	$COCH_3$	24-Norchol-5-en-3β-ol-22-one	Acetate	—	156

[a] Abbreviations used: M = mesophase, not identified; Ch = cholesteric mesophase; c = cholesteric colors; S = smectic mesophase; — = not observed.

4. Position of Double Bonds in the Sterane Skeleton

Wiegand [173] determined the melting and clearing points of the benzoates of the then-known cholesten-3β-ols and cholestadien-3β-ols, but unfortunately did not identify the respective mesophases. Only two of his original seven compounds have since been recognized as being cholesteric. Although the number of sterol olefins has expanded considerably in the past 25 years, only a few have been investigated with respect to mesophases. The known cholestadien- and trien-3β-ols and cholesten-3β-ols and their derivatives are presented in Tables V and VI.

TABLE V

DERIVATIVES OF CHOLESTADIENES AND CHOLESTATRIENES

No.	Compound	X	Derivatives	Mesophase[a]	References
53	5α-Cholesta-1,3-dien-3-ol	OH	Acetate	NR	162
54	Cholesta-3,5-dien-3-ol	OH	Acetate	M	161, 163
55	3-Chlorocholesta-3,5-diene	Cl	—	M	164
56	Cholesta-4,6-dien-3β-ol	OH	Acetate, benzoate	NR	165, 166
57	Cholesta-4,7-dien-3β-ol	OH	Acetate, benzoate	NR	167
58	Cholesta-5,7-dien-3β-ol	OH	Alkanoates	S	168
			Acetate	Ch	116
			Benzoate, substituted benzoates	Ch	98
59	Cholesta-5,7-dien-3β-thiol	SH	Acetate, benzoate	M	169
60	3β-Chlorocholesta-5,7-diene	Cl	—	Ch	89
61	Cholesta-5,8-dien-3β-ol	OH	Acetate, benzoate	NR	170
62	Cholesta-5,16-dien-3β-ol	OH	Acetate, benzoate	NR	171
63	5α-Cholesta-6,8-dien-3β-ol	OH	Acetate, benzoate	M	172
64	5α-Cholesta-7,9(11)-dien-3β-ol	OH	Benzoate	M	173, 174
65	5α-Cholesta-7,14-dien-3β-ol	OH	Benzoate	NR	175, 176
66	5α-Cholesta-8,14-dien-3β-ol	OH	Benzoate	NR	176, 177
67	5α-Cholesta-6,8(14),9(11)-trien-3β-ol	OH	p-Nitrobenzoate	M	178

[a] Abbreviations used: M = unidentified mesophase; S = smectic; Ch = cholesteric; NR = not reported; — = not observed.

TABLE VI

Derivatives of Cholestenes

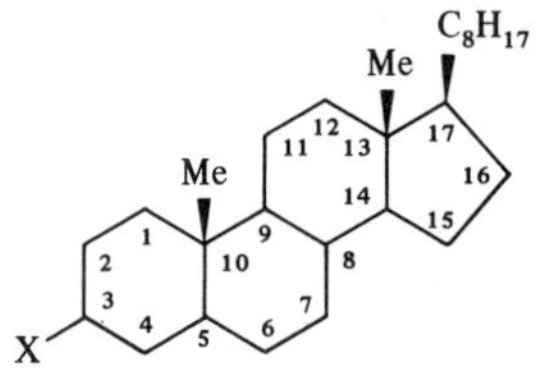

No.	Compound	X	Derivatives	Mesophase[a]	References
68	5α-Cholest-1-en-3β-ol	OH	Octadecanoate	—	159
69	3-Phenyl-5α-cholest-2-ene	C_6H_5	p-Substituted	Ch	179
70	5α-Cholest-3-en-3-ol	OH	Acetate	NR	180
71	Cholest-4-en-3β-ol	OH	Octadecanoate	—	159
72	5α-Cholest-6-en-3β-ol	OH	Benzoate	NR	181
73	5α-Cholest-7-en-3β-ol (lathosterol)	OH	Alkanoates	S	168
			Benzoate	M	174
74	5α-Cholest-8-en-3β-ol	OH	Benzoate	M	173, 175
75	5α-Cholest-8(14)-en-3β-ol	OH	Alkanoates	Ch, S, c	116, 117
			Benzoate	Ch	116
	3β-Chloro-5α-cholest-8(14)-ene	Cl	—	Ch	116
76	5α-Cholest-9(11)-en-3β-ol	OH	Acetate	NR	163, 182
77	5α-Cholest-14-en-3β-ol	OH	Benzoate	NR	173

[a] Abbreviations used: Ch = cholesteric; S = smectic; M = unidentified mesophase; c = cholesteric colors; NR = not reported; — = not observed.

5. Double Bonds in 17β-Side Chain and Sterane Moiety

Only a few representatives of this group have been investigated. Neither the acetate nor the benzoate of 5α-cholesta-7,22-dien-3β-ol (**78**) are reported to be mesomorphic [*183*]. A clearing point has been reported [*173, 184, 185*] for the benzoate of 5α-cholesta-8,24-dien-3β-ol (zymosterol, **79**) but neither the acetate nor the hexadecanoate appear to be mesomorphic [*186*]. No mesophase has been observed in the benzoate of 5α-cholesta-14,24-dien-3β-ol (**80**) [*173*].

6. Triterpenoids

In the past few years Nicholas and co-workers have investigated derivatives of triterpenoids for their mesomorphic properties [*187*]. Triterpenoids form a very large group of naturally occurring substances, widely distributed throughout the plant kingdom [*188, 189*]. A small but important group, which includes lanosterol, it is of animal origin. The compounds investigated are classified as derivatives of lanostane and cycloartane [*188, 190*].

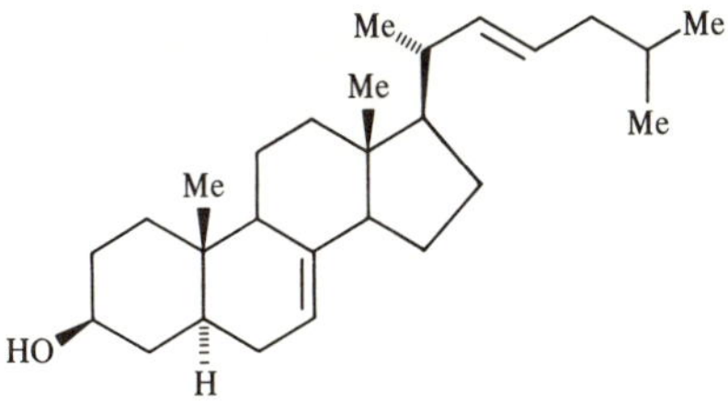

78

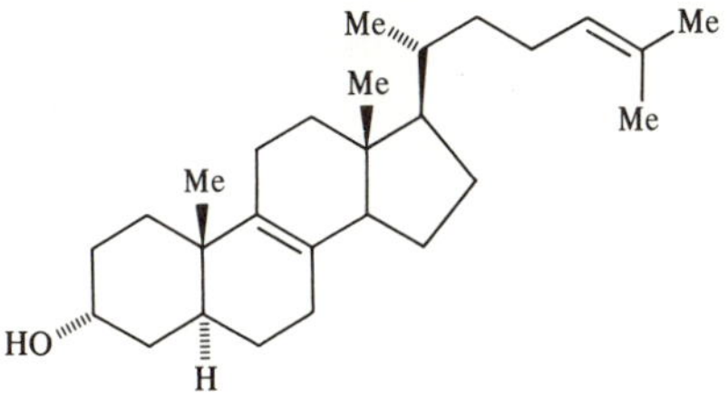

79

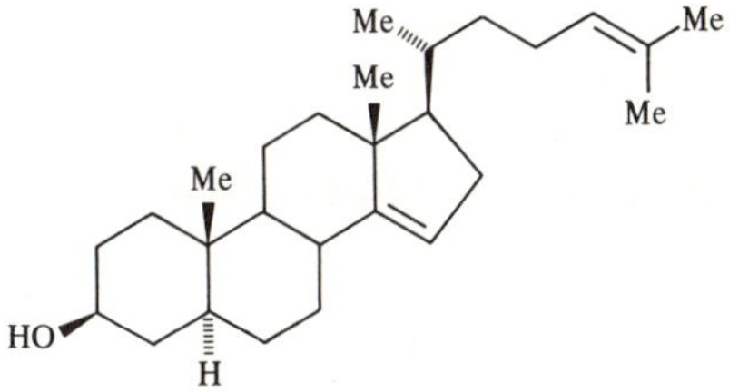

80

a. *Lanostane Group.* Smectic mesophases were observed [*191*] in alkanoates of 4α-methyl-5α-cholest-7-en-3β-ol (lophenol, **81**), but a recent calorimetric investigation [*192*] of its hexadecanoate also revealed an enantiotropic cholesteric mesophase which exists only within a narrow temperature range. Alkanoates of 4,4-dimethyl-5α-cholest-7-en-3β-ol (**82**) are also reported to be smectic [*168*]. The acetate of 4,4,14α-trimethyl-5α-cholest-7-en-3β-ol (**83**) is not mesomorphic [*168*]. Neither the acetate of 4α,14α-dimethyl-5α-cholest-8-en-3β-ol (**84**) nor a homologous series of alkanoates of lanost-8-en-3β-ol (dihydrolanosterol, **85**) exhibited any mesophases [*186*].

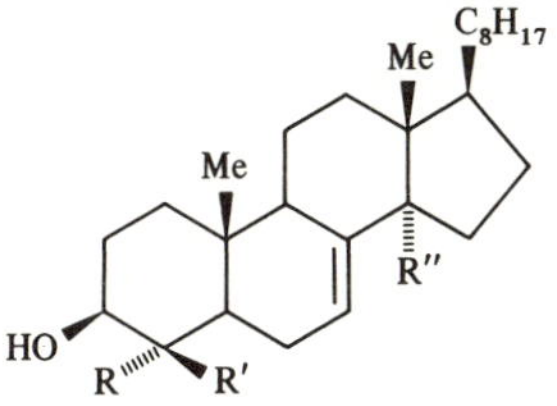

81 R' = R" = H; R = Me
82 R = R' = Me; R" = H
83 R = R' = R" = Me

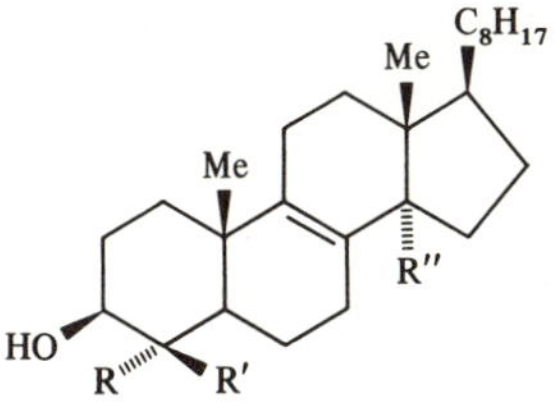

84 R' = H; R = R" = Me
85 R = R' = R" = Me

No mesophases were observed in alkanoates of lanosta-8,24-dien-3β-ol (lanosterol, **86**) and of its epimer 13α,14β,17β$_H$-lanosta-8,24-dien-3β-ol (tirucallol, **87**) [*151*, *186*].

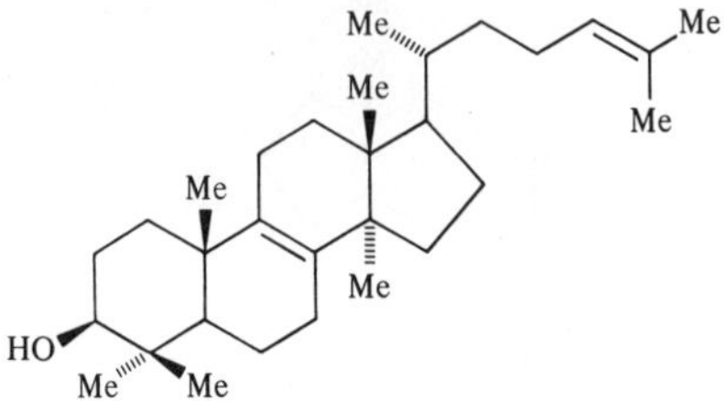

86

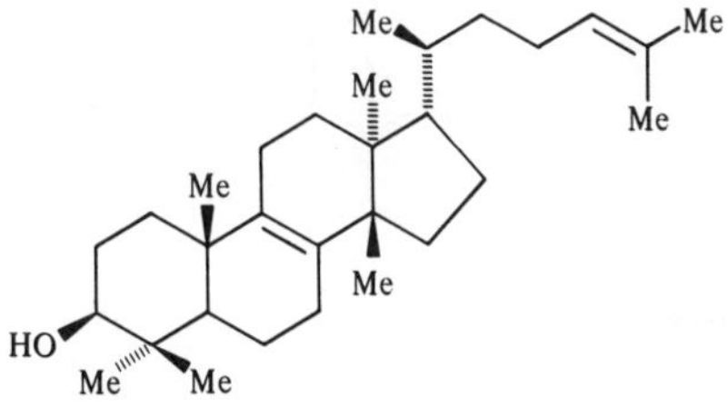

87

No mesophase has been reported for the benzoate of 4,4-dimethyl-5α-cholest-8(14)-en-3β-ol (**88**) [*193*]. The acetate of 4α,14α-dimethyl-5α-cholest-9(11)-en-3β-ol (**89**) is monotropic cholesteric and exhibits the visible spectrum of selectively reflected light [*186*]. No mesophase has been reported for the benzoate of 4,4-dimethyl-5α-cholesta-8,14-dien-3β-ol (**90**) [*193*], in line with Wiegand's observation that a double bond in the 14-position apparently prevents the formation of a mesophase.

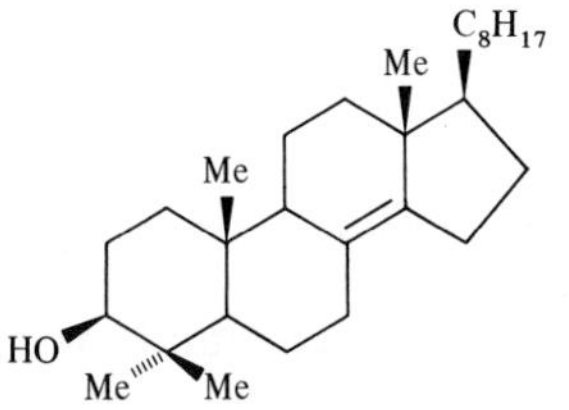

88

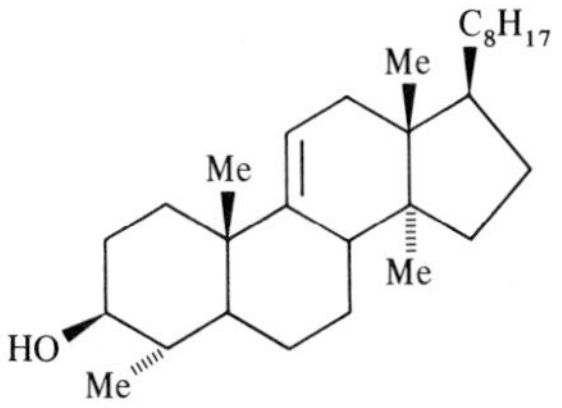

89

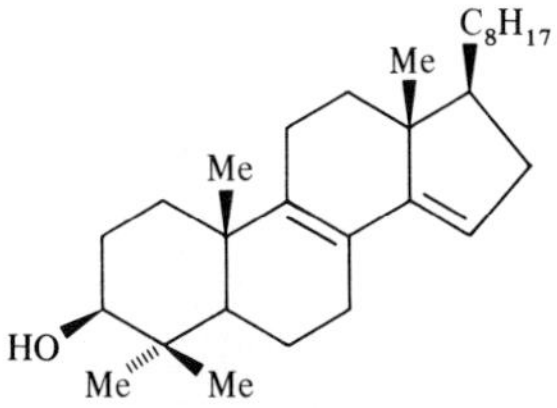

90

b. *Cycloartane Group.* Smectic and cholesteric mesophases as well as cholesteric colors have been observed in alkanoates of 14α-methyl-9,19-cyclo-5α,9β-cholestan-3β-ol (pollinastanol, **91**) [*194*], of 4α,14α-dimethyl-9,19-cyclo-5α,9β-cholestan-3β-ol (31-norcycloartanol, **92**) [*195*], and of 9,19-cyclo-9β-lanostan-3β-ol (cycloartanol, **93**) [*196, 197*]. Only a monotropic cholesteric mesophase was found in 9,19-cyclo-9β-lanost-24-en-3β-ol (cyclo-

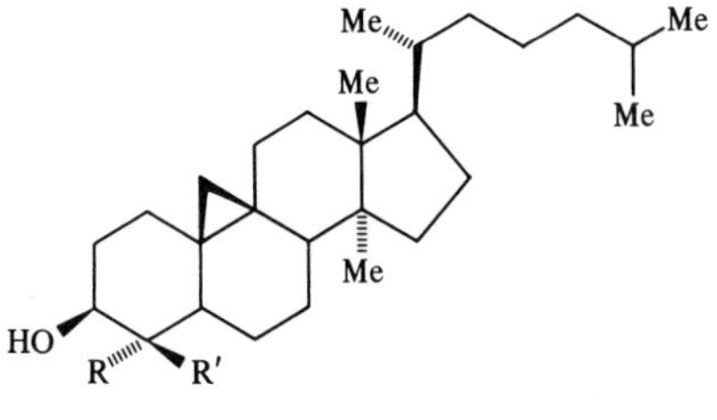

91 R = R' = H
92 R = Me; R' = H
93 R = R' = Me

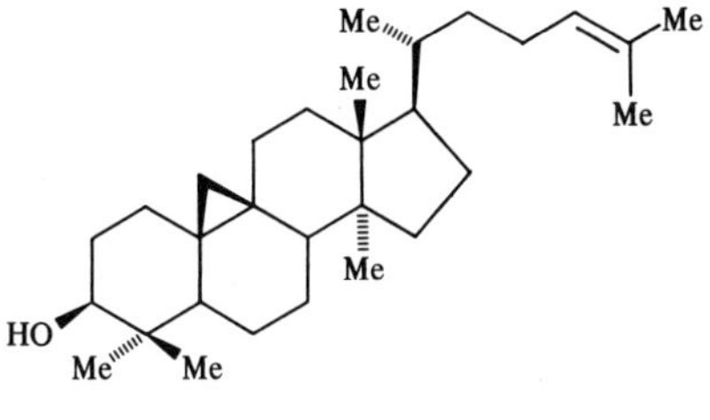

94

artenol, **94**) [*151, 197, 198*]. Smectic mesophases have been observed in alkanoates of 4α, 14α-dimethyl-9,19-cyclo-5α,9β-ergost-24(28)-en-3β-ol (cycloeucalenol, **95**) [*151, 195*] and of 4,4,14α-trimethyl-9,19-cyclo-5α,9β-ergost-24(28)-en-3β-ol (24-methylenecycloartanol, **96**) [*151, 196*]. The reduction of the 24-methylene group leads to cholesteric compounds. Cholesteric colors have been observed in derivatives of 4α,14α,24ξ-trimethyl-9,19-cyclo-5α,9β-cholestan-3β-ol (cycloeucalanol, **97**) [*151, 196*] and of 24ξ-methyl-9,19-cyclolanostan-3β-ol (**98**) [*196, 199*].

95 R = Me; R' = H
96 R = R' = Me

97 R = Me; R' = H
98 R = R' = Me

A comparison of the terpenoids and the respective sterols shows that methyl substituents in the 4- and 14α-positions and the 9,19-cyclopropane ring do not drastically influence the mesomorphic properties.

C. Nonsteroidal Cholesteric Liquid Crystals

Although cholesteric behavior was already observed in 1910 by Vorländer's school with (S)-(−)-2-methylbutyl p-[(p-cyanobenzylidene)amino]cinnamate (**99**) and related derivatives [*200, 1*], only recently have several additional optically active compounds been investigated. Since hardly any measurements of their respective selective reflection have been reported, we will only briefly discuss some representative examples (Table VII).

Compounds investigated are derivatives of optically active p-(2-methylbutyl)benzoic acid (**100**) [*201*], optically active 4-alkoxybiphenyl-4'-carboxylic acids (**101**) [*202, 203*], and alkyl p-[(p-arylidene)amino]cinnamates (**102**) [*203*]; (S)-(−)-2-methylbutyl p-[(p-subst. benzylidene)amino]cinnamates (**103**) [*204*], p-(2-methylbutyl)-N-(p-alkoxybenzylidene)anilines (**104**) [*205*], N-p-[(p-alkoxybenzylidene)amino]phenyl acylates (**105**) [*206*], N-p-[(p-alkoxybenzylidene)amino]benzonitriles (**106**) [*206*], optically active bis-Schiff bases (**107**) [*207*], and a homologous series of N-(p-alkoxybenzylidene)-p-methylalkylanilines (**108**) [*208*]. The racemates of these compounds exhibit

TABLE VII

Nonsteroidal Cholesteric Liquid Crystals

$NC-\langle\text{ph}\rangle-CH=N-\langle\text{ph}\rangle-CH=CH-CO_2CH_2\overset{*}{C}H(CH_3)CH_2CH_3$ **99**

$CH_3CH_2\overset{*}{C}H(CH_3)CH_2-\langle\text{ph}\rangle-CO_2R$ **100**

$HO_2C-\langle\text{ph}\rangle\langle\text{ph}\rangle-OCH_2\overset{*}{C}H(CH_3)R$ **101**

$X-\langle\text{ph}\rangle-CH=N-\langle\text{ph}\rangle-Y$ **102–109**

X	Y	No.
R	$CH=CH\ CO_2CH_2\ \overset{*}{C}H(CH_3)R'$	102
R	$CH=CH\ CO_2CH_2\ \overset{*}{C}H(CH_3)CH_2CH_3$	103
$CH_3(CH_2)_nO$	$CH_2\ \overset{*}{C}H(CH_3)CH_2CH_3$	104
$CH_3CH_2\overset{*}{C}H(CH_3)(CH_2)_mO$	$CO_2(CH_2)_nCH_3$	105
$CH_3CH_2\overset{*}{C}H(CH_3)(CH_2)_nO$	CN	106
$CH_3CH_2\overset{*}{C}H(CH_3)CH_2CO_2-\langle\text{ph}\rangle-N=CH$	$CO_2CH_2\overset{*}{C}H(CH_3)CH_2CH_3$	107
$CH_3(CH_2)_mO$	$(CH_2)_nCH_2\overset{*}{C}H(CH_3)CH_2CH_3$	108
NC	$CH=CH\ CO_2\overset{*}{C}H(D)(CH_2)_2CH_3$	109

nematic and smectic mesophases, depending on the length of the alkyl chain. The enantiomers exhibit cholesteric mesophases, with polymesomorphism in certain cases (the enantiomers of **101**, $R = n-C_4H_9, C_6H_5CH_2$, have two cholesteric and the racemates two nematic mesophases each [*203*]). Dolphin et al. [*205*] therefore proposed the term "chiral nematic" for this type of cholesteric mesophase in contrast to "cholesteric" for derivatives of sterols. Coates and Gray [*209*] have recently demonstrated with 1-butyl-1-D p-[(p-cyanobenzylidene)amino]cinnamate (**109**) that the H–D asymmetry is sufficient to produce a cholesteric mesophase. Both the measured pitch, which is very much longer than that of common cholesteric compounds—about 10–1000 times—, and the observed Grandjean terraces prove the existence of a cholesteric mesophase.

D. Stability

Cholesterol deteriorates in the presence of air in the dark, a process which is accelerated under the influence of heat and light. The complexity of the autoxidation of cholesterol was recognized only with the advent of modern analytical methods, such as thin-layer and gas–liquid chromatography [210]. Commercial cholesterol contains several autoxidation products which can be removed only by the bromination–debromination method of Fieser [211].

Aliphatic esters, alkyl carbonates, and S-alkyl thiocarbonates of cholesterol, especially the long-chain derivatives, also deteriorate over a period of several months, while aromatic esters appear to be more stable. Derivatives of cholest-5-en-3β-thiol are definitely more stable than their cholesteryl analogs. We observed that S-cholesteryl alkyl thiocarbonates (VII) and S-cholesteryl ω-phenylalkanethioates (IV) are stable for a period of several years if they are protected from light.

Scala and Dixon [212, 213] investigated the stabilities of cholesteryl nonanoate, cholesteryl oleyl carbonate, 3β-chlorocholest-5-ene, and their respective mixtures, which are frequently used as temperature indicators. They found that aging in air and in the dark does not significantly affect the temperature ranges of the visible spectra of selectively reflected light, but sharply reduces the reflectance. The presence of both oxygen and UV radiation (0.254 μm) accelerated the deterioration, and thin films and solutions deteriorated even faster.

Cholesteryl oleyl carbonate decomposes gradually at room temperature with the evolution of carbon dioxide. A decomposition rate of the order of 10^{-5} mol %/day *in vacuo* at 39° was obtained by mass spectroscopic analysis. In the presence of oxygen, oxidative decomposition occurs with the evolution of carbon monoxide. This presumably free-radical process was not observed in the presence of the inhibitor nitric oxide [214].

Our studies on the stability of high-purity cholesteryl oleyl carbonate [215], mainly by chromatographic methods, support these investigations. We found that it is reasonably stable if it is kept under an inert gas in the dark. As a matter of fact, thin films hermetically sealed between 6-μm thick Mylar* membranes had about the same selective reflectance in approximately the same temperature region after over five years of storage in the dark. But exposure to air and light eventually results in a material that can no longer be repurified by chromatographic methods. As one of the many oxidation products we identified dicholesteryl carbonate.

The addition of UV absorbers to cholesteryl oleyl carbonate greatly extends

---

* Mylar is the trademark of E. I. DuPont de Nemours & Co. for poly(ethylene terephthalate).

its stability. A few percent of 4-(4′-ethoxyphenylazo)phenyl hexanoate do not significantly influence the peak reflectance temperature of mixtures containing cholesteryl oleyl carbonate. The combination of this absorber and a small amount of silicone resulted in mixtures which changed very little over a period of several months under UV radiation (0.366 μm) if they were sandwiched between Mylar films [213]. Other UV absorbers studied include 4-(phenylazo)phenol and 4-(4′-ethoxyphenylazo)phenyl 10-undecenoate. Sandwiched films, claimed to be stable to about 4000 hours of UV radiation, are obtained by adding about 5% of these azo dyes to the cholesteric mixtures [216].

E. Summary and Conclusion

The study of the influence of substitutions on the mesomorphism of cholesteric liquid crystals has resulted in some interesting correlations.

1. The replacement of the hydroxyl groups of cholesterol by small electronegative groups can induce cholesteric behavior.

2. 3β-Substitutions of cholesterol, which enlarge the lateral dimensions of the molecule, apparently prevent the formation of mesophases.

3. 3β-Substitutions of cholesterol derivatives, which extend and modify the chain, lead to the following:

   a. In alkanoates and carbonates the cholesteric–isotropic transition temperatures decrease with chain length while the smectic–cholesteric transition temperatures first rise and then fall.

   b. The ω-phenylalkanoates exhibit a pronounced odd–even effect of the cholesteric–isotropic transition temperatures. ω-Phenylalkyl carbonates should show similar behavior.

   c. The substitution of oxygen by sulfur next to the carbonyl group increases both the cholesteric–isotropic and the smectic–cholesteric transition temperatures of the alkanethioates and alkyl thiocarbonates and accentuates the odd–even effect of the ω-phenylalkanethioates. The latter display, in addition, a small odd–even effect of the smectic–cholesteric transition temperatures, which is in phase with the cholesteric–isotropic transition temperatures.

   d. Hydrogenation of the 5,6-double bond of the cholesterol to the 5α-cholestan-3β-ol moiety lowers the cholesteric–isotropic transition temperatures of the alkanoates, alkyl carbonates, and S-alkyl thiocarbonates.

   e. The known alkoxy esters of cholesterol have much higher smectic–cholesteric and cholesteric–isotropic transition temperatures than the corresponding cholesterol alkanoates. The substantial increase of the smectic stability may be explained by the higher lateral polarizability of the alkoxy functions.

The information about the effect of variations of the 17β-side chain, the sterane moiety, and of both is insufficient to correlate with the same degree of confidence changes of the molecular features with those of the mesomorphic state. To achieve the latter requires the investigation of sequences of corresponding homologous series which could allow separation of the effects of chain length within a specific homologous series from the effects of those molecular differences that distinguish the various homologous series.

IV. TEMPERATURE DEPENDENCE OF SELECTIVE REFLECTION

A. Single Compounds

As we have seen, impurities can strongly influence the temperature dependence of the selective reflection. They can suppress this effect in the visible and change temperature interval and temperature region of the color band. Furthermore, certain impurities can introduce a hysteresis effect, as indicated by the different temperatures of the maximum spectral reflectance observed for heating and cooling [36, 217]. For these reasons it is necessary to reinvestigate the cholesteric properties of those compounds which were prepared before modern methods of purifications were known.

1. Temperature Regions of Color Bands

Table VIII lists the temperature intervals associated with the color bands of the homologous series which were purified by chromatographic methods. Nevertheless, one should expect some inaccuracies that are larger than those implied by the last digit of the temperature values, because in many instances the width of the color band was subjectively determined by the human eye. Although most of the compounds listed are cholesteric, we reported only relatively stable color bands and did not give the estimated temperatures of those color regions which could be observed only temporarily on rapid undercooling. As a consequence, these tables provide an overview of the compounds that might be useful for thermometric applications.

2. Dependence of Selective Reflectance on Temperature and Wavelength

The temperature and wavelength dependence of the selective reflection was investigated for relatively few compounds [36, 47]. The samples were films about 20 μm thick which exhibited a polydomain structure of planar texture elements. It was estimated that the helical axes of the domains were aligned to within a few degrees of the surface normal (see Section II, B, 2). The incident and selectively reflected light rays filled a narrow cone perpendicular to the

TABLE VIII
Temperature Intervals of Visible Spectrum of Selective Reflection[a]

n	I	II	III	IV	V	VI	VII	XIV	XV	XVI	XVII
5	—	—	—	—	—	—	—	—	—	—	—
6	—	—	—	—	—	—	56.3-56.0	—	—	—	—
7	—	72.7-71.5	83.0-82.0	—	—	56.0-53.0	68.5-68.0	—	—	—	—
8	—	74.0-72.9	—	—	—	68.0-67.0	64.3-64.1	—	—	—	34.3-31.3
9	77.9-76.5	—	—	—	41.0-37.3	—	67.3-66.7	—	—	—	47.8-47.0
10	77.1-76.1	—	—	50.3-48.1	50.0-45.5	—	70.2-70.1	—	—	—	52.4-51.9
11	—	—	39.5-39.3	—	51.0-49.4	—	70.5	—	—	—	55.4-55.2
12	—	—	47.0-44.2	49.9-48.7	52.3-51.5	—	70.5	—	—	—	—
13	—	—	44.6-44.3	—	53.0-52.1	—	68.7	61.9-61.1	—	39.5-39.2	—
14	—	—	50.0-48.1	51.1-49.8	53.1-52.5	—	67.8	63.6-63.1	26.6-26.1	43.5-42.1	—
15	—	—	48.7-48.3	47.35	53.9-52.9	—	66.5	—	—	43.6-42.6	—
16	—	—	53.0-52.0	52.4-51.6	54.9-54.1	—	65.4	—	36.0-35.6	—	—
17	—	—	51.7-51.3	49.6-49.4	55.7-55.0	—	64.5	—	—	41.4-40.3	—
18	—	—	—	—	56.2-55.6	57.8-57.4	63.2-63.1	—	44.0-41.7	—	—
19	—	—	—	—	56.4-55.8	—	62.5-62.3	—	—	—	—
20	—	—	—	—	55.1-49.9	57.0-56.0	60.6-60.4	—	—	—	—

[a] The temperature values were obtained by optical microscopy. n = alkyl or acyl chain length, respectively; for structures see Table II. I, cholesteryl alkanoates; II, S-cholesteryl alkanethioates; III, cholesteryl alkanethioates; IV, S-cholesteryl ω-phenylalkanethioates; V, cholesteryl alkyl carbonates; VI, cholesteryl S-alkyl thiocarbonates; VII, S-cholesteryl alkyl thiocarbonates; XIV, S-cholestan-3β-yl alkanoates; XV, 5α-cholestan-3β-yl ω-phenylalkanoates; XVI, 5α-cholestan-3β-yl alkyl carbonates; XVII, 5α-cholestan-3β-yl S-alkyl thiocarbonates.

sample surface. Because of the small angular dependence of the selective reflectance for light incident and reflected close to the average direction of the helical axes (see Fig. 7), the measured data should be representative of perpendicular incidence and perpendicular selective reflection.

Figures 3, 5, and 6 show the wavelength dependence of the selective reflection at various temperatures as well as the temperature dependence of the peak wavelength and of the selective reflectance of monochromatic light of cholesteryl oleyl carbonate [36]. S-Cholesteryl 14-phenyltetradecanethioate [47], cholesteryl nonanoate, and cholesteryl erucyl carbonate (containing 5% of cholesteryl chloride to achieve wetting of the glass substrate and thus a good sample) are the only other compounds for which a set of data has been reported [36]. As indicated by the influence of the thermal history on the selective reflectance, the degree of alignment within these polydomain samples was insufficient for the determination of the selective reflectivity. Consequently, only gross variations of the intensity observed under identical experimental conditions may be meaningful.

3. Chain Length Dependence of the Color Band

The influence of the acyl chain length on the temperature interval of the color band was investigated with cholesteryl alkanoates [218] and S-cholesteryl ω-phenylalkanethioates [219]. The former series is reported to exhibit color bands from the propionate to the hexanoate and from the nonanoate to the octadecanoate which, except for minor variations, decrease from about 30° to 3° for the octadecanoate. Considering that these materials were of unspecified origin and that in this homologous series of pure alkanoates only the nonanoate and the decanoate have well-defined color bands (see Table VIII), we conclude that impurity effects must be responsible for this discrepancy. This is confirmed by Kassubek and Meier [45] who have obtained a maximum pitch p of 0.310 μm (i.e., $\lambda_{p_{\max}} = np \approx 1.5 \times 0.310\,\mu m = 0.465\,\mu m$) at 88° for the pure propionate and of about 0.300 μm (i.e., $\lambda_{p_{\max}} \sim 0.450\,\mu m$) at 86° for the pure pentanoate. Furthermore, they found only a linear decrease of these pitches with temperatures (100° range) and did not observe the nonlinear dependence characteristic of pretransitional effects.

In the S-cholesteryl ω-phenylalkanethioates the relation between the temperature width of the color band and the chain length dependence indicates an odd–even effect (see Fig. 22).

For odd acyl chain lengths the temperature interval decreases with increasing chain length at the rate of 0.16° per methylene group, while it increases with even chain length at about half that rate. Considering the linearity of the odd branch and assuming linearity for the even branch, one should observe a crossover at about the 16-phenylhexadecanethioate and no color bands for the 19-phenylnonadecanethioate and higher odd members. By the

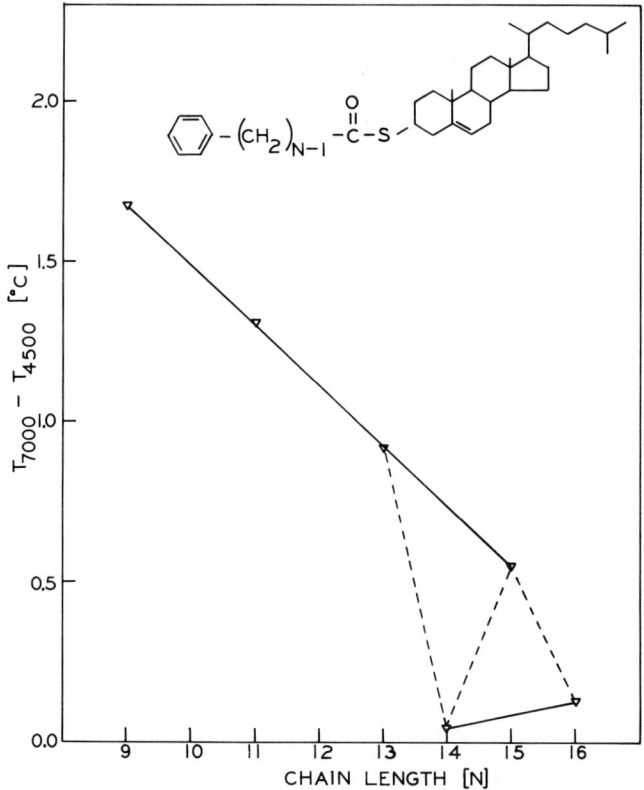

Fig. 22. Width of the cclor band (4500–7000 Å) of S-cholesteryl ω-phenylalkanethioates as a function of chain length.

same token, extrapolation of the even branch toward shorter chain length implies that no even member lower than 14-phenyltetradecanethioate should exhibit a color band. Experimental experience supports the latter speculation. This odd–even effect is in phase with that observed for the cholesteric–smectic and the cholesteric–isotropic transition temperatures [*121*]. Since the addition of a methylene group to higher ω-phenylalkanethioates does not significantly change lateral polarizability and size of the molecule, these drastic odd–even effects may be caused by differences of the molecular conformation [*118*].

4. Extremely Temperature-Sensitive Selective Reflectance

S-Cholesteryl 14-phenyltetradecanethioate exhibits the highest temperature sensitivity of the selective reflection reported to date, because its color band occupies a temperature interval of 0.07° [*47*]. But in spite of this high temperature dependence, the measurements obtained during a given run (see Fig. 23) fit the theoretical relation derived by Keating [*80*]

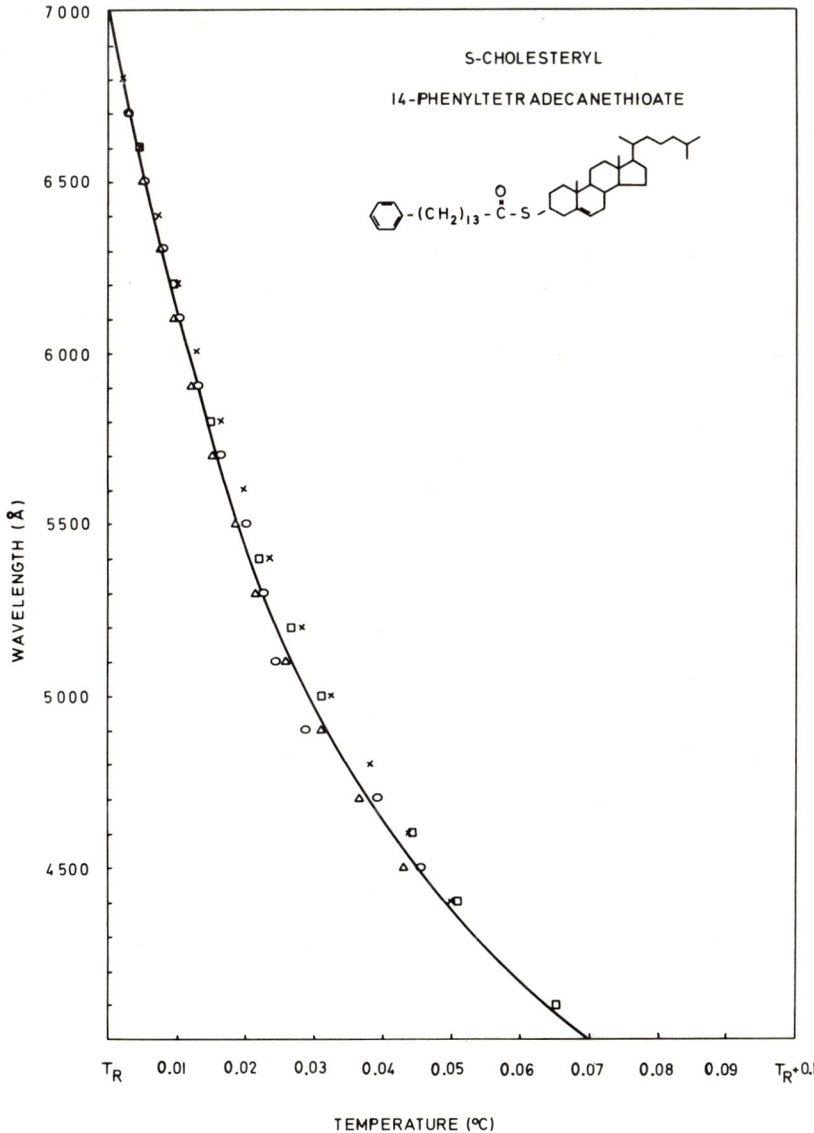

Fig. 23. Wavelength of maximum selective reflection as a function of temperature. Material: *S*-cholesteryl 14-phenyltetradecanethioate. ×, Data set I; □, data set II, measured on same day on same sample as set I; △, data set III, second sample, cooled at 2 m°/min, measured 8 days after set I; ○, data set IV, third sample, cooled at 5 m°/min, measured 9 days after set I (cf. Table IX). [Reproduced from Ennulat *et al.*, *Mol. Cryst. Liquid Cryst.* **26**, 245 (1974), with permission.]

$$\lambda_p = \frac{A}{T}\left(1 + \frac{B}{T-T_0}\right)^2 \tag{5}$$

as well as the curve postulated by Alben [76]

$$\lambda_p = \lambda_0 + \frac{A}{(T-T_0)^2} \tag{6}$$

within the relative measurement uncertainty of ± 0.2 m°C.

As Table IX indicates, the characteristic temperatures T_0 differ within the test runs for each formula by less than 0.03°. Furthermore, the T_0 values of both formulas coincide within 0.03°. This can be explained by the fact that the expansion of Keating's formula contains $1/(T-T_0)^2$ as the dominant

TABLE IX

PARAMETERS OF CURVE FITS $\lambda_p(T)$ TO FIGURE 23

		Data set			
		I	II	III	IV
$\lambda_p = A/T\{1+[B/(T-T_0)]\}^2$	T_0 (°K)	320.070	320.093	320.097	320.085
	A (Å °K)	479319	699985	693706	607294
	B (°K)	0.0930	0.0455	0.0422	0.0601
$\lambda_p = \lambda_0 + [A/(T-T_0)^2]$	T_0 (°K)	320.048	320.071	320.080	320.065
	λ_0 (Å)	2287.5	2863.4	2933.5	2695.5
	A (Å °K^2)	48.903	26.112	19.970	30.889

term. However, the other parameters of the equations vary strongly with the test run. We were unable to determine whether these discrepancies were caused by measurement errors or by small differences of the thermal history of the sample.

Figure 24 shows the peak wavelength curves of the other members of the series. It would be interesting to know how closely these experimental results can be represented by the above equations with the same set of parameters A, B, and λ_0 and whether or not these parameters depend on chain length. We believe that the observed unwinding of the helical structure is a pretransitional effect. Consequently, the closer the reference temperature T_0 is to the cholesteric–smectic transition temperature, the narrower is the width of the color band.

The temperature dependence of the selectively reflected light intensity at a

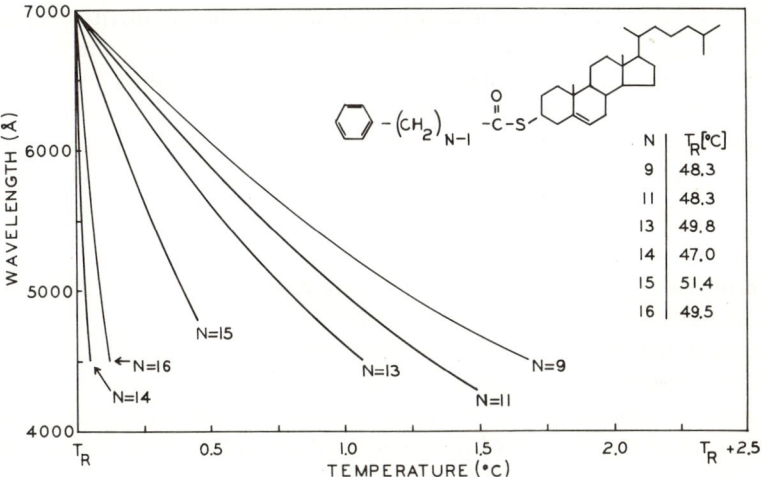

Fig. 24. Wavelength of maximum selective reflection as a function of temperature. Materials: S-cholesteryl ω-phenylalkanethioates.

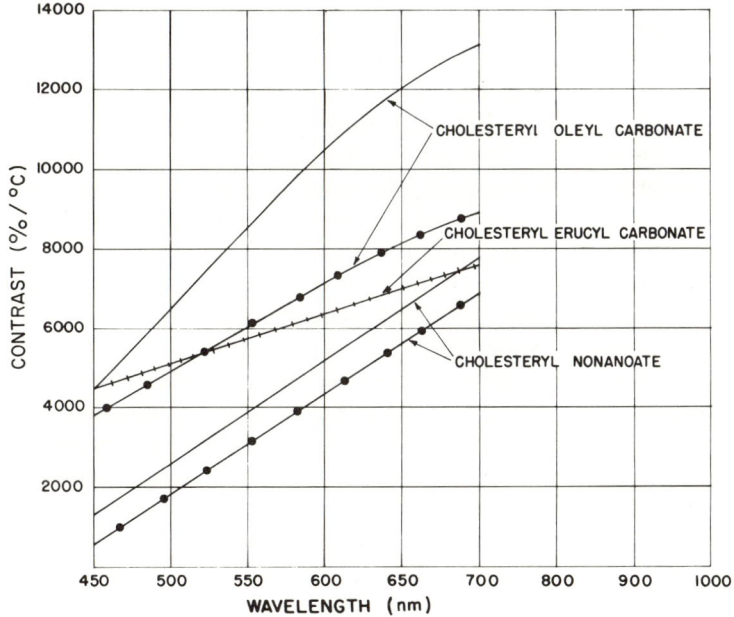

Fig. 25. Maximum temperature coefficients of selective reflection as function of wavelength. Solid line = low temperature side; broken line = high temperature side of intensity curve at constant wavelength. [Reproduced from Ennulat, *Mol. Cryst. Liquid Cryst.* **13**, 337 (1971), with permission.]

given wavelength is another measure of the temperature sensitivity of selective reflection. The intensity curve measured at a wavelength of 0.5 μm has a half-width of 7.6 m°C [47], which is about one-third of that reported for cholesteryl oleyl carbonate (see Fig. 6). Furthermore, we found that the maximum relative intensity change per degree (that is, the maximum contrast per degree Celsius) amounts to about 50,000% per degree Celsius [47]. The latter exceeds previously reported data, summarized in Fig. 25, by almost a factor of five.

B. Mixed Liquid Crystals

In mixed liquid crystals the temperature region of the color band can be varied by changing type and concentration of the constituents. Since phase separation may be noticeable only after a prolonged period of time, it is difficult to establish the equilibrium concentration without knowing the phase diagram in sufficient detail. Although an extensive literature deals with the concentration dependence of the color band, we did not find much information about the associated phase diagrams. Because of this and possible impurity effects, we doubt that the long-term stability of the temperature interval of the color band can presently be predicted for the known temperature-sensitive cholesteric mixtures.

1. Cholesteric–Cholesteric Mixtures

The phase diagrams of 16 binary mixtures of cholesteryl alkanoates were determined by differential scanning calorimetry in conjunction with microscopic investigations [220–222, 102]. Since this work was devoted to thermodynamic studies, it does not address the concentration dependence of the selective reflection. But we expect color bands at least for compositions which contain a dominant amount of a constituent with a color band, if the mixture exhibits smectic–cholesteric transition temperatures. This is the case for binary mixtures of cholesteryl derivatives containing a high concentration of nonanoate and tetradecanoate [221], pentanoate, isopentanoate, 2-methylbutyrate, or 2-ethylbutyrate [102].

Dilatometric measurements near the smectic–cholesteric phase transition of the cholesteryl oleyl carbonate-cholesteryl chloride system revealed first-order phase transitions for a cholesteryl oleyl carbonate mole fraction greater than 0.63 and second-order phase transitions for smaller mole fractions [223]. This result agrees with the theoretical prediction based on the theories of McMillan and Kobayashi. Since these mixtures exhibit a very temperature-sensitive color band, it appears that the pretransitional selective reflection can occur near first- and second-order phase transitions.

The pitch of a mixture can be predicted from pitch and concentration of the components if the effective rotary power follows the linear superposition law [see Eq. (1), Section II,A,5]. As we have seen, the latter should hold for compounds of similar molecules when pretransitional effects are absent.

Since the high temperature sensitivity of the selective reflection is directly related to the pretransitional unwinding of the helical structure, it is impossible to determine the temperature region of the color band in mixtures from our present theoretical knowledge.

The experimental results indicate a nonlinear relationship between the temperature region of the color band and concentration [224]. For example, in the system cholesteryl nonanoate-cholesteryl oleyl carbonate, the color band shifts monotonically toward higher temperatures with increasing concentration of cholesteryl nonanoate [49, 225]. This effect is accompanied by a gradual expansion of the temperature interval of the color band. However, the system cholesteryl decanoate-cholesteryl hexanoate behaves quite differently [226, 227]. The temperature level of the color band is substantially depressed in the mid-concentration region to below the values characteristic of either component. Concurrently the temperature width of the color band assumes a maximum. Note that the former system consisting of substantially different component molecules exhibits less nonlinearity than the latter system, which is composed of similar homologs.

Other binary systems reported in the literature consist of the following combinations of cholesteryl derivatives: decanoate-chloride, decanoate-acetate, decanoate-cinnamate, decanoate-oleate, octanoate-oleate, and acetate-cinnamate [226, 227]. They behave essentially like the binary mixture cholesteryl decanoate-cholesteryl hexanoate regardless of whether the temperature width of the color band is wide or narrow for both components or wide for one and narrow for the other. Moreover, the smallest temperature interval of the color bands exhibited by these mixtures is not smaller than the smallest one of the two respective components. The system cholesteryl nonanoate-cholesteryl butyrate seems to be an exception to the latter [218], possibly because of the influence of impurities.

A few binary systems were investigated over a limited range of concentration. They include mixtures of cholesteryl nonanoate with the cholesteryl derivatives propionate, butyrate, isobutyrate, crotonate, 3-phenylpropionate, methyl carbonate [228], acetate [228, 229], and chloride [228, 229, 224]; binary mixtures of cholesteryl decanoate with several cholesteryl β-alkoxy-propionates [128]; and a 1:1 mixture of cholesteryl decanoate and cholesteryl hexanoate [230]. For the last system, selective reflection data, such as spectral profiles at various temperatures, the angular dependence of peak wavelength and selective reflectance, and the peak wavelength as a function of temperature have been reported.

The dependence of the pitch on temperature and composition has been discussed for mixtures of cholesteryl chloride and the cholesteryl alkanoates nonanoate, decanoate, and dodecanoate, respectively [224].

Ternary mixtures of cholesteric materials were developed to achieve a wider choice of temperature region and temperature width of the color band. Woodmansee [*231*] selected the constituents as follows: a base compound that exhibits a narrow color band of relatively low temperature (cholesteryl oleate), a second component that reduces the width of the color band (cholesteryl octanoate, nonanoate, decanoate, *p*-nitrobenzoate), and a third component that increases the temperature level of the color band without significantly widening it (cholesteryl acetate, propionate). By properly selecting the constituents and their respective concentrations, he was able to shift the color band from about 21°–45° and from 50°–58°, respectively, while maintaining the width within the range of about 0.3°–0.4°. Figure 26 shows some of those results.

Magne and Pinard [*50*] thoroughly investigated ternary mixtures of the cholesteryl alkanoates oleate, nonanoate, and propionate. They measured the peak wavelength versus temperature relations for compositions which

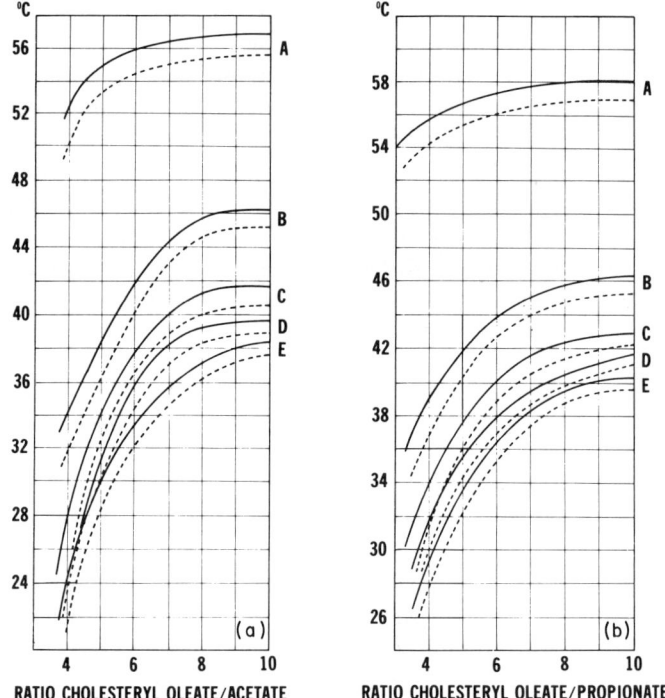

Fig. 26. Temperature range (red to blue) of selective reflection of ternary mixtures. Basic binary system: cholesteryl oleate–cholesteryl propionate. (a) % nonanoate, octanoate, or decanoate for curve A: 68.25, curve B: 35.8, curve C: 20.8; curve D: 11.4, curve E: 0.0. (b) % nonanoate, octanoate, or decanoate for curve A: 68.2, curve B: 35.8, curve C: 23.7, curve D: 11.4, curve E: 0.0. Redrawn from Woodmansee, U.S. Patent 3,441,513 (1969).

selectively reflect the visible within a temperature interval of 1°–3° over a temperature range from 5° to 75°.

Melamed and Rubin [232] studied relatively temperature-insensitive ternary mixtures of cholesteryl chloride (22–45 wt %) with a 40:60 wt % composition of cholesteryl nonanoate-cholesteryl oleyl carbonate. Although the peak wavelength was shifted by only about 50 Å/°C over the temperature range from 20° to 35°, the 50-μm thick sample containing 32% cholesteryl chloride changed its optical rotation up to the rate of 60°/°C. This means that the readily measurable variation of rotation of 0.2° corresponds to a temperature change of 3 m°C.

Obviously, even with the limited number of temperature-sensitive compounds known to date, it is possible to prepare a large number of mixtures. Since most of the newer temperature-sensitive materials, such as the S-cholesteryl 14-phenyltetradecanethioate, are not commercially available, cholesteryl oleyl carbonate and cholesteryl oleate are still the basic materials for close-to-room-temperature operation, and cholesteryl nonanoate or decanoate for higher temperature regions. Because many calibrated mixtures are commercially available, we list only a few representative compositions in Table X covering the temperature range from 0 to 250°. For special mixtures,

TABLE X

TEMPERATURE INTERVALS OF COLOR BANDS OF MIXTURES

Components: cholesteryl	Ratio	Temp. range (°C)	Reference
Oleyl carbonate, acetate	80:20	0–4	234
Oleyl carbonate, phenyl carbonate	80:20	14–16	239
Oleyl carbonate, acetate	95:5	16–18	234
Oleyl carbonate, nonanoate, benzoate	65:25:10	17–23	234
Oleyl carbonate, nonanoate, benzoate	70:10:20	20–25	239
Methyl carbonate, nonanoate	20:80	22–47	234
Nonanoate, oleate, crotonate	25:55:20	22–25	239
Nonanoate, oleate, crotonate	10:70:20	24–26	239
Oleyl carbonate, nonanoate, benzoate	45:45:10	26.5–30.5	240
Oleyl carbonate, nonanoate, benzoate	43:47:10	29–32	239
Oleyl carbonate, nonanoate, benzoate	44:46:10	30–33	239
Oleyl carbonate, nonanoate, benzoate	38:52:10	33–36	239
Oleyl carbonate, nonanoate, benzoate	32:58:10	36–39	239
Nonanoate, oleate, crotonate	30:60:10	40–42	239
Nonanoate, propionate	80:20	45–65	234
Nonanoate, butyrate	80:20	55–57	234
3-Phenylpropionate, nonanoate	20:80	64–67	234
Cinnamate, nonanoate	90:10	140–250	234

adjusted to temperatures from 30° to 40°, see Tables XII and XIII in Section V,D.

2. Hysteretic Cholesteric Mixtures

Binary and ternary mixtures have been developed which exhibit a certain selective reflection color over a wide temperature region. These materials lose their color if an upper temperature limit is exceeded and reflect the color only minutes, hours, or weeks after returning to the lower original temperature. Some hysteretic compositions remain colorless as long as the lower temperature is maintained. Depending on the composition, binary mixtures containing a cholesteryl halide, such as the chloride, and another component, such as cholesteryl oleyl or erucyl carbonate, cholesteryl oleate or erucate, display the color with a delay of 3–30 minutes [233]. About 15–40% of the halide is required to practically eliminate the temperature sensitivity of the other component and to establish the desired reflection color. A third component can be added to either raise or lower the clearing temperatures. Certain cholesteryl esters (cholesteryl nonanoate) induce the former and certain fatty acid derivatives (oleic acid, triolein) induce the latter effect.

The addition of a few percent of oil-soluble nigrosine or indulene dyes to binary cholesteric mixtures containing one cholesteryl halide results in a hysteretic time delay of the order of weeks [234]. Because of the small dye concentration and the fact that mechanical disturbance restores the color of the cholesteric mesophase, we surmise that surface effects caused by the dye molecules control the alignment and thus the texture of the cholesteric mesophase.

The largest hysteresis effects can be achieved by glassy cholesteric liquid crystals [235]. Fast cooling of the cholesteric mesophase reflecting a certain color to below the glass transition point results in a frozen-in planar texture. As long as the temperature is kept below the glass transition temperature and as long as crystallization does not occur, this solid cholesteric mesophase will selectively reflect at a peak wavelength characteristic of the frozen-in pitch. Once the temperature exceeds that of the glass transition, the frozen-in color is eliminated and the sample behaves like an ordinary mesophase or an isotropic liquid. These glassy liquid crystals consist of a mixture of cholesteryl oleyl carbonate and cholesteryl nonanoate (20–80% each) and up to 45% of a viscosity-increasing cholesteric compound. The latter can be cholesteryl p-nonylphenylcarbonate or dicholesteryl esters of α,ω-alkanedicarboxylic acids, such as dicholesteryl succinate, adipate or sebacate.

3. Cholesteric–Nematic Mixtures

The properties of mixtures consisting of cholesteric and nematic compounds depend strongly on concentration. For example, small concentrations of p-butyl-N-(p-methoxybenzylidene)aniline (MBBA) in cholesteryl oleyl carbonate reduce the smectic–cholesteric transition temperature and consequently shift the color band toward lower temperatures without affecting the temperature sensitivity [236]. However, high concentrations (40%) result in visible selective reflection over a wide temperature range which is practically temperature-insensitive [37].

Thin films of dilute solutions of cholesteryl oleate in MBBA exhibit a well-behaved temperature dependence of the optical rotatory power in the vicinity of the clearing point. To eliminate the disturbing visible selective reflection, a concentration of 90% MBBA had to be used to shift the minimum peak wavelength into the near infrared wavelength region. Depending on the concentration and the proximity to the clearing point, the optical rotatory power of the visible light had a temperature coefficient between 5 and 25% per degree Celsius. This effect can be utilized in a fiberoptic thermometer capable of resolving temperature differences of 0.02° or better over a range of several degrees Celsius [237].

Labes *et al.* [238] found that high concentrations of certain nematic materials in cholesteric mixtures reduce substantially the angular dependence of the selective reflection. Since high concentrations of nematic materials also increase substantially the selectively reflected light intensity and reduce the response time of the field-induced cholesteric–nematic phase transitions, these mixtures are being considered for display applications.

4. NEMATIC–CHIRALIC MIXTURES

Cholesteric mesophases obtained by adding a chiralic compound [34] to a nematic material exhibit a temperature dependence of the peak wavelength that is characteristic of cholesteric mesophases in the pretransitional region. However, the materials reported to date have a relatively small temperature dependence of the selective reflection. The temperature dependence of the optical rotation was recently measured for binary mixtures of MBBA and *d*- and *l*-menthol [34].

V. APPLICATION

A. Methods of Temperature Sensing

Three temperature-sensitive effects associated with the selective reflection are utilized for the detection, measurement, and imaging of thermal, mechanical, and electromagnetic energy. In the following we will discuss the methods of the optical observation, describe the structure of liquid crystal sensing layers proposed to date, and address means of coupling the energy into the liquid crystal.

1. Optical Read-Out

We will discuss liquid crystal temperature sensors under ideal conditions. By that we mean that the liquid crystal film consists of one planar texture domain whose helical axis is perpendicular to the surface and that collimated light is used to avoid complications arising from the angular dependence of the selective reflection. Furthermore, we prescribe that regions of different temperatures are sufficiently separated to eliminate the degradation of the optically displayed temperature information due to lateral heat conduction.

a. *Peak Wavelength Shift.* In the simplest method the liquid crystal is illuminated by white light and the temperature differences are indicated as color differences of the selectively reflected light. For a given resolvable wavelength difference $(\Delta\lambda)_R$ (which, in the worst case, is the half width of the spectral profile) one obtains the following minimum detectable temperature difference

$$\Delta T_{min} = \left(\frac{\Delta\lambda(T)}{\Delta T}\right)^{-1} (\Delta\lambda)_R$$

Because of the pretransitional effect, the highest values of $\Delta\lambda(T)/\Delta T$ are achieved by those liquid crystals for which the color band is close to the cholesteric–smectic transition temperature.

b. *Selective Reflectance.* The second method is based on the selective reflection of monochromatic light. As shown in Fig. 27, the minimum detectable temperature difference obeys the relation

$$\Delta T_{min} = \left[\frac{\Delta I}{I\Delta T}\right]^{-1} \left(\frac{\Delta I}{I}\right)_R$$

where $(\Delta I/I)_R$ is the resolvable relative brightness change (i.e., the resolvable contrast). The factor $\Delta I/I\Delta T$ is proportional to the function $\Delta\lambda/\Delta T$ and the inverse of the halfwidth of the spectral selective reflectance [47].

c. *Optical Rotation.* The third method relies on the temperature dependence of the optical rotatory power in the wavelength region outside but in the

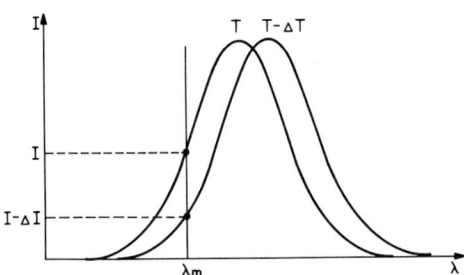

Fig. 27. Temperature modulation of selective reflectance.

vicinity of the color band. For a liquid crystal film sandwiched between polarizers one obtains

$$\Delta T_{min} = \left[\frac{\Delta\phi(T)d}{\Delta T}\right]^{-1}(\Delta\psi)_R$$

if one measures a resolvable rotation $(\Delta\psi)_R$, and

$$\Delta T_{min} = \frac{1}{\sin 2[\psi(T)+\alpha]}\left[\frac{\Delta\phi(T)d}{\Delta T}\right]^{-1}\left[\frac{\Delta I}{I}\right]_R$$

if one perceives the contrast $(\Delta I/I)_R$ of the transmitted light. $\Delta\phi(T)/\Delta T$ is the change of the specific optical rotation per degree Celsius and per micrometer thickness of the liquid crystal, d the thickness in micrometers, and $\psi(T) = \phi(T)d$ and α are the optical rotation and angle between the polarizers, respectively. For a given operating temperature T, the denominator $\sin 2[\psi(T)+\alpha]$ can be made equal to unity by setting $\alpha = \pi/4 - \psi(T)$.

d. *Comparison of Methods.* In Table XI we compare the performance of these methods. The temperature resolution of method 1a is primarily limited by the spectral halfwidth of the selectively reflected light, ranging from about 250 Å for short to about 500 Å for long wavelengths. Regardless of whether or not this spectral profile is square (perfectly aligned planar texture, see Fig. 2) or bell-shaped (polydomain sample, see Fig. 3), one can discriminate in the worst case only about ten different colors. Consequently, the human observer can see a minimum temperature difference of about 50 m°C in cholesteryl oleyl carbonate $[\Delta\lambda/\Delta T \equiv 10^4 \text{ Å}/°C]$ [36] or 5 m°C in S-cholesteryl 14-phenyltetradecanethioate $[\Delta\lambda/\Delta T \equiv 10^5 \text{ Å}/°C]$ [47].

In method 1b the limiting factor is the minimum contrast perceivable by the human observer. Since this quantity depends on the observer, the background luminance, the resolution, and on other parameters, we use a conservative threshold contrast of 0.1. For cholesteryl oleyl carbonate with $[\Delta I/I\Delta T]_{max} \simeq 10^4 \%/°C$, we expected a temperature difference ΔT_m of 1 m°C but observed about 3 m°C [241]. Because of the polydomain structure of the liquid crystal, the use of uncollimated light and the uncertainty of the threshold contrast, we believe that this discrepancy is insignificant.

Method 1c yields a respectable temperature resolution over a relatively wide temperature region. For example, the mixture of 6.4% cholesteryl oleate and MBBA exhibits a sensitivity of $\Delta\phi/\Delta T = 0.44$ deg/°C μm for temperatures between 26° and 31°, which increases to $\Delta\phi/\Delta T \simeq 8$ deg/°C μm just below the clearing point at 36° [237]. However, thickness variations Δd across the liquid crystal film cause variations $(\Delta\psi)_v$ of the optical rotation which can mask those induced by temperature differences. If we consider

$$(\Delta\psi)_v = (\Delta\psi)_R \tag{7}$$

TABLE XI

Discernible Temperature Differences by Peak Wavelength Shift (1a), Selective Reflectance (1b), and Optical Rotation (1c)

Method	Physical effect	Illumination	Liquid crystal	Limitation of observable	$\Delta T_{\min}$ (m°C) Calculated	$\Delta T_{\min}$ (m°C) Observed	Ref.
1a	$\lambda_p(T)$	White	$\Delta\lambda/\Delta T \approx 10^4 - 10^5$ Å/°C^a	$(\Delta\lambda)_R \approx 500$ Å^a	5–50	100	36, 47
1b	$I(T, \lambda_0)$	λ_0	$\Delta I/\Delta TI \approx 10^4 - 5 \times 10^4$ %/°C	$(\Delta I/I)_R \approx 10\%$	0.2–1	3	241
1c	$\psi(T, \lambda_0)$	λ_0	$\Delta\phi/\Delta T \approx 0.44 - 8$ deg/°C μm	$(\Delta I/I)_R \approx 10\%$	0.3–6^b	20	237

^{a} Data pertain to peak wavelength of 6500 Å.
^{b} Thickness of liquid crystal 36 μm.

as a practical limit of the temperature resolution, we obtain

$$\Delta d = (\Delta\psi)_R/\phi \tag{8}$$

and

$$\Delta d = [\Delta I/I]_R \, 1/\phi \tag{9}$$

where $\phi = \phi(T)$ is the specific optical rotation in degrees per micrometer of the liquid crystal. Since ϕ is usually larger than one, which is a reasonable assumption unless the operating temperature is extremely close to the clearing point, we conclude from Eq. (8) that it is not sensible to utilize the full angular resolution of high performance polarimeters because of the prohibitive requirements on thickness uniformity. For example, an angular resolution $(\Delta\psi)_R = 0.01°$ could be utilized only if the thickness variations are kept below 100 Å. It should be pointed out that the mixture discussed above is designed to exploit the strong temperature dependence of the optical rotatory power in the pretransitional region of the clearing point and that ϕ tends toward zero as the clearing point is approached. This is not the case for the optical rotatory power in the vicinity of the selective reflection region of an ordinary cholesteric material [242] or close to the nematic temperature of a compensated material [243]. In either case the quantities $\Delta\phi/\Delta T$ and ϕ both increase strongly with temperature as the reflection band or the nematic temperature is approached. As a consequence, the thickness requirement rather than the attainable resolution $(\Delta\psi)_R$ or $[\Delta I/I]_R$ determines the minimum detectable temperature difference.

2. STRUCTURE OF LIQUID CRYSTAL SENSING LAYERS

a. *Continuous Films.* The direct deposition of liquid crystal films on the test surface provides the best thermal contact for temperature sensing. Films of uniform thickness can be obtained either by spreading and brushing the melted compound or by precipitating the liquid crystal, dissolved in a volatile solvent, by slow evaporation. In one method a certain amount of solution deposited in the center of a horizontal surface rapidly spreads in radial directions, losing solvent during this process [241]. The remaining liquid crystal film can be sufficiently uniform over extended regions if one selects properly the solvent and the environmental temperature. In the other method the solution is sprayed, for example, with an ordinary artist's air brush. We found that the spraying method allows a better control of the film thickness and provides a higher thickness uniformity [241].

For many applications it is necessary to sandwich the liquid crystal film between thin sheets of materials [241, 244]. Usually the transparent top layer protects the liquid crystal from dust and prevents or at least reduces degrading effects that are caused by exposure to air. We found that cholesteryl oleyl carbonate films, sandwiched between Mylar membranes, which were hermetically sealed at the edges, maintained the same temperature region of the color band for many years when stored in the dark. We also noticed that the surface alignment of the Mylar induced by calendering improves the orientation of the planar texture and thus the selective reflectance. Depending on the application, these composite films can be attached to curved surfaces with the aid of electrostatic charges, a contact grease, or by vacuum action [245, 246].

In most sensing layers the substrate of the liquid crystal is blackened to absorb the transmitted component of the incident light and thus to increase the contrast. The latter can also be achieved by transmitting the light through the sensing layer and by preventing the return of this light to the other side of the liquid crystal [247].

b. *Encapsulated Liquid Crystals.* Microencapsulation [248] is a unique approach to protect the liquid crystal from the environment. In this process minute droplets of liquid crystals are dispersed in an aqueous mixture of gelatin and gum arabic. The latter compounds cover the droplets with a thin film which can be hardened by a diol (e.g., pentanediol [249, 250]), or a dialdehyde, such as glutardialdehyde [251, 252], or formaldehyde [253]. A more recent patent [254] claims poly(urethanes) as an encapsulating material and a subsequent modification utilizes gelatin and poly(vinyl alcohol) as the wall material which is hardened by potassium dichromate and UV irradiation [255]. The diameter of the microcapsules is restricted to the range from 10 to 50 μm [249], because larger capsules reduce the spatial resolution of temperature sensors while smaller ones exhibit a low selective reflectance.

The slurry of encapsulated liquid crystals can be used in various ways. It can be directly deposited on blackened surfaces by spraying, by precipitating from a suspension, or with the aid of a mechanical applicator. A polymeric binder such as poly(vinyl alcohol) [249] or cellulose ethers [256] can be used to glue the capsules to each other and to the substrate. Based on this principle, several "liquid crystal inks" have been developed which can be applied to white paper directly [256] or through a patterned 150-mesh silk screen [257]. The binder can be made water soluble, thus permitting recovery of the capsules. On the other hand, glued encapsulated films can be stripped from the substrate in order to obtain a self-supporting film consisting essentially only of capsules.

Encapsulated liquid crystals exhibit in general a lower selective reflectance than a continuous liquid crystal film of equivalent thickness. This is explained by the fact that the selective reflectance depends strongly on the angle of incidence and that, for a given direction of incident light with respect to the film, only a relatively small area on the liquid crystal droplet is within the optimum angle of incidence. However, a top coating of a suitable polymer can reduce the angle of incident light and thus increase the selective reflectance of the film [*251*].

Various temperature-sensitive liquid crystals have been successfully encapsulated. Among them are mixtures of cholesteryl oleyl carbonate and cholesteryl nonanoate exhibiting color bands within temperature intervals of 1°–3° [*251*]. Furthermore, mixtures of different encapsulated liquid crystals have been prepared which display color bands in separated temperature regions [*249*]. The selective reflectance of cholesteric mixtures was increased by the addition of certain room-temperature nematic liquid crystals (such as MBBA or similar Schiff bases). This phenomenon may be explained by the fact that such mixtures exhibit higher selective reflectances at higher angles of incidence than ordinary cholesteric materials (see Section IV, B, 3, and [*238*]). Certain mixtures of MBBA and cholesteryl derivatives exhibit visible selective reflection only after encapsulation [*236, 258*].

We have not found any information comparing the operating life of encapsulated and sandwiched liquid crystal sensors. It is our experience that the former appear to lose the color band sooner than liquid crystals sealed between Mylar sheets. One patent [*256*] claims that an encapsulated composition having a relatively wide temperature interval of the color band (8°) was unaffected during about five months of storage.

c. *Imbedded Liquid Crystals.* Several methods have been developed by which volume elements of liquid crystals can be imbedded into a homogeneous matrix. For example, a "liquid crystal sheet" was obtained by hot-pressing of microencapsulated liquid crystals which had been coated on a support film and then covered with a thermoplastic polymer [*259*]. Cholesteric liquid crystals can also be dispersed in aqueous solutions of gelatin [*260*], poly(vinyl alcohol) [*260, 261, 262*], dispersions of poly(vinyl acetate) [*263*], and polymer films such as poly(acrylates), poly(styrene) and poly(esters) [*264*]. These films selectively reflect light several degrees below the neat cholesteric mixtures. Carbon black can be directly incorporated into the polymer matrix to enhance contrast [*264*]. Buttons with cholesteric liquid crystals dispersed in melamine-formaldehyde [*265*], twill fabric coated with cholesteric liquid crystals dispersed in poly(urethanes) [*266*], and thermoplastic hollow fibers containing liquid crystals [*267*] complete the cholesteric wardrobe. A sprayable lacquer consists of a solution of cholesteric liquid crystals and a binder, such as poly(vinyl alcohol), shellac, or collodion, in a suitable solvent. The addition of poly(vinylpyrrolidone) as an emulsifier and of sorbitol esters as a surfactant permits the regulation of the droplet size [*268*].

The use of heat-curable monomers, such as styrene and epoxy systems, etc., results in films which no longer display cholesteric colors, but change from opaque to clear with temperature [*264*].

Additional formulations, tailored for the measurement of skin temperatures in man, will be briefly discussed under Medical Applications, Section V, D.

3. Energy Coupling

Applications exploiting the temperature sensitivity of liquid crystals require the conversion of the input energy into an equivalent temperature change in the liquid crystal film that can be optically read out by one of the three methods described above. This technique can be superior to other temperature sensors in terms of cost and performance because liquid crystal transducers can readily perform this conversion simultaneously in the two-dimensional spatial domain. Of course, this presupposes that the transfer function relating the energy input with the optical signal output does not cause intolerable distortions.

In some applications the liquid crystal film is directly deposited on the test surface to transfer the temperature pattern of the latter to the former by heat conduction. The resolution limit of such a configuration was tested with electrically heated thin film resistors deposited on silicon substrates [269]. The selective reflection of either white or monochromatic light converted the temperature pattern into the visible. The results indicate a limiting resolution of about 100 line pairs/millimeter (lp/mm). Furthermore, a 1-μm wide resistor strip was displayed as a color line of the same width. It has been reported [49] that the shift of the peak wavelength in response to a sudden change in temperature level occurs with a time constant of 0.1 sec in cholesteryl nonanoate and 0.2 sec in cholesteryl oleyl carbonate. Of course, if a liquid crystal film sandwiched between polymer sheets or protected by microencapsulation is in thermal contact with the test surface, the resulting increase of lateral heat conduction and of the specific heat will degrade the spatial resolution, reduce the attainable temperature differences, and lengthen the thermal time constant of the liquid crystal transducer [270].

Radiation energy has to be converted into heat before it can be indicated by liquid crystal films. Unless the latter itself absorbs the energy, an absorber has to be added to the liquid crystal [245] or to the substrate. For example, viscoelastic materials are used for the absorption of ultrasound; carbon black, gold black, and other materials for electromagnetic radiation ranging from the ultraviolet to the long wavelength infrared; and space-matched metal films for microwave radiation.

The thermal coupling of the radiation sensor to the environment has a pronounced influence on spatial resolution of the temperature pattern, temperature sensitivity, and time response of the transducer. In the case of ultrasound the liquid crystal is directly coupled to a liquid or solid medium to achieve good ultrasonic propagation. Consequently, thermal conduction through this medium dominates, and spatial and temperature resolution are greatly affected.

On the other hand, electromagnetic radiation can be imaged by transducers that are radiation coupled to the environment. In this case the temperature resolution is limited primarily by the lateral heat conduction within the liquid crystal sensor.

If a sinusoidal, one-dimensional intensity pattern

$$P(x) = P_0 \cos[2\pi(X/L)]$$

of pattern wavelength L is absorbed in the liquid crystal, one obtains an amplitude modulation of the temperature (i.e., the modulation transfer function [241])

$$\frac{T}{T_0} = \frac{1}{1+[2\pi(L_0/L)]^2}$$

where

$$L_0 = \left(\frac{\Sigma K_v d_v}{8\varepsilon\sigma T_0^3}\right)^{1/2}$$

is the characteristic length, and K_v and d_v are thermal conductivity and thickness of the Mylar substrate ($v = 1$) and liquid crystal ($v = 2$), respectively. ε is the emissivity of the film, σ the Stefan–Boltzmann constant, and T_0 the average film temperature. Using typical values [241] we obtain

$$\frac{T}{T_0} = \frac{1}{1+39.5(L_0/L)^2} \quad \text{and} \quad L_0 = 0.41 \text{ mm}$$

This result illustrates the spatial resolution limitations of radiation-coupled liquid crystal films.

The thermal time response t follows a similar expression:

$$t = \frac{Cd}{8\varepsilon\sigma T_0^3} \frac{1}{1+[2\pi(L_0/L)]^2}$$

where C is the heat capacity of the film. In the worst case (i.e., $L > L_0$) we obtain a response time of the order of several seconds [241]. Any mass added to the liquid crystal film, be it a thicker substrate or absorbing particles suspended in the liquid crystal, tends to increase both lateral heat conduction and heat capacity and thus degenerates spatial resolution and time response.

B. Liquid Crystal Devices

In this section we will describe classes of liquid crystal devices that fulfill functions not limited to nondestructive testing and medical applications. Examples are thermometers, pressure indicators, voltage and current meters, and radiation sensors.

1. Indicators and Measuring Devices

a. *Thermal Devices.* Liquid crystal thermometers represent the most direct application of the temperature-dependent selective reflection. Special mixtures of cholesteric liquid crystals are commercially available that allow the user the design of thermometric films that can be adjusted for a large variety of temperature intervals and operating temperatures (see Section IV, B, 1, Table X, and Fig. 26). The applications range from encapsulated liquid crystal thermometers designed for desks, swimming pools, and fish tanks [271, 272] to temperature-limit indicators [273]. The latter devices utilize hysteretic mixtures discussed in Section IV, B, 2 to visibly display, for example, whether or not frozen food has been temporarily thawed [233–235]. A thin film device has been developed which can be attached by vacuum action to a variety of shaped surfaces for the purpose of thermal mapping [246]. An electrothermal analog temperature indicator was proposed to display temperatures or temperature regions of electrical heating appliances [273]. This approach can be used for warning purposes or for the qualitative measurement of temperatures. A patent discloses a thermal exposure meter which optimizes the thermographical reproduction of documents [274].

b. *Mechanical Devices.* A liquid crystal pressure sensor was described that indicates pressures [275] from 500–5000 bars as a red shift of the selective reflectance maximum at constant temperature (see also [35, 35a]).

c. *Electrical Devices.* Voltage and rms current meters have been patented which consist of resistive films supporting a liquid crystal layer. The measuring current shifts the temperature distribution and thus the location of the color band. In one case a specific peak wavelength is selected by a filter [276], while in the other one the shift of a narrow color band is used for the indication of the electrical input quantity [277]. The latter device contains in addition a temperature-irreversible hysteretic liquid crystal to record peak rms values.

d. *Optical Devices.* The optical features characteristic of planar textures can be exploited for optical filtering. Combinations of such films with polarizers and retardation plates result in low-cost, bandpass and notch filters [244, 278, 279]. Mixtures of certain cholesteric and nematic materials were suggested because of their wide optical bandwidth and their small temperature dependence of selective reflection and associated optical properties [37]. On the other hand, the temperature dependence can be used to modulate the filter action.

Thermally addressed light valves have been built in which an infrared laser beam locally heats regions of a planar texture film to or beyond the clearing point. Subsequent cooling converts those regions into focal conic structures which scatter the viewing light while the surrounding planar texture remains transparent. Light valves, xy addressed by an infrared laser, display high quality black and white images of graphic and alphanumeric information with a resolution of 50–66 lines/mm [280, 281].

The temperature dependence of the selective reflectance is utilized in a variety of displays. They consist of an electrically xy-addressed panel of dissipative elements which are in thermal contact with the liquid crystal film. These dissipative elements can be thin film resistors [282], magnetic memory elements (transfluxors) [283], or microtransistors [284]. Other displays are addressed by input radiation (UV, visible, or IR), which modulates the electrical conductivity of a biased photoconductive layer sandwiched with a liquid crystal film. The resulting pattern of heat dissipation in the photoconductor is converted into the visible either by selective reflection [285] or by the optical rotation of the plane of polarization of the incident viewing light [286]. In other displays the information is directly inscribed on the liquid crystal by an electron beam, causing local heating. Again this display can operate either with reflected or transmitted viewing light [286].

2. RADIATION SENSORS

Radiation transducers, consisting of an absorber sandwiched with a liquid crystal film or suspended in the liquid crystal and of the appropriate optical components, may be the most versatile and inexpensive devices for the conversion of invisible to visible radiation. They can perform this task with virtually any radiation for which a practical absorber exists. So it is not surprising to find applications ranging from thermal imaging for night vision to imaging and holography of infrared, microwave, and ultrasonic radiation.

a. *Ionizing Radiation.* It is suggested that films of cholesteric liquid crystals may be suitable for the registration of elementary particle tracks [287]. Furthermore, theoretical investigations show that fast, charged particles moving in a cholesteric planar texture should generate observable, Cerenkov-type radiation [288, 289]. It is also estimated that heavy moving ions of a charge as low as six should produce in a planar texture heat tracks which could be visualized by selective reflection of light [290]. As in certain displays described above [286], electrons impinging on a cholesteric liquid crystal can be detected due to local heating. A patent describes an X-ray dosimeter utilizing the temperature shift of the color band in response to X-ray exposure [291]. This effect can be enhanced by adding iodine-containing compounds (e.g., cholesteryl iodide) to the liquid crystal.

b. *Ultraviolet Radiation.* This radiation can change the composition of the liquid crystal and thus lower the temperature region of the color band [292–294]. For example, irradiation of a 1:1 mixture of cholesteryl iodide and nonanoate shifts the color band by 1000 Å. This effect is utilized in an imaging system for UV radiation [295]. Extremely short pulses of UV radiation can be recorded by a film of a glassy cholesteric liquid crystal which has a frozen-in selective reflection color. Exposed areas of this film are rendered either transparent or in a different color [296].

c. *Infrared Radiation.* Thermal imaging systems have been built which can convert the heat radiation emanating from persons and objects in the scene into the visible [297–300, 241]. These devices consist of an optics focusing the infrared image on the sensing layer. The latter usually consists of a thin Mylar substrate carrying an infrared absorber on one side and the liquid crystal film on the other side. The temperature distribution associated with the absorbed infrared image is displayed by the intensity of selectively reflected monochromatic light. Such a system [241] detected a temperature difference of 3 m°C on the sensing layer at a resolution of 0.1 lp/mm. However, it displayed 26 m°C at 1 lp/mm and had a time constant of several seconds. These values are typical for 15–25-μm thick films of cholesteryl oleyl carbonate sandwiched between Mylar membranes (each about 6 μm thick) which are radiation-coupled to the ambient environment. This imaging approach was used to observe the mode pattern of a 3.39 μm He–Ne laser [301, 302]. A similar imaging device utilizes white instead of monochromatic illumination of the liquid crystal and relies on the direct observation of the temperature-induced color change [303]. This approach should exhibit lower responsivity. One patent describes a method to reduce fixed patterns in liquid crystal imagers caused by the thermal environment of the sensing layer [304].

Much higher spatial resolution has been obtained with special viewers for laser radiation. The device automatically photographs the laser-induced temperature pattern in the liquid crystal during the transient state. This is accomplished by exposing the sensor to a laser pulse of sufficient energy and by delaying the illumination furnished by a flash lamp until the evolving temperature pattern is within the region of the color band. As a result the local heating imposed on the liquid crystal occurs faster than the process of lateral heat conduction. This approach was used to photograph single pulses of a Nd laser [305, 306] and the modes, the diffraction, and the interference patterns generated by the radiation of HCN lasers [307–310]. A similar device converted into the visible the infrared Schlieren images formed by a plasma which was illuminated by a CO_2 laser through an interferometer. Such experiments were performed to determine the index of refraction associated with the electron density in an argon plasma jet [311] and in electrical spark discharges [312, 313]. Others increased the spatial resolution and decreased the thermal time constant of liquid crystal sensing layers by convection cooling [314] or by heatsinking it to a water-cooled substrate [315].

Cholesteric liquid crystal films were used as holographic recording media for CO_2 laser radiation. The color pattern associated with the CO_2 laser radiation is photographed. The illumination of this hologram by visible He–Ne laser radiation resulted in successful reconstruction of the original image [*316, 317*]. Real-time holography was achieved by producing the temperature hologram in the liquid crystal with the aid of two mutually inclined beams of CO_2 laser radiation [*318*]. In this case the temperature dependence of the optical rotatory power of the liquid crystal was used to modulate the intensity of a He–Ne laser beam and thus to reconstruct the hologram in the visible.

d. *Microwave Radiation.* Microwave sensors consist of a liquid crystal which is thermally coupled to a thin metal film [*319, 320*]. The incident radiation induces electrical currents in the metal which in turn causes the dissipation of heat. This process results in a maximum absorptance of 0.5, if the film resistance per square is equal to one-half of the characteristic impedance of vacuum (i.e., $\frac{1}{2}(\mu_0/\varepsilon_0)^{1/2} = 188.5 \ \Omega$ [*321*]). The temperature-induced color change of the liquid crystal was either directly observed or photographed. A simplified theory supported by experimental evidence [*321*] shows that typical, large-area microwave sensors, which are thermally coupled to the environment by free convection and radiation, have a sensitivity of about 4–6 mW/cm² °C. These devices can resolve large microwave spots, separated by less than 1 mm, with a thermal time constant of several seconds. The responsivity of the sensors can be substantially increased by selecting liquid crystals having a color bandwidth much smaller than that customarily used (i.e., 2–4°) and by observing the temperature-dependent selective reflectance of monochromatic light. Microwave converters were used to study the near-field pattern of antennas, the radiation pattern of an open X-band waveguide, and the interference patterns caused by standing waves [*322, 319, 320*]. A patent describes an active microwave system for the detection of concealed weapons [*323*]. A microwave fluoroscope was proposed [*322*] that can detect density variations and flaws in microwave-transmitting materials. It consists of two opposing antennas, which generate an extended field of standing planar waves, and of a liquid crystal sensor, positioned in the plane of a standing-wave trough. The field strength is adjusted to achieve a desirable selective reflection color. If a sample is inserted between one of the antennas and the sensor, any resulting change in loss of energy or phase will cause an equivalent color pattern in the liquid crystal. These variations can be measured quantitatively with the aid of a phase shifter and an attenuator. Others have found that microwave sensors with absorbers of high resistance per square display the electrical field [*324–326*], while those having sufficiently low resistance per square indicate magnetic field patterns [*325, 326*]. Other sensor configurations were designed to record the mode patterns of waveguides operating in the millimeter and submillimeter region [*327*]. Microwave holograms were generated in liquid crystal films, recorded by photographing the resulting color pattern, and reconstructed in the visible by He–Ne laser radiation. This approach, proposed in 1968 [*322, 328*], was demonstrated by several investigators [*329–332*] and applied to detect metal objects concealed in a purse [*333*].

e. *Ultrasonic Radiation.* Temperature-sensitive liquid crystal films were used to visualize ultrasonic radiation ranging in frequency from about 34 kHz to 1 GHz. A recent paper describes some of these devices in the context of a review on the properties of liquid crystals in sound fields [*334*].

In the first demonstration of this conversion technique, ultrasonic waves of 1 GHz were coupled from a sapphire rod through a reflection-reducing layer directly into a liquid crystal [*335*]. It was estimated that an absorbed acoustic power density of less than 2.5 W/cm^2 generated a perceivable color pattern. The liquid crystal displayed the color band within a temperature interval of 0.2°. The radiation of an electroacoustic horn was visualized by contacting the end face of the horn with a 0.5-mil thick blackened Mylar membrane which supported the liquid crystal film [*336*]. In this case the Mylar membrane served as the acoustic absorber. The vibrational modes of a lead zirconate titanate vibrator were converted into the visible by loosely coupling the supporting membrane of the liquid crystal to the vibrator surface via a thin layer of viscous grease [*337*]. The motion of the vibrator surface elements relative to the membrane caused a frictional heat pattern in the grease layer and thus an equivalent color pattern in the liquid crystal. One patent applies ultrasonic liquid crystal transducers to nondestructive testing of solids [*338*]. The test piece is uniformly insonified and the cholesteric liquid crystal, which is in mechanical contact with the test piece, displays the color pattern associated with the transmitted ultrasound. Any deviation from the unperturbed sound pattern indicates defects such as flaws and variations of density and thickness in the test piece. The inventors claimed to control the thermal and optical environment of the converter film by a hermetic inclosure.

More significant applications deal with the conversion of ultrasound in liquid media. An acoustographic imaging system was built in which the liquid crystal film is separated from the liquid medium by a steel membrane capable of withstanding the hydrostatic pressure [*339*]. The acoustic absorber is attached to the side facing the liquid, while the liquid crystal, deposited on the other side of the steel membrane, is maintained in the proper thermal and optical environment by a hermetic enclosure. The major drawbacks of this sensor are the high power density (100 W/cm^2 for a color band of 2°) required for forming the converted image within 0.5 sec and the degradation of the spatial resolution after about 10 sec because of lateral heat conduction. A patent proposed the use of this sensor in an underwater camera for ultrasonic imaging and for nondestructive testing of solids [*338*]. In the latter application the sensor is mounted into the side wall of a water tank containing the ultrasonic source and the test object. Others submerge the sensor in the water tank and observe it through a window in the wall [*337*]. The sensing layer consists of encapsulated liquid crystals attached to a blackened poly-(ester) film which is sandwiched with the ultrasound-absorbing poly(vinyl chloride) sheet. A coating of grease protects this assembly against adverse effects of the water. Depending on the peak wavelength of the selective reflection, sound pressures between 1.2 and 3.9 N/cm^2 caused perceivable color changes in a liquid crystal with a 1.5°-wide color band. The spatial resolution of this converter is limited by diffraction effects because the ultrasonic waves have a wavelength of 3.5 mm.

Liquid crystal transducers can also be placed at the water–air interface and insonified from underneath. This method yields higher order, near-field mode patterns of a round and a square-shaped, x-cut quartz transducer (air-backed), which agree reasonably well with the results derived from theoretical and other experimental evidence [*340*]. A sound intensity of 0.25 W/cm^2 was sufficient to produce these images with a liquid crystal having a color band of 1.5° width and being supported by a 0.135-mm thick sheet of poly(ethylene). The threshold intensity was 0.03 W/cm^2. The image built up within 3–5 seconds after the start of the insonification and degraded after about 10 seconds due to lateral heat conduction.

Acoustic holographs of hole patterns in an opaque brass plate were made with a liquid crystal transducer located at the air–water interface. The ultrasonic field established by the interfering object and reference beams caused a temperature pattern in the liquid crystal that was recorded by photographing selectively reflected monochromatic light [*247, 341*]. The resulting transparency is the amplitude holograph of the object, which can be readily reconstructed by conventional optical methods. Holes of 1.5 mm diameter were reasonably well reproduced. This performance is close to the estimated resolution limit of the device. Although the liquid crystal displays the visible colors within a 2° temperature interval, each ultrasonic source had to provide an output power of about 6 W/cm^2 at a frequency of 5 MHz.

A patent describes a device capable of real-time reconstruction of ultrasonic holographs [*342*]. This system differs from the previous ones in the read-out method of the "temperature hologram" established in the liquid crystal by the ultrasonic reference and object beams. The liquid crystal is located between two polarizers and is illuminated by collimated light. The intensity of the transmitted light is modulated by the temperature pattern in the liquid crystal because of the temperature dependence of the optical rotatory power. The transmitted light is focused on a spatial filter which passes only one diffracted beam of first order to the viewing optics. The observer perceives a three-dimensional reconstructed image of the insonified object.

Experiments were conducted to determine whether or not ultrasonic energy can directly affect the pitch of cholesteric mesophases and thus the related optical properties. An ultrasonic transducer based on such an effect would not have the severe limitations on time response and resolution that is characteristic of the thermal approach to ultrasonic imaging. One experimental investigation failed, apparently because of the smallness of the effect [*339*]. However, the measurement of the spectral shift of one of the side lobes of the selective transmission profile measured between linear polarizers shows that pressure variations of 1500 N/cm^2 change the pitch by 0.1 nm [*343*]. This experiment was conducted on a ternary mixture of cholesteric and nematic materials which exhibits a rather temperature-insensitive selective reflectance. Since the relaxation frequency of this liquid crystal should be about 15 MHz, it is expected that the variations of the light intensity caused by the pitch changes should follow those ultrasonic frequencies that are below the relaxation limit. An ultrasonic imaging converter based on this effect was proposed [*343*], in which a pulsed tunable laser samples the ultrasonically induced pitch changes at the rate of several megahertz. If practical, this device would permit truly real-time ultrasonic imaging.

C. Nondestructive Testing

Thermal mapping, conducted with temperature-sensing films, which are in direct contact with the test item, may be the most important application of cholesteric liquid crystals. This technique can solve many problems because both temperature level and temperature interval of the color band can be adjusted over a wide range by selecting mixtures of appropriate compounds. Furthermore, the thermal impedance of the sensing film is usually so high that it does not affect significantly the temperature distribution on the test surface.

This temperature pattern can be thermally mapped in the steady and in the transient state. The steady-state approach is limited by the lateral heat conduction, which tends to wash out those characteristic details of the pattern that are associated with tiny cracks, minute inhomogeneities, and other small structural deviations in the test piece. These small defects are enhanced by the transient-state approach because heat waves of short wavelengths propagate faster than those of long wavelengths. Consequently, the fine detail of a temperature pattern appears before the establishment of the cruder, steady-state pattern. Nevertheless, the steady-state method is favored because the testing is simpler and less expensive.

One of the problems is the determination of the detection limit of flaws, structural deviations, and other features that render temperature differences when a uniform heat flux is applied to the sample. Since the theoretical approach is cumbersome even for simple test specimens, most often empirical criteria are used to assess the outcome of nondestructive testing.

In the following we give only representative examples of nondestructive testing with cholesteric liquid crystals because a bibliography of this subject [344] was recently published. Furthermore, most of the pertinent literature describes essentially the same approach applied to a large variety of different test items.

1. COMPOSITE STRUCTURES

Flaws and bonding faults were detected in metal composites and in laminated and honeycomb structures [245, 218, 226, 345–351]. Potential fracture sites in composite materials were determined with an axial fatigue machine [352].

2. METALLOGRAPHICAL APPLICATIONS

Cholesteric liquid crystals have been used to indicate cracks in marine propellers [345], imperfections in the welding of axisymmetric vessels [353], and cracks and voids close to the surface of a weld [354]. They have further been used to detect laminations in sheet metal caused by hot or cold rolling [355], to locate stress areas and potential fracture sites in metals [346], and defects in springs [350, 13d], and to visualize Lüder bands (i.e., regions of unstable plastic flow) of aluminum alloy specimens [345]. Other applications address the detection of shrinkage cavities in metal castings [356], the testing of the thermal isolation of aluminum rivets [351], and the temperature distribution on the surface of heating coils [355].

3. QUALITY CONTROL OF COMPONENTS

Thermal mapping with liquid crystals has been used to test ordinary printed circuits [349, 13d, 357–359] and multilayer printed circuit boards [345, 360] for electrical shorts, and to inspect high-resistance connections on circuit boards [354]. Moreover, oxide breakdowns and gate oxide shorts in field effect transistors [361] and shorts between the windings of small toroidal transformers [360] have been detected, and the temperature pattern of resistors [269, 351, 354] and transistors [269, 362] has been visualized. Other applications address the determination of the uniformity of chromium depositions in chromium-on-glass thin-film resistors [360], the detection of inhomogeneities in sapphire substrates of $Hg_{1-x}Cd_xTe$ infrared detectors [363], and the observation of switching phenomena in Zr–ZrO_2–Au junctions [364] and in discontinuous gold films [365].

Liquid crystals displayed the heat pattern generated by the vibrations of piezoelectric transducers [366, 367] and by miniaturized active ultrahigh-frequency devices [368]. They also indicated discontinuities of electrical conductors imbedded in automobile windshields [369] and heat leaks of refrigerator doors [370].

4. AERONAUTICAL AND SPACE APPLICATIONS

Temperature differences visualized by liquid crystals showed restricted coolant channels of aerospace components [245, 345], flaws in coalescers used for the filtration of jet fuel [371], nonuniformities of the resistive defrosting coatings of aircraft windows [345, 357] and of a Skylab window [372], and hot spots of miniature heaters for spacecraft [360]. Several papers describe the mapping of laminar and turbulent boundary layers of airplane models and other objects located in wind tunnels [373–379]. Other applications involve the measurement of local convective heat transfer coefficients for a plate exposed to a jet of air [380, 381] and the determination of the temperature shift of the color band caused by the shearing action of the air flow [382]. The latter effect has to be considered if exposed liquid crystals films are used for quantitative thermal mapping in wind tunnels [373]. Cholesteric liquid crystals have also been used aboard Apollo 14 [383–385] and Apollo 17 [386–388] as temperature indicators in heat flow and convection experiments of gases and liquids in an environment of less than $10^{-6}\ g$ gravity.

D. Medical Applications*

In medicine, thermal mapping is used to detect those defects and malfunctions in human beings that can generate temperature patterns on the skin. The correct interpretation of these patterns can represent a formidable medical problem because the exact correlation between medical cause and observed temperature effect is difficult to quantify. Furthermore, a number of different medical causes can result in temperature patterns of the same type. Expensive infrared thermographs, equipped with sophisticated electronic signal processors, are used for quantitative thermal mapping, while the rather inexpensive liquid crystals are primarily applied to qualitative and comparative thermal mapping. In addition, liquid crystals serve as temperature monitors of medical equipment.

Thermography with cholesteric liquid crystals reported so far relies on the conductive heat exchange between sensing film and test surface. In one method the liquid crystals are directly deposited on the skin. The usual procedure is first to blacken the skin with a water-soluble black dye or with carbon black dispersed in either water or a water-soluble polymer such as poly(vinyl alcohol), methyl cellulose, and poly(vinylpyrrolidone) in order to reduce reflection of light. Subsequently, a solution of a suitable liquid crystal in chloroform or in a volatile hydrocarbon is brushed or sprayed on the dry,

---

* In this section, an asterisk after a reference number indicates that the article contains photographs.

TABLE XII

Temperature Range of Color Band of Mixtures[a]

Temperature range (°C)	wt % Cholesteryl		
	Oleyl carbonate	Nonanoate	Benzoate
30–33	44	46	10
31–34	42	48	10
32–35	40	50	10
33–36	38	52	10
34–37	36	54	10
35–38	34	56	10
36–39	32	58	10
37–40	30	60	10

[a] From Goldberg and Fergason [239].

blackened skin [239]. Thermal mapping can commence as soon as the solvent has evaporated and the patient is in thermal equilibrium with the environment. Table XII lists typical mixtures covering the temperature range from 30° to 40° with a color band width of 3°.

Alternate methods employ sandwiched, microencapsulated, and imbedded liquid crystals to simplify and shorten the deposition process. Patents claim silk-screened liquid crystals covered with a protective acrylic layer [240], the use of black-pigmented latices [389] and black polymer films [390] as substrates, and the addition of bacteriostatic agents [391] to liquid crystal mixtures. A "liquid crystal cream" consisting of microencapsulated liquid crystals, carbon black, and a surfactant in poly(vinyl alcohol) or poly(vinylpyrrolidone) has been described in a patent. Thin layers of this cream form a tough film after drying that is suitable for thermal mapping and can be readily peeled off after use [392]. Liquid crystals imbedded in a self-adhesive polymer film are marketed as "liquid crystal tape." It contains a yellow dye to enhance the selective reflectance in the green and an antioxidant to extend its shelf life [393, 394]. Studies address the effects of ambient temperature [395] and exposure to light [396] on these tapes. Some of the liquid crystal mixtures used in these films and their respective color bands [393] are listed in Table XIII.

Several patents describe the use of cholesteric liquid crystals in clinical thermometry. The temperature range of liquid crystal thermometers can be extended without a loss of sensitivity by imbedding different liquid crystals in close proximity [272]. For example, a plastic sensing patch has been proposed in which each liquid crystal, contained in a radial segment, selectively reflects in a different temperature region [397]. Mixtures of cholesteryl oleyl carbonate,

TABLE XIII

SELECTIVE REFLECTION IN THE GREEN OF IMBEDDED LIQUID CRYSTALS

Green color (°C)	Parts of cholesteryl (g)			
	Nonanoate	Oleyl carbonate	Benzoate	Chloride
31	4.2	1.4	—	0.6
32	4.0	1.1	1.0	—
34	4.4	1.1	—	0.6
35	4.45	1.2	—	0.43
37	3.34	2.42	0.24	—
40	4.8	0.7	—	0.6

cholesteryl nonanoate, and carbon black confined to small pockets in a special wrist strap have been proposed as optical fever indicators. It is claimed that the replacement of carbon black by small glass beads or translucent poly(urethane) foam improves the performance of this device [398]. Imbedded hysteretic liquid crystals (see Section IV,B,2) have been used to indicate whether or not the temperature has exceeded a prescribed limit. Three geometric configurations have been claimed in which different, physically separated, hysteretic mixtures cover a set of limit temperatures [399].

The photographic recording of the color patterns has been improved by using polarizers to suppress specular reflection [400*] and interference filters to quantify the temperature pattern [401*].

Although several brief review articles describe the applications of cholesteric liquid crystals in various medical disciplines [13a, 402–407], a complete survey has not been published yet. Therefore, a chemist and a physicist, with the assistance of friends in the medical profession, will attempt to summarize the clinical applications of cholesteric liquid crystals as temperature sensors.

1. CUTANEOUS THERMOGRAPHY

The usefulness of cholesteric liquid crystals as an indicator of skin temperature patterns was first demonstrated by Crissey et al. [401*], who also discussed potential applications such as the localization and diagnosis of neoplasms and pharmacological studies of drugs. Subsequent investigations address the influence of temperature variations of the surrounding air [408*], high-velocity air streams [409*], and thermal shock [410] on liquid crystal films. Other applications include the differentiation of embolic, inflammatory, and atelectatic pulmonary lesions [411*], as well as the diagnosis of pulmonary thromboembolism [412], angina pectoris [413], thyroidal dysfunctions [414], acute cholecystitis and, in combination with ultrasound, cholelithiasis [407]. Liquid crystal thermography assisted in the rather difficult diagnosis of retroperitoneal and other intra-abdominal abscesses. It was also used to detect osteomyelitis of the bones of the pelvis, acute appendicitis and localized peritonitis of appendicularian origin [407], and to study patients with rheumatic arthritis, including involvement of the metacarpal bones, osteoarthritis, and hemoarthritis [415*].

Other applications are the visual inspection of algodystrophic syndromes, ankylosing spondylitis [402*], psoriasis of the abdomen [416*], hemangioma, Wegener's disease [417*], angiopapilloma, fibroadenoma, gynecomastia [418], and hemorrhage, especially hemorrhage of hemophiliacs [405*].

The study of surface temperature patterns of the human foot and ankle suggests widespread applications for thermography in podiatry [419].

Dynamic liquid crystal thermography has been used to enhance temperature patterns in the patient. In this method a temperature gradient (i.e., a thermal stress) is imposed across the region of investigation by external means (e.g., by heating or cooling of the surface) and the return of this area to thermal equilibrium is continuously monitored. The latter is possible only with fast-scanning infrared thermographs or with two-dimensional temperature indicators such as liquid crystal sensors. Varicose veins were visualized by cooling the surface of the affected region and by reducing the blood flow of superficial veins and tributaries with tourniquets. Thus the heat flow through incompetent perforators from deep to superficial veins resulted in dynamic thermograms which clearly outlined varicose veins [411*]. Other studies concerned obstructions of the iliac, femoral [414], and the carotid arteries [420]. The segments of arteries with a pronounced inflammatory reaction were recognized due to their higher temperature.

Several papers report the use of liquid crystals in the investigation of rapid temperature changes due to vasoconstriction and vasodilation induced by thermal stress [404*, 408*, 421*, 422*]. Intravenous administration of papaverine (a vasodilator) and methoxamine (a vasoconstrictor) to rabbits resulted in perceivable temperature changes in response to their pharmacological action [423*]. Nicotine [417*] or a combination of histamine and epinephrine [408*] induced similar effects in man. Selective reflection visualized temperature changes associated with Buerger's disease [415*], endarteritis obliterans and atherosclerosis [407].

2. DENTISTRY

Liquid crystal thermography supplemented common diagnostic techniques used to determine pulp vitality. Preliminary results indicate that a nonvital tooth containing necrotic pulp tissue has a lower surface temperature than a normal tooth with an intact internal blood supply, and that the surrounding oral tissues do not significantly affect the surface temperature of a nonvital tooth [424]. Cholesteric liquid crystals indicated the surface temperature of teeth during cavity preparation. Temperature increases greater than 75° were recorded at a distance of about 1 mm from a mechanically driven dental drill [425*].

3. GYNECOLOGY

The abdominal temperatures in 108 pregnant women were measured with cholesteric liquid crystals sandwiched between polymer sheets. The placental site was correctly indicated by an elevated skin temperature in 84 patients. However, this method failed for very early pregnancies [426*]. In a series of investigations conducted during the late stages of pregnancy, the implantation sites were correctly identified with an overall accuracy of 85%. Thermographic location of anterior, frontal, and lateral implantation sites was very accurate, but the identification of posterior implantation was rather uncertain [427*].

Vascular patterns of the breast have been studied extensively and typical thermal patterns were obtained as a function of age, past pregnancies, previous lactation, and breast size [428*]. In women with apparently normal menstrual cycle, the coldest breast temperatures were recorded during the menstrual period, and the warmest breast temperatures during ovulation. These cyclic changes were absent in post-menopause women or women taking oral contraceptives [429].

4. NEUROLOGY

Patients with disorders of the cranial and spinal marrow, with syringomelia, and with spinal trauma from injuries exhibited correlated temperature patterns on the respective skin areas [430]. Liquid crystal thermography was also used for the diagnosis of spina bifida [416*, 431*].

5. ONCOLOGY

A preliminary study [417*] indicated that liquid crystal thermography can detect subcutaneous and intracutaneous malignant tumors, which are typically 0.9°–3.3° warmer than the apparently normal surrounding skin. This method was therefore recommended as an aid in the detection of early cancer of the female breast, despite some false-negative results in poorly vascularized carcinomas and false-positive results in patients with inflammatory lesions. A comparison with infrared thermography [432*] also favors the application of cholesteric liquid crystals because of the simplicity of the method and the real-size presentation of the temperature patterns on the patient. The reliability of liquid crystal thermography in breast cancer detection was evaluated. In a study over a period of 28–45 days [429] it was found that daily variations of the temperature pattern associated with normal activity and the menstrual cycle should not affect the reliability of this method. However, alcohol consumption and cigarette smoking prior to the thermographic examination can drastically change the normal thermogram. Therefore, information about prior activity level, smoking history, alcohol consumption, and time relative to the menstrual cycle are prerequisites of a definitive interpretation of the thermogram.

In the thermographic examination of 105 women with breast lesions that required a biopsy, only one of 17 histologically proven malignancies gave no thermographic evidence of malignancy. False-positive thermograms occurred only in cases of fibrocystic diseases [428*]. The authors also found that liquid crystal thermography was more reliable than mammography and physical diagnosis. During routine examinations of 836 patients in 1972, palpable tumors of the breast were detected in 56 cases; the liquid crystal thermograms of another 21 (2.7%) of the remaining 780 patients indicated a miniature breast tumor which could not be diagnosed clinically. Histological examination confirmed the diagnoses. The authors report the removal of a small (1 × 2 mm) adenocarcinoma in one case and therefore suggest the use of liquid crystals for routine detection of tumors of the breast [433*, 434].

Extended thermographic investigations showed that tumors of the head and the neck (83 neoplasms) could also be identified [435, 436], as well as vertebral metastases [414] and carcinomas of the lung [412], malignant and benign tumors of the mammary gland [407], tumors of the testicle [437], and vertebral tumors [430].

Results similar to those in man were also obtained with tumors in mice [438*].

6. Ophthalmology

The corneal temperature was measured *in vivo* with a thermally sensitive contact lens consisting of a laminated cholesteric liquid crystal layer [439].

7. Pediatrics

Clinical studies of neonates have shown that liquid crystal thermography can be used to estimate accurately the body temperature. The skin temperature over the right upper abdominal quadrant corresponds closely to the rectal temperature. Liquid crystal monitor tapes applied to the abdominal skin of new-born infants yielded a satisfactory low number of false readings in a total of 375 observations made in 66 infants [440].

Measuring the surface temperature of neonates in incubators with cholesteric liquid crystals can be useful for the early detection of illnesses. Out of 100 cases studied, the extremity temperatures of six infants were lower than the standard deviation attributable to incubator temperature. This indicated peripheral vasoconstriction. Subsequent clinical investigation revealed neonatal sepsis and meningitis, congestive heart failure, and perinatal asphyxia [441].

8. RADIOLOGY

In a study on possible biological effects caused by the exposure of the human head to microwave radiation (2.45 GHz), the energy absorbed by a full-scale phantom model of the human head was determined by measuring the associated temperature pattern with the aid of cholesteric liquid crystals [442]. For other radiological applications, such as X-ray dosimeters, the reader is referred to the previous section on Radiation Sensors (Section V, B, 2).

9. SURGERY

Cholesteric liquid crystals were used simultaneously with angiography, transcutaneous blood flow velocity monitors, and surface thermistor thermometers on patients scheduled for vascular surgery and lumbar sympathectomy for peripheral vascular occlusive diseases [420]. Postoperative monitoring of a femoropopliteal bypass allows evaluation of graft patency [420]. In a thermographic investigation of patients with occlusive endarteritis or atherosclerosis, vessel patency was evaluated prior to amputation of the extremity [443, 444]. It has been reported [420] that the entire leg of a potential amputee was saved by evidence obtained from liquid crystal thermography. In chronic gunshot and hematogenic osteomyelitis (18 patients), an extensive zone of inflammatory reaction was thermographically determined even in places where osseous pathology could not be obtained by roentgenological means [407]. Valuable evidence was obtained in the localization of suppurative diseases of the bone and fingers (33 patients). The surgeon was able to utilize the thermographic information for the correction of the initial phases of paronychia [407]. Liquid crystal thermography, used in a study of flap circulation in rabbits and humans, showed that complete viability is achieved consistently only if the major dorsal vessels are included in the flap [445*].

Resistively heated surgical probes are used to destroy tissue by raising the temperature of the surgical target above 55°. The temperature distribution around such a probe was visually monitored with cholesteric liquid crystals in a test medium which closely simulates the thermal properties of tissue [446]. The temperature distribution was also measured around radio-frequency surgical probes [447] and cryoprobes [448] imbedded in their respective test media.

10. UROLOGY

Patients with acute epididymitis exhibited an increased skin temperature on the infected side of the scrotum, while patients with unilateral and bilateral acute pyelonephrosis had a reduced skin temperature directly over the afflicted kidney [437].

REFERENCES

1. G. Friedel, *Ann. Phys.* (*Paris*) **18**, 273 (1922).
2. G. H. Brown and W. G. Shaw, *Chem. Rev.* **57**, 1049 (1957).
3. G. W. Gray, "Molecular Structure and the Properties of Liquid Crystals." Academic Press, New York, 1962.
3a. G. W. Gray and P. A. Winsor, eds., "Liquid Crystals and Plastic Crystals," Vols. 1 and 2. Ellis Horwood, Chichester, England, 1974.
4. G. H. Brown, G. J. Dienes, and M. M. Labes, "Liquid Crystals I," Int. Liquid Crystal Conf., (Pap.), 1st, 1965, Gordon & Breach, New York, 1967.
5. G. H. Brown, "Liquid Crystals II," Int. Liquid Cryst. Conf., (Pap.), 2nd, 1968. Gordon & Breach, New York, 1970.
6. G. H. Brown and M. M. Labes, "Liquid Crystals III," Int. Liquid Cryst. Conf., (Pap.), 3rd, 1970. Gordon & Breach, New York, 1972.
7. Papers presented at the *4th Int. Liquid Cryst. Conf., 1972*, have been published in *Mol. Cryst. Liquid Cryst.* **21–25**, (1973–1975).
8. R. S. Porter and J. F. Johnson, "Ordered Fluids and Liquid Crystals," Advan. Chem. Ser. 63. Amer. Chem. Soc., Washington, D.C., 1967.
9. J. F. Johnson and R. S. Porter, "Liquid Crystals and Ordered Fluids." Plenum, New York, 1970.
10. J. F. Johnson and R. S. Porter, "Liquid Crystals and Ordered Fluids II." Plenum, New York, 1974.
11. P. G. de Gennes, "The Physics of Liquid Crystals." Oxford Univ. Press (Clarendon), London and New York, 1974.
12. For example: (a) A. Saupe, *Mol. Cryst. Liquid Cryst.* **7**, 59 (1969); (b) H. Baessler, *Festkörperprobleme* **11**, 99 (1971); (c) I. G. Chistyakov, "Liquid Crystals" (In Russian). Sci. Publ. House, Moscow, 1966; (d) J. L. Fergason and G. H. Brown, *J. Amer. Oil Chem. Soc.* **45**, 120 (1968); (e) R. Steinsträsser, *Chem.-Ztg.* **95**, 661 (1971); (f) W. H. Toliver, C. G. Roach, R. W. Roundy, and P. E. Hoffman, *Aerosp. Med.* **40**, 35 (1969); (g) G. H. Brown and J. W. Doane, *Appl. Phys.* **4**, 1 (1974); (h) R. Steinsträsser and L. Pohl, *Angew. Chem., Int. Ed. Engl.* **12**, 617 (1973); *Angew. Chem.* **85**, 706 (1973).
13. For example: (a) H. Kelker and R. Hatz, *Chem.-Ing.-Tech.* **45**, 1005 (1973); (b) M. Hareng, *Rev. Tech. Thomson-CSF* **5**, 319 (1973); (c) J. L. Fergason, T. R. Taylor, and T. B. Harsch, *Electro-Technol.* (*New York*) **85**, 41 (1970); (d) B. Hampel, *Z. Werkstofftech.* **3**, 149 (1972).
14. O. Lehmann, *Z. Phys. Chem.* (*Leipzig*) **4**, 462 (1889).
15. R. D. Ennulat, *Mol. Cryst. Liquid Cryst.* **3**, 405 (1968).
16. H. Baessler, P. A. G. Malya, W. R. Nes, and M. M. Labes, *Mol. Cryst. Liquid Cryst.* **6**, 329 (1970).
17. D. Coates and G. W. Gray, *Phys. Lett. A* **45**, 115 (1973).
18. W. Elser, J. L. W. Pohlmann, and P. R. Boyd, *Mol. Cryst. Liquid Cryst.* **20**, 77 (1973).
19. H. Baessler and M. M. Labes, *J. Chem. Phys.* **52**, 631 (1970).
20. J. E. Adams, W. Haas, and J. J. Wysocki, *in* "Liquid Crystals and Ordered Fluids" (J. F. Johnson and R. S. Porter, eds.), p. 463. Plenum, New York, 1970.
21. J. E. Adams and L. B. Leder, *Chem. Phys. Lett.* **6**, 90 (1970).
22. K. Ko, I. Teucher, and M. M. Labes, *Mol. Cryst. Liquid Cryst.* **22**, 203 (1973).
23. H. Stegemeyer and H. Finkelmann, *Chem. Phys. Lett.* **23**, 227 (1973); *Ber. Bunsenges. Phys. Chem.* **78**, 860 (1974).
24. J. E. Adams and W. Haas, *Mol. Cryst. Liquid Cryst.* **15**, 27 (1971).
25. L. B. Leder, *Chem. Phys. Lett.* **6**, 285 (1970).

26. R. Cano, *Bull. Soc. Fr. Mineral.* **90**, 333 (1967).
26a. J. M. Pochan and D. D. Hinman, *J. Phys. Chem.* **78**, 1206 (1974).
27. H. Hakemi and M. M. Labes, *J. Chem. Phys.* **58**, 1318 (1973).
28. E. Sackmann, *J. Amer. Chem. Soc.* **93**, 7088 (1971).
29. J. Voss and E. Sackmann, *Z. Naturforsch. A* **28**, 1496 (1973).
30. F. D. Saeva and J. J. Wysocki, *J. Amer. Chem. Soc.* **93**, 5928 (1971).
31. E. Sackmann and J. Voss, *Chem. Phys. Lett.* **14**, 528 (1972).
32. F. D. Saeva, *J. Amer. Chem. Soc.* **94**, 5135 (1972).
33. K.-J. Mainusch and H. Stegemeyer, *Z. Phys. Chem. (Frankfurt)* **77**, 210 (1972).
34. H. Stegemeyer and K.-J. Mainusch, *Naturwissenschaften* **58**, 599 (1971).
35. P. H. Keyes, H. T. Weston, and W. B. Daniels, *Phys. Rev. Lett.* **31**, 628 (1973).
35a. P. Pollmann and H. Stegemeyer, *Ber. Bunsenges. Phys. Chem.* **78**, 843 (1974).
36. R. D. Ennulat, *Mol. Cryst. Liquid Cryst.* **13**, 337 (1971).
37. F. J. Kahn, *Appl. Phys. Lett.* **18**, 231 (1971).
38. J. L. Fergason, *U.S. Nat. Tech. Inform. Serv., Rep. AD 741 898* (1971); see also Fergason *et al.* [*88*]; W. H. Toliver, J. L. Fergason, E. Sharpless, and P. E. Hoffman, *Aerosp. Med.* **41**, 18 (1970), and references cited therein.
39. For example, we observed that the outgassing of a certain epoxy glue destroyed the color band of cholesteryl oleyl carbonate.
40. C. J. Gerritsma and P. van Zanten, *Phys. Lett. A* **42**, 329 (1972).
41. R. Dreher, Ph.D. Dissertation, Freiburg, 1971.
42. J. E. Adams, W. Haas, and J. Wysocki, *J. Chem. Phys.* **50**, 2458 (1969).
43. D. W. Berreman and T. J. Scheffer, *Phys. Rev. Lett.* **25**, 577 (1970); *Mol. Cryst. Liquid Cryst.* **11**, 395 (1970).
44. R. Dreher, G. Meier, and A. Saupe, *Mol. Cryst. Liquid Cryst.* **13**, 17 (1971).
45. P. Kassubek and G. Meier, *Mol. Cryst. Liquid Cryst.* **8**, 305 (1969).
46. P. Chatelain, *Bull. Soc. Fr. Mineral.* **66**, 105 (1943); for a different explanation see de Gennes [*11*], p. 77.
47. R. D. Ennulat, L. E. Garn, and J. D. White, *Mol. Cryst. Liquid Cryst.* **26**, 245 (1974).
48. J. L. Fergason, *Mol. Cryst. Liquid Cryst.* **1**, 293 (1966).
49. J. L. Fergason, *Appl. Opt.* **7**, 1729 (1968).
50. M. Magne and P. Pinard, *J. Phys. (Paris)* **30**, C4, 117 (1969).
51. See Friedel [*1*], pp. 447–448.
52. C. W. Oseen, *Trans. Faraday Soc.* **29**, 883 (1933).
53. H. de Vries, *Acta Crystallogr.* **4**, 219 (1951).
54. S. Chandrasekhar and K. N. Srinivasa Rao, *Acta Crystallogr., Sect. A* **24**, 445 (1968).
55. G. H. Connors, *J. Opt. Soc. Amer.* **58**, 875 (1968).
56. A. S. Marathay, *J. Opt. Soc. Amer.* **61**, 1363 (1971).
57. R. M. A. Azzam and N. M. Bashara, *J. Opt. Soc. Amer.* **62**, 1252 (1972).
58. R. M. A. Azzam, B. E. Merrill, and N. M. Bashara, *Appl. Opt.* **12**, 764 (1973).
59. S. Chandrasekhar and J. S. Prasad, *Mol. Cryst. Liquid Cryst.* **14**, 115 (1971).
60. M. Aihara and H. Inaba, *Opt. Commun.* **3**, 77 (1971).
61. M. Aihara and H. Inaba, *Oyo Butsuri* **41**, 338 (1972); *Chem. Abstr.* **77**, 145689 (1972).
62. M. Aihara and H. Inaba, *Oyo Butsuri* **41**, 345 (1972); *Chem. Abstr.* **77**, 145691 (1972).
63. G. Holzwarth and N. A. W. Holzwarth, *J. Opt. Soc. Amer.* **63**, 324 (1973).
64. G. S. Ranganath, S. Chandrasekhar, U. D. Kini, K. A. Suresh, and S. Ramaseshan, *Chem. Phys. Lett.* **19**, 556 (1973).
65. S. Chandrasekhar and J. S. Prasad, *in* "Physics of the Solid State" (S. Balakrishna, M. Krishnamurthi, and B. R. Rao, eds.), p. 77. Academic Press, New York, 1969.
66. E. Sackmann, S. Meiboom, L. C. Snyder, A. E. Meixner, and R. E. Dietz, *J. Amer. Chem. Soc.* **90**, 3567 (1968).

67. D. Taupin, *J. Phys. (Paris)* **30**, C4, 32 (1969).
68. E. I. Kats, *Sov. Phys.—JETP* **32**, 1004 (1971).
69. B. Böttcher, *Materialpruefung* **13**, 11 (1971).
70. V. A. Belyakov and V. D. Dmitrienko, *Fiz. Tverd. Tela* **15**, 2724 (1973); *Sov. Phys.—Solid State* **15**, 1811 (1974).
71. I. G. Chistyakov, *Usp. Fiz. Nauk* **89**, 563 (1966); *Sov. Phys.—Usp.* **9**, 551 (1967).
72. R. Dreher and G. Meier, *Phys. Rev. A* **8**, 1616 (1973).
73. F. C. Frank, *Discuss. Faraday Soc.* **25**, 19 (1958).
74. W. Maier and A. Saupe, *Z. Naturforsch. A* **14**, 882 (1959).
75. W. J. A. Goossens, *Phys. Lett. A* **31**, 413 (1970); *Mol. Cryst. Liquid Cryst.* **12**, 237 (1971).
76. R. Alben, *Mol. Cryst. Liquid Cryst.* **20**, 231 (1973).
77. T. Harada and P. P. Crooker, *Bull. Amer. Phys. Soc.* **18**, 436 (1973).
78. In this argument we assume that the materials used in Baessler et al. [16] and Goossens [75] are approximately of the same purity, which is reasonable because both samples were purified by chromatographic methods.
79. J. Cheng and R. B. Meyer, *Phys. Rev. Lett.* **29**, 1240 (1972).
80. P. N. Keating, *Mol. Cryst. Liquid Cryst.* **8**, 315 (1969).
81. B. Böttcher, *Chem.-Ztg.* **96**, 214 (1972).
82. R. S. Porter and J. F. Johnson, *J. Appl. Phys.* **34**, 55 (1963).
83. D. B. Davies and A. J. Matheson, *Trans. Faraday Soc.* **63**, 596 (1967).
84. B. Böttcher, Ph.D. Dissertation, Techn. Univ. Berlin, 1971.
85. R. S. Pindak, C. C. Huang, and J. T. Ho, *Solid State Commun.* **14**, 821 (1974); *Phys. Rev. Lett.* **32**, 43 (1974).
86. P. G. de Gennes, *Solid State Commun.* **10**, 753 (1972); *Mol. Cryst. Liquid Cryst.* **21**, 49 (1973).
87. J. Frenkel, "Kinetic Theory of Liquids." Dover, New York, 1955.
88. J. L. Fergason, N. N. Goldberg, C. H. Jones, R. S. Rush, and L. C. Scala, *U.S. Nat. Tech. Inform. Serv., Rep. AD 620 940* (1965).
89. L. B. Leder, *J. Chem. Phys.* **55**, 2649 (1971).
90. W. Elser, R. D. Ennulat, and J. L. W. Pohlmann, *Mol. Cryst. Liquid Cryst.* **27**, 375 (1974).
91. G. L. O'Connor and H. R. Nace, *J. Amer. Chem. Soc.* **74**, 5454 (1952).
92. W. Elser, *U.S. Nat. Tech. Inform. Serv., Rep. AD 449 684* (1964).
93. W. R. Young, E. M. Barrall, and A. Aviram, in "Analytical Calorimetry II" (R. S. Porter, ed.), p. 113. Plenum, New York, 1970.
94. E. M. Barrall, private communication.
95. C. W. Shoppee and R. J. Stephenson, *J. Chem. Soc.* 2230 (1954).
96. L. B. Leder, *J. Phys. Chem.* **54**, 4671 (1971).
97. J. L. W. Pohlmann, R. D. Ennulat, W. Elser, and P. R. Boyd, to be published.
98. G. H. Brown, *U.S. Nat. Tech. Inform. Serv., Rep. AD 822 208* (1967).
99. J. S. Dave and R. A. Vora, *Indian J. Chem.* **11**, 19 (1973).
100. E. M. Barrall, K. E. Bredfeldt, and M. J. Vogel, *Mol. Cryst. Liquid Cryst.* **18**, 195 (1972).
101. D. Heydenhauss, G. Jaenecke, and H. Schubert, *Z. Chem.* **13**, 295 (1973).
102. H. W. Gibson and J. M. Pochan, *J. Phys. Chem.* **77**, 837 (1973).
103. D. Gross, *Z. Naturforsch. B* **27**, 472 (1972).
104. K. Tsuji, M. Sorai, and S. Seki, *Bull. Chem. Soc. Jap.* **44**, 1452 (1971).
105. W. Mahler and M. Panar, *J. Amer. Chem. Soc.* **94**, 7195 (1972).
106. W. Mahler and M. Panar, Ger. Offen. 2,245,924 (1974); *Chem. Abstr.* **81**, 31903 (1974).
107. R. D. Ennulat, *Mol. Cryst. Liquid Cryst.* **8**, 247 (1969).

108. P.-J. Sell and A. W. Neumann, *Z. Phys. Chem. (Frankfurt)* **65**, 13 (1969).
109. H. Arnold, D. Demus, H.-J. Koch, A. Nelles, and H. Sackmann, *Z. Phys. Chem. (Leipzig)* **240**, 185 (1969).
110. G. J. Davis and R. S. Porter, *J. Therm. Anal.* **1**, 449 (1969).
111. M. Leclerq, J. Billard, and J. Jacques, *C. R. Acad. Sci., Ser. C* **264**, 1789 (1967).
112. L. M. Cameron, R. E. Callender, and A. J. Kramer, *Mol. Cryst. Liquid Cryst.* **16**, 75 (1972).
113. R. D. Ennulat, *in* "Analytical Calorimetry I" (R. S. Porter and J. F. Johnson, eds.), p. 219. Plenum, New York, 1968.
114. W. Elser, J. L. W. Pohlmann, and P. R. Boyd, *Mol. Cryst. Liquid Cryst.* **11**, 279 (1970).
115. W. Elser, J. L. W. Pohlmann, P. R. Boyd, and A. J. Brown, to be published.
116. L. B. Leder, *J. Chem. Phys.* **58**, 1118 (1973); U. S. Patents 3,888,892 and 3,907,406 (1975).
117. J. Y. C. Chu, *J. Chem. Soc., Chem. Commun.* 374 (1974); *J. Phys. Chem.* **79**, 119 (1975).
118. R. D. Ennulat and A. J. Brown, *Mol. Cryst. Liquid Cryst.* **12**, 367 (1971).
119. J. L. W. Pohlmann, W. Elser, and P. R. Boyd, *Mol. Cryst. Liquid Cryst.* **20**, 87 (1973).
120. W. Elser, J. L. W. Pohlmann, and P. R. Boyd, *Mol. Cryst. Liquid Cryst.* **15**, 175 (1971).
121. W. Elser, J. L. W. Pohlmann, and P. R. Boyd, *Mol. Cryst. Liquid Cryst.* **27**, 325 (1974).
122. J. L. W. Pohlmann, W. Elser, and P. R. Boyd, *Mol. Cryst. Liquid Cryst.* **26**, 59 (1974).
123. W. Elser, *Mol. Cryst. Liquid Cryst.* **2**, 1 (1966).
124. W. Elser, *Mol. Cryst. Liquid Cryst.* **8**, 219 (1969).
125. W. Elser and R. D. Ennulat, *J. Phys. Chem.* **74**, 1545 (1970).
126. J. L. W. Pohlmann and W. Elser, *Mol. Cryst. Liquid Cryst.* **8**, 427 (1969).
127. J. L. W. Pohlmann, R. D. Ennulat, and W. Elser, to be published.
128. D. Gross and B. Böttcher, *Z. Naturforsch.* **B 25**, 1099 (1970).
129. J. S. Dave and R. A. Vora, *in* "Liquid Crystals and Ordered Fluids" (J. F. Johnson and R. S. Porter, eds.), p. 477. Plenum, New York, 1970.
130. J. S. Dave and R. A. Vora, *Mol. Cryst. Liquid Cryst.* **14**, 319 (1971).
131. J. S. Dave and G. Kurian, *Indian J. Chem.* **11**, 833 (1973).
132. J. S. Dave and G. Kurian, *Curr. Sci.* **42**, 200 (1973).
133. J. S. Dave and G. Kurian, *Mol. Cryst. Liquid Cryst.* **24**, 347 (1973).
134. W. Elser, J. L. W. Pohlmann, and P. R. Boyd, *Mol. Cryst. Liquid Cryst.* **13**, 255 (1971).
135. E. Fernholz and W. L. Ruigh, *J. Amer. Chem. Soc.* **62**, 3346 (1940).
136. H. I. Bolker, *Nature (London)* **213**, 905 (1967).
137. For a recent review of marine sterols, see P. J. Scheuer, "Chemistry of Marine Natural Products," Chapter 2. Academic Press, New York, 1973.
138. D. H. R. Barton and C. H. Robinson, *J. Chem. Soc.* 3045 (1954).
139. W. Bergmann and J. P. Dusza, *Justus Liebigs Ann. Chem.* **603**, 36 (1957).
140. W. Cole and P. L. Julian, *J. Amer. Chem. Soc.* **67**, 1369 (1945).
141. E. P. Burrows, G. M. Hornby, and E. Caspi, *J. Org. Chem.* **34**, 103 (1969).
142. B. Riegel and I. A. Kaye, *J. Amer. Chem. Soc.* **66**, 723 (1944).
143. G. G. Maidachenko and I. G. Chistyakov, *Zh. Obshch. Khim.* **37**, 1730 (1967); *J. Gen. Chem. USSR* **37**, 1649 (1967).
144. G. G. Maidachenko and I. G. Chistyakov, *Uch. Zap., Ivanov. Gos. Pedagog. Inst.*, **77**, 61 (1970); *Chem. Abstr.* **76**, 18886 (1972).
145. J. L. W. Pohlmann, *Mol. Cryst. Liquid Cryst.* **2**, 15 (1966).
146. L. Verbit and G. A. Lorenzo, *Mol. Cryst. Liquid Cryst.* **30**, 87 (1975).
147. J. L. W. Pohlmann, unpublished observations.
148. A. Tanaka, Y. Satomi, and A. Kato, Japan. Kokai 73 10,061 (1973); *Chem. Abstr.*

78, 136542 (1973); for the corresponding cholesteryl analogs, see Y. Satomi, Japan. Kokai 73 68,560 (1973); *Chem. Abstr.* **80**, 48246 (1974).
149. F. R. Valentine and W. Bergmann, *J. Org. Chem.* **6**, 452 (1941).
150. J. L. W. Pohlmann, *Mol. Cryst. Liquid Cryst.* **8**, 417 (1969).
151. F. F. Knapp and H. J. Nicholas, in "Liquid Crystals and Ordered Fluids" (J. F. Johnson and R. S. Porter, eds.), p. 147. Plenum, New York, 1970.
152. F. F. Knapp and H. J. Nicholas, *Mol. Cryst. Liquid Cryst.* **10**, 173 (1970).
153. W. Bergmann and J. P. Dusza, *J. Org. Chem.* **23**, 1245 (1958).
154. J. Kučera and F. Šorm, *Coll. Czech. Chem. Commun.* **23**, 116 (1958).
155. H. de Vries and H. J. Backer, *Rec. Trav. Chim. Pays-Bas* **69**, 1252 (1950).
156. W. Elser, J. L. W. Pohlmann, and P. R. Boyd, *Mol. Cryst. Liquid Cryst.* **13**, 271 (1971).
157. W. Elser, J. L. W. Pohlmann, and P. R. Boyd, to be published.
158. J. Malthete, J. Billard, and J. Jacques, *Bull. Soc. Chim. Fr., Ser. I*, 1199 (1974).
159. J. L. W. Pohlmann, W. Elser, and P. R. Boyd, *Mol. Cryst. Liquid Cryst.* **13**, 243 (1971).
160. F. C. Chang, *J. Chin. Chem. Soc.* [II] **9**, 53 (1962).
161. H. Reich and A. Lardon, *Helv. Chim. Acta* **29**, 671 (1946).
162. A. J. Bellamy and G. H. Whitham, *J. Chem. Soc., C* 215 (1967).
163. J. T. Edward and N. E. Lawson, *J. Org. Chem.* **35**, 1426 (1970).
164. J. A. Ross and M. D. Martz, *J. Org. Chem.* **29**, 2784 (1964).
165. V. A. Petrow, *J. Chem. Soc.* 66 (1940).
166. F. S. Spring and G. Swain, *J. Chem. Soc.* 320 (1941).
167. C. F. Cohen, S. L. Louloudes, and M. J. Thompson, *Steroids* **9**, 591 (1967).
168. A. M. Atallah and H. J. Nicholas, *Mol. Cryst. Liquid Cryst.* **19**, 217 (1973).
169. S. Bernstein and K. J. Sax, *J. Org. Chem.* **16**, 685 (1951).
170. K. Tsuda, K. Arima, and R. Hayatsu, *J. Amer. Chem. Soc.* **76**, 2933 (1954).
171. N. K. Chaudhuri, R. Nickolson, J. G. Williams, and M. Gut, *J. Org. Chem.* **34**, 3767 (1969).
172. A. Windaus, O. Linsert, and H. J. Eckhardt, *Justus Liebigs Ann. Chem.* **534**, 22 (1938).
173. C. Wiegand, *Z. Naturforsch. B* **4**, 249 (1949).
174. C. F. Hammer and R. Stevenson, *Steroids* **5**, 637 (1965).
175. F. Schenk, K. Buchholz, and O. Wiese, *Ber. Deut. Chem. Ges.* **69**, 2696 (1936).
176. L. F. Fieser and G. Ourisson, *J. Amer. Chem. Soc.* **75**, 4404 (1953).
177. W. J. Adams, V. Petrow, and R. Royer, *J. Chem. Soc.* 678 (1951).
178. A. Windaus, U. Riemann, H. H. Rüggeberg, and G. Zühlsdorff, *Justus Liebigs Ann. Chem.* **552**, 142 (1942).
179. L. Verbit and J. Hudec, private communication.
180. G. Storck, M. Nussim, and B. August, *Tetrahedron, Suppl.* **8**, Pt. 1, 105 (1966).
181. O. Wintersteiner and M. Moore, *J. Amer. Chem. Soc.* **72**, 1923 (1950).
182. L. F. Fieser and W.-Y. Huang, *J. Amer. Chem. Soc.* **75**, 5356 (1953).
183. K. Sakai and K. Tsuda, *Chem. Pharm. Bull. (Tokyo)* **11**, 529 (1963).
184. H. Wieland, F. Rath, and W. Benend, *Justus Liebigs Ann. Chem.* **548**, 19 (1941).
185. H. Wieland and L. Görnhardt, *Justus Liebigs Ann. Chem.* **557**, 248 (1947).
186. A. M. Atallah and H. J. Nicholas, *Mol. Cryst. Liquid Cryst.* **18**, 339 (1972).
187. Derivatives of several triterpenes discussed are covered under U.S. Patent 3,686,235 (1972) [H. J. Nicholas and F. F. Knapp].
188. J. D. Connolly and K. H. Overton, in "Chemistry of Terpenes and Terpenoids" (A. A. Newman, ed.), Chapter 5. Academic Press, New York, 1972.
189. A. M. Atallah and H. J. Nicholas, *Chem. Tech.* **2**, 486 (1972); *Lipids* **9**, 613 (1974).
190. For a brief review of naturally occurring triterpenes with a cyclopropane ring, see

S. Beckmann and H. Geiger, *in* "Houben-Weyl, Methoden der Organischen Chemie" (E. Müller, ed.), 4th ed., Vol. IV/4, p. 472. Thieme, Stuttgart, 1971.
191. A. M. Atallah and H. J. Nicholas, *Mol. Cryst. Liquid Cryst.* **18**, 321 (1972).
192. R. J. Krzewki and R. S. Porter *in* "Thermal Analysis: Comparative Studies on Materials" (H. Kambe and P. D. Garn, eds.), p. 1. Kodansha, Tokyo and Wiley, New York, 1974.
193. F. Gautschi and K. Bloch, *J. Biol. Chem.* **233**, 1343 (1958).
194. A. M. Atallah and H. J. Nicholas, *Mol. Cryst. Liquid Cryst.* **24**, 213 (1973).
195. A. M. Atallah and H. J. Nicholas, *Mol. Cryst. Liquid Cryst.* **17**, 1 (1972).
196. F. F. Knapp, H. J. Nicholas, and J. P. Schroeder, *J. Org. Chem.* **34**, 3328 (1969).
197. R. J. Krzewki, R. S. Porter, A. M. Atallah, and H. J. Nicholas, *Mol. Cryst. Liquid Cryst.* **29**, 127 (1974).
198. F. F. Knapp and H. J. Nicholas, *Mol. Cryst. Liquid Cryst.* **6**, 319 (1970).
199. F. F. Knapp and H. J. Nicholas, *J. Org. Chem.* **33**, 3995 (1968).
200. F. Stumpf, *Phys. Z.* **11**, 780 (1910).
201. B. H. Klanderman and T. R. Criswell, *J. Amer. Chem. Soc.* **97**, 1585 (1975).
202. M. Leclerq, J. Billard, and J. Jacques, *C. R. Acad. Sci., Ser. C* **266**, 654 (1968).
203. M. Leclerq, J. Billard, and J. Jacques, *Mol. Cryst. Liquid Cryst.* **8**, 367 (1969).
204. G. W. Gray, *Mol. Cryst. Liquid Cryst.* **7**, 127 (1969).
205. D. Dolphin, Z. Muljiani, J. Cheng, and R. B. Meyer, *J. Chem. Phys.* **58**, 413 (1973).
206. J. A. Castellano, C. S. Oh, and M. T. McCaffrey, *Mol. Cryst. Liquid Cryst.* **27**, 417 (1974).
207. D. Coates, K. J. Harrison, and G. W. Gray, *Mol. Cryst. Liquid Cryst.* **22**, 99 (1973).
208. Y. Y. Hsu and D. Dolphin, *in* "Liquid Crystals and Ordered Fluids II" (J. F. Johnson and R. S. Porter, eds.), p. 461. Plenum, New York, 1974.
209. D. Coates and G. W. Gray, *Mol. Cryst. Liquid Cryst.* **24**, 163 (1973).
210. J. E. van Lier and L. L. Smith, *J. Org. Chem.* **35**, 2627 (1970) and references cited therein.
211. L. F. Fieser, *Org. Syn., Collect. Vol.* **4**, 195 (1963).
212. L. C. Scala and G. D. Dixon, *Mol. Cryst. Liquid Cryst.* **7**, 443 (1969).
213. L. C. Scala and G. D. Dixon, *Mol. Cryst. Liquid Cryst.* **10**, 411 (1970).
214. L. C. Scala, G. D. Dixon, and D. F. Ciliberti, *U.S. Nat. Tech. Inform. Serv., Rep. AD 716 003* (1970).
215. W. Elser, J. L. W. Pohlmann, and P. R. Boyd, *Mol. Cryst. Liquid Cryst.*, to be published.
216. G. D. Dixon and L. C. Scala, U.S. Patent 3,656,909 (1972).
217. G. D. Dixon and L. C. Scala, *Mol. Cryst. Liquid Cryst.* **10**, 317 (1970).
218. S. P. Brown, *Mater. Eval.* **26**, 163 (1968).
219. R. D. Ennulat, to be published.
220. A. V. Galanti and R. S. Porter, *J. Phys. Chem.* **76**, 3089 (1972).
221. C. W. Griffen and R. S. Porter, *Mol. Cryst. Liquid Cryst.* **21**, 77 (1973).
222. R. J. Krzewki and R. S. Porter, *Mol. Cryst. Liquid Cryst.* **21**, 99 (1973).
223. W. U. Müller and H. Stegemeyer, *Chem. Phys. Lett.* **27**, 130 (1974).
224. J. Adams, W. Haas, and J. Wysocki, *Phys. Rev. Lett.* **22**, 92 (1969).
225. J. H. J. Vogelzangs, *Ned. Tijdschr. Natuurk.* **38**, 18 (1972).
226. B. Böttcher, D. Gross, and E. Mundry, *Materialpruefung* **11**, 156 (1969).
227. B. Böttcher, *Mater. Constr. (Paris)* **4**, 241 (1971).
228. J. L. Fergason, N. N. Goldberg, and R. J. Nadalin, *Mol. Cryst. Liquid Cryst.* **1**, 309 (1966).
229. T. Ishikawa, T. Mizuno, F. Kato, Y. Yashiro, and H. Nagao, *Bull. Nagoya Inst. Technol.* **22**, 243 (1970).

230. B. Böttcher, *Chem.-Ztg.*, *Chem.-Techn.* **1**, 195 (1972).
231. W. E. Woodmansee, U.S. Patent 3,441,513 (1969).
232. L. Melamed and D. Rubin, *Appl. Opt.* **10**, 1103 (1971).
233. J. L. Fergason and N. N. Goldberg, British Patent 1,153,959 (1969).
234. F. Davis, U.S. Patent 3,576,761 (1971).
235. J. L. Fergason and N. N. Goldberg, U.S. Patent 3,594,126 (1971).
236. H. Kelker and B. Scheurle, *Angew. Chem.* **81**, 903 (1969); *Angew. Chem., Int. Ed. Engl.* **8**, 884 (1969).
237. C. W. Smith, D. G. Gisser, M. Young, and S. R. Powers, *Appl. Phys. Lett.* **24**, 453 (1974).
238. N. Oron, J. L. Yu, and M. M. Labes, *Appl. Phys. Lett.* **23**, 217 (1973).
239. N. N. Goldberg and J. L. Fergason, U.S. Patent 3,533,399 (1970).
240. F. Davis, U.S. Patent 3,619,254 (1971).
241. R. D. Ennulat and J. L. Fergason, *Mol. Cryst. Liquid Cryst.* **13**, 149 (1971).
242. J. E. Adams and W. E. L. Haas, U.S. Patent 3,726,584 (1973).
243. H. Baessler, T. M. Laronge, and M. M. Labes, *J. Chem. Phys.* **51**, 3213 (1969).
244. J. E. Adams, L. B. Leder, and W. E. L. Haas, U.S. Patent 3,679,290 (1972).
245. W. E. Woodmansee and H. L. Southworth, *Proc. Int. Conf. Nondestr. Test., 5th, 1967* p. 81 (1969); *Chem. Abstr.* **73**, 81746 (1970).
246. G. L. Waterman and W. E. Woodmansee, U.S. Patent 3,439,525 (1969).
247. M. J. Intlekofer and D. C. Auth, *Appl. Phys. Lett.* **20**, 151 (1972).
248. J. E. Vandegaer, "Microencapsulation—Processes and Applications." Plenum, New York, 1974.
249. D. Churchill, J. V. Cartmell, and R. E. Miller, U.S. Patent 3,697,297 (1972).
250. D. Churchill, J. V. Cartmell, and R. E. Miller, U.S. Patent 3,732,119 (1973).
251. T. L. Hodson, J. V. Cartmell, D. Churchill, and J. W. Jones, U.S. Patent 3,585,381 (1971).
252. D. Churchill and J. V. Cartmell, U.S. Patent 3,578,844 (1971).
253. S. Kubo and H. Arai, *Chiba Daigaku Kogakubu Kenkyu Hohohu* **21**, 163 (1970); *Chem. Abstr.* **77**, 10696 (1972).
254. M. Kano, T. Ninomiya, and Y. Nishimura, Japan. Kokai 73 71,377 (1973); *Chem. Abstr.* **80**, 15612 (1974).
255. H. Arimoto, T. Kakishita, and M. Takada, Japan. Kokai 73 44,177 (1973); *Chem. Abstr.* **79**, 93136 (1973).
256. F. Hori, B. Kato, and N. Arima, Japan. Kokai 73 49,502 (1973); *Chem. Abstr.* **79**, 116435 (1973).
257. M. Ueda, F. Hori, B. Kato, and N. Arima, Japan. Kokai 73 83,903 (1973); *Chem. Abstr.* **80**, 97526 (1974).
258. J. V. Cartmell and D. Churchill, U.S. Patent 3,720,623 (1973).
259. K. Goto, Japan. Kokai 73 19,488 (1973); *Chem. Abstr.* **79**, 85685 (1973).
260. D. Churchill, J. V. Cartmell, R. E. Miller, and P. D. Bouffard, British Patent 1,161,039 (1969).
261. D. Churchill and J. V. Cartmell, U.S. Patent 3,600,060 (1971).
262. D. Churchill and J. V. Cartmell, U.S. Patent 3,734,597 (1973).
263. M. Sagane, Japan. Kokai 74 35,427 (1974); *Chem. Abstr.* **81**, 137702 (1974).
264. D. H. Baltzer, U.S. Patent 3,620,889 (1971); Ger. Offen. 1,929,256 (1970).
265. M. Ono, T. Ito, T. Sawa, and N. Tokuyama, Japan. Patent 74 07,595 (1974); *Chem. Abstr.* **81**, 121900 (1974).
266. M. Ono, T. Ito, T. Sawa, and T. Kato, Japan. Patent 74 01,676 (1974); *Chem. Abstr.* **81**, 92920 (1974).

267. H. Arimoto, T. Kakishita, and M. Takada, Japan. Kokai 73 44,522 (1973); *Chem. Abstr.* **79**, 116286 (1973).
268. R. Hesse, G. Edler, and H. Keller, Ger. Offen. 2,201,121 (1973); *Chem. Abstr.* **80**, 5037 (1974).
269. G. V. Lukianoff, *Mol. Cryst. Liquid Cryst.* **8**, 389 (1969).
270. R. Parker, *Mol. Cryst. Liquid Cryst.* **20**, 99 (1973).
271. R. Parker, U.S. Patent 3,861,213 (1975).
272. H. Seto, M. Ueda, and H. Segawa, U.S. Patent 3,704,625 (1972).
273. R. Parker, U.S. Patent 3,822,594 (1974).
274. J. Gaynor and T. G. Anderson, U.S. Patent 3,733,485 (1973).
275. P. Pollmann, *Ber. Bunsenges. Phys. Chem.* **78**, 374 (1974); *J. Phys. E* **7**, 490 (1974).
276. P. U. Schulthess, Swiss Patent 520,939 (1972); *Chem. Abstr.* **77**, 106805 (1972).
277. R. Parker, *IEEE Trans. Power App. Syst.* **92**, 104 (1973).
278. J. Adams, W. Haas, and J. Dailey, *J. Appl. Phys.* **42**, 4096 (1971).
279. J. E. Adams and L. B. Leder, U.S. Patent 3,697,152 (1972).
280. H. Melchior, F. J. Kahn, D. Maydan, and D. B. Fraser, *Appl. Phys. Lett.* **21**, 392 (1972).
281. D. Maydan, *Proc. IEEE* **61**, 1007 (1973).
282. J. L. Fergason and A. E. Anderson, U.S. Patent 3,410,999 (1968).
283. J. A. Asars, British Patent 1,119,253 (1968).
284. S. A. Hadjistavros, *U.S. Nat. Tech. Inform. Serv., Rep. AD 734 442* (1971).
285. J. L. Fergason and A. E. Anderson, U.S. Patent 3,401,262 (1968).
286. A. E. Anderson and J. L. Fergason, British Patent 1,120,093 (1968).
287. A. K. Jalaluddin and H. Husain, *Proc. Nucl. Phys. Solid State Phys. Symp., 15th, 1970* **2**, 527 (1971); *Chem. Abstr.* **75**, 146947 (1971).
288. E. I. Kats, *Zh. Eksp. Teor. Fiz.* **61**, 1686 (1971); *Sov. Phys.—JETP* **34**, 899 (1972).
289. V. A. Belyakov and V. P. Orlov, *Phys. Lett. A* **42**, 3 (1972).
290. D. E. Nagle, J. W. Doane, R. Madey, and A. Saupe, *Mol. Cryst. Liquid Cryst.* **26**, 71 (1974).
291. J. L. Fergason and N. N. Goldberg, U.S. Patent 3,663,390 (1972).
292. W. Haas, J. Adams, and J. Wysocki, *Mol. Cryst. Liquid Cryst.* **7**, 371 (1969).
293. J. Adams and W. Haas, *J. Electrochem. Soc.* **118**, 2026 (1971).
294. Westinghouse Electric Co., British Patent 1,309,558 (1973).
295. W. Haas, J. Adams, and J. Wysocki, *Appl. Opt., Suppl.* **3**, 196 (1969).
296. L. B. Leder, U.S. Patent 3,789,225 (1974).
297. J. L. Fergason, T. P. Vogl, and M. Garbuny, U.S. Patent 3,114,836 (1963).
298. J. L. Fergason, *Westinghouse Res. Lab., Rep. 912-J904-R4-X* (1961).
299. G. Jankowitz and J. Axelrod, *Barnes Eng. Co., Rep. 67/53* (1967).
300. G. Jankowitz and D. Pearsall, *Barnes Eng. Co., Rep. 69/57* (1969).
301. J. R. Hansen, J. L. Fergason, and A. Okaya, *Appl. Opt.* **3**, 987 (1964).
302. M. Ohi, Y. Akimoto, and T. Tako, *Oyo Butsuri* **41**, 363 (1972); *Chem. Abstr.* **77**, 146087 (1971).
303. M. R. Wank, U.S. Patent 3,569,709 (1971).
304. G. Jankowitz, U.S. Patent 3,527,945 (1970).
305. V. M. Ginzburg, V. I. Smirnov, A. S. Sonin, B. M. Stepanov, and I. G. Chistyakov, *Pribory Tekn. Eksp.* **2**, 206 (1969); *Instr. Exp. Techn. USSR* **2**, 549 (1970).
306. A. V. Tolmachev and V. M. Kuz'michev, *Pis'ma Zh. Eksp. Teor. Fiz.* **14**, 220 (1971); *J. Exp. Theor. Phys. USSR, Lett.* **14**, 144 (1971).
307. F. Keilmann, *Max-Planck Inst. Plasmaphys., Rep. IPP IV/18* (1971); *U.S. Nat. Tech. Inform. Serv., Rep. N 72-16360* (1972).
308. F. Keilmann and K. F. Renk, *Appl. Phys. Lett.* **18**, 452 (1971).

309. J. P. Lesieur, M. C. Sexton, and D. Véron, *J. Phys. D* **5**, 1212 (1972).
310. D. Véron, *C. R. Acad. Sci., Ser. B* **274**, 1013 (1972).
311. F. Keilmann, *Appl. Opt.* **9**, 1319 (1970).
312. M. Hugenschmidt and K. Vollrath, *C. R. Acad. Sci., Ser. B* **274**, 1221 (1972).
313. M. Hugenschmidt and K. Vollrath, *Int. Congr. High-Speed Photogr. Cinematogr., 10th, 1972.*
314. H. J. Stocker, *U.S. Nat. Tech. Inform. Serv., Rep. AD 752 201* (1972).
315. R. F. Morton, *Proc. Spec. Meet. Unconventional Infrared Detectors, 1971* p. 21; *U.S. Nat. Tech. Inform. Serv., Rep. AD 726 224* p. 21 (1971).
316. S. Kobayashi and T. Sakusabe, *Proc. Electro-Opt. Syst. Design Conf., 1971* p. 280.
317. T. Sakusabe and S. Kobayashi, *Jap. J. Appl. Phys.* **10**, 758 (1971).
318. W. A. Simpson and W. E. Deeds, *Appl. Opt.* **9**, 499 (1970).
319. Bendix Corp., *U.S. Dep. Health, Educ., Welf., Rep. BRH/DEP 70-8* (1970).
320. C. F. Augustine, U.S. Patent 3,693,084 (1972).
321. K. Magura, *Nachrichtentechn. Z.* **23**, 440 (1970).
322. C. F. Augustine, *Electronics* **41**, 118 (24 June 1968).
323. R. G. Pothier, U.S. Patent 3,713,156 (1973).
324. J. C. Sethares and M. R. Stiglitz, *Appl. Opt.* **8**, 2560 (1969).
325. J. C. Sethares and S. Gulaya, *Appl. Opt.* **9**, 2795 (1970).
326. J. Puyhaubert, *Onde Electr.* **52**, 213 (1972).
327. A. V. Tolmachev, E. Ya. Govorun, and V. M. Kuz'michev, *Zh. Eksp. Teor. Fiz.* **63**, 583 (1972); *Sov. Phys.—JETP* **36**, 309 (1973).
328. H. E. Stockman and B. Zarwyn, *IEEE Proc.* **56**, 763 (1968).
329. C. F. Augustine and W. E. Kock, *IEEE Proc.* **57**, 354 (1969).
330. C. F. Augustine, C. Deutsch, D. Fritzler, and E. Marom, *IEEE Proc.* **57**, 1333 (1969).
331. D. Fritzler, C. F. Augustine, C. Deutsch, and E. Marom, *Bendix Tech. J.* **2**, 83 (1969).
332. H. E. Stockman, *Electronics* **42**, 110 (24 Nov. 1969).
333. K. Iizuka, *Electron. Lett.* **5**, 26 (1969).
334. O. A. Kapustina, *Akust. Zh.* **20**, 1 (1974); *Sov. Phys.—Acoust.* **20**, 1 (1974).
335. J. F. Havlice, *Electron. Lett.* **5**, 477 (1969).
336. K. Iizuka, *IEEE Proc.* **58**, 288 (1970).
337. Y. Kagawa, T. Hatakeyama, and Y. Tanaka, *J. Sound Vib.* **36**, 407 (1974).
338. S. E. Cohen and W. H. Sproat, U.S. Patent 3,599,477 (1971).
339. W. H. Sproat and S. E. Cohen, *Mater. Eval.* **28**, 73 (1970).
340. B. D. Cook and R. E. Werchan, *Ultrasonics* **9**, 101 (1971).
341. M. J. Intlekofer, D. C. Auth, and M. E. Fourney, *Proc. Soc. Photo-Opt. Instrum. Eng.* **29**, 83 (1972).
342. B. B. Brenden and H. R. Curtin, British Patent 1,194,544 (1970).
343. J.-L. Dion and M. Bader, *Proc. Soc. Photo-Opt. Instrum. Eng.* **38**, 43 (1973).
344. M. A. Wall, *U.K., At. Energy Res. Establ., Bibliogr. AERE-Bib 181* (1972).
345. W. E. Woodmansee, *Appl. Opt.* **7**, 1721 (1968).
346. C. M. Forman, *U.S. Nat. Tech. Inform. Serv., Rep. AD 878 790* (1970).
347. S. P. Brown, *Appl. Polym. Symp.* **19**, 463 (1972).
348. B. Böttcher and D. Gross, *Umschau* 574 (1969).
349. M. Magne, P. Pinard, P. Thome, and N. Chretien, *Colloq. Met.* **12**, 241 (1968); *Bull. Inform. Sci. Techn., Comm. Energ. At. (Fr.)*, **136**, 45 (1969).
350. A. J. Parisi, *Prod. Eng.* **39**, No. 16, 19 (1968).
351. T. A. Simcox, *Techn. Inform. Serv., Amer. Inst. Aeronaut. Astronaut., Rep. A 69-25293* (1969).
352. L. J. Broutman, T. Kobayashi, and D. Carrillo, *J. Compos. Mater.* **3**, 702 (1969).

353. J. C. Manaranche, *J. Phys. D* **5**, 1120 (1972).
354. M. H. Perry, *NASA, Rep. N 70-35 867* (1970).
355. W.-U. Kopp, *Prakt. Metallogr.* **9**, 370 (1972).
356. M. Pazdur, *Hutnik* **37**, 205 (1970); *Chem. Abstr.* **73**, 58691 (1970).
357. W. E. Woodmansee and H. L. Southworth, *Mater. Eval.* **26**, 149 (1968).
358. E. Grzejdziak, A. Rogowski, R. Szylhabel, A. Szymański, and J. Hejwowski, *Elektronika (Warsaw)* **13**, 234 (1972).
359. D. L. Uhls, *IBM Tech. Discl. Bull.* **15**, 1670 (1972).
360. L. C. Mizell, *Amer. Inst. Aeronaut. Astronaut., Rep. A 71-40 738* (1971).
361. P. L. Garbarino and R. D. Sandison, *J. Electrochem. Soc.* **120**, 834 (1973); *IBM Tech. Discl. Bull.* **15**, 1738 (1972).
362. G. V. Lukianoff in Brown [5], p. 219.
363. R. S. Ziernicki and W. F. Leonard, *Rev. Sci. Instrum.* **43**, 479 (1972).
364. K. C. Park and S. Basavaiah, *J. Non-Cryst. Solids* **2**, 284 (1970).
365. S. K. Dey and G. D. Dick, *J. Vac. Sci. Technol.* **11**, 97 (1974).
366. R. D. Maple, *U.S. Nat. Tech. Inform. Serv., Rep. AD 744 211* (1972).
367. P. M. Kendig, *Meet. Acoust. Soc. Amer., 84th, 1972*; *Electr. Electron. Abstr.* **76**, 4973 (1973).
368. G. N. Glazkov, A. M. Zhmud, G. G. Molokov, and Yu. K. Nepochatov, *Coll. Rep., All-Union Sci. Conf. Liquid Cryst., Acad. Sci. USSR, 1st, 1970* p. 296. (1972).
369. H. E. Shaw, U.S. Patent 3,590,371 (1971).
370. E. Sprow, *Mach. Des.* **41**, No. 2, 37 (1969).
371. A. P. Pontello, *U.S. Nat. Tech. Inform. Serv., Rep. AD 886 071* (1971); *AD 772 099/8GA* (1973).
372. R. A. Champa, *U.S. Nat. Tech. Inform. Serv., Rep. AD 755 831* (1972).
373. E. J. Klein, *Amer. Inst. Aeronaut. Astronaut., Aerodyn. Testing Conf., 3rd, 1968, Paper 68-376*.
374. E. J. Klein, *Astronaut. Aeronaut.* p. 70 (July 1968).
375. G. M. Zharkova and A. P. Kapustin, *Izv. Sib. Otd. Akad. Nauk SSSR, Ser. Tekh. Nauk* **13**, 65 (1970); *Chem. Abstr.* **74**, 80809 (1971).
376. E. D. McElderry, *Air Force Flight Dynamics Lab., Wright-Patterson Air Force Base, Tech. Rep. FDMG-TM-70-3* (1970).
377. A. Szymański, *Post. Astronaut.* **5**, 27 (1971).
378. D. Vennemann and K.-A. Bütefisch, *Deut. Luft-Raumfahrt, Rep. DLR-FB 73-121* (1973).
379. T. E. Cooper, R. J. Field, and J. F. Meyer, *U.S. Nat. Tech. Inform. Serv., Rep. AD/A-002 458* (1974).
380. C. den Ouden, *Delft Progr. Rep., Ser. A* **1**, 33 (1973).
381. C. den Ouden and C. J. Hoogendoorn, *Proc. Int. Heat Transfer Conf., 5th, 1974* p. 293.
382. G. M. Zharkova and A. V. Lokotko, *Coll. Rep., All-Union Sci. Conf. Liquid Cryst., Acad. Sci. USSR, 2nd, 1972* p. 271 (1973).
383. T. C. Bannister, *NASA, Marshall Space Flight Center, Summary Rep. I and II, Apollo 14* (1971).
384. P. G. Grodzka, C. Fan, and R. O. Hedden, *Lockheed Missiles and Space Co., Rep. LMSC-HREC D-225,333* (1971).
385. P. G. Grodzka and T. C. Bannister, *Science* **176**, 506 (1972).
386. T. C. Bannister, *NASA, Marshall Space Flight Center, Res. Technol. Rev. S & E-SSL-T 453-3090* (1973).
387. T. C. Bannister, P. G. Grodzka, L. W. Spradley, S. V. Bourgeois, R. O. Hedden, and B. R. Facemire, *NASA, Marshall Space Flight Center, Rep. TM X-64772* (1973).

388. P. G. Grodzka and T. C. Bannister, *Science* **187**, 165 (1975).
389. J. Tricoire, French Patent 2,110,505 (1972); *Chem. Abstr.* **78**, 78173 (1973).
390. F. A. Mina, Canadian Patent 912,806 (1972).
391. A. Koff, Fr. Demande 2,141,907 (1973); *Chem. Abstr.* **79**, 24519 (1973).
392. P. G. Pick, Canadian Patent 899,610 (1972).
393. A. Koff, Ger. Offen. 2,018,028 (1970); *Chem. Abstr.* **74**, 105716 (1971).
394. A. Koff, Ger. Offen. 2,059,789 (1971); *Chem. Abstr.* **75**, 146260 (1971); Canadian Patent 901,277 (1972).
395. P. G. Pick and J. Fabijanic, *Mol. Cryst. Liquid Cryst.* **15**, 371 (1972).
396. P. G. Pick, J. Fabijanic, and A. Stewart, *Mol. Cryst. Liquid Cryst.* **20**, 47 (1973).
397. E. Flam, U.S. Patent 3,661,142 (1970).
398. R. A. Sanford, U.S. Patent 3,633,425 (1972).
399. W. J. Jones, U.S. Patent 3,440,882 (1969).
400. A. Benjamin, *Photogr. Appl. Sci., Technol., Med.* **6**, No. 22, 30 (1971).
401. J. T. Crissey, E. Gordy, J. L. Fergason, and R. B. Lyman, *J. Invest. Dermatol.* **43**, 89 (1964).
402. C. Gros, M. Gautherie, P. Bourjat, and F. Archer, *Ann. Radiol.* **13**, 333 (1970).
403. W. M. Portnoy, *J. Ass. Adv. Med. Instrum.* **5**, 176 (1970).
404. S. Friberg, *Svensk Kem. Tidskr.* **83**, No. 4, 26 (1971).
405. Anonymous, *Med. World News* **12**, 71 (24 Sept. 1971).
406. Anonymous, *Nikkei Med.* **1**, 68 (1972).
407. Yu. M. Gerusov, in Glazkov *et al.* [*368*], p. 304.
408. J. T. Crissey, J. L. Fergason, and J. M. Bettenhausen, *J. Invest. Dermatol.* **45**, 329 (1965).
409. A. Balla, A. Romanin, and P. Rota, *Riv. Med. Aeronaut. Spaz.* **35**, 51 (1972).
410. Y. Grall and J. Tricoire, *C. R. Acad. Biol.* **16**, 1309 (1967).
411. Anonymous, *Geriatrics* **27**, 60 (1972).
412. C. Potanin, *Chest* **58**, 491 (1970).
413. C. Potanin, D. Hunt, and L. T. Sheffield, *Circulation* **42**, 199 (1970).
414. J. Tricoire, *Presse Med.* **78**, 2481 (1970).
415. A. Benjamin, *J. Biol. Photogr. Ass.* **41**, 13 (1973).
416. J. T. Crissey, E. Gordy, J. L. Fergason, and R. B. Lyman, *Roche Med. Image Comment.* **11**, 6 (1969).
417. O. S. Selawry, H. S. Selawry, and J. F. Holland, *Mol. Cryst. Liquid Cryst.* **1**, 175 (1966).
418. A. I. Kubarko and E. P. Demidchik, in Zharkova and Lokotko [*382*], p. 235.
419. P. L. Tai, *J. Amer. Pod. Ass.* **63**, 119 (1973).
420. B. Y. Lee, F. S. Trainor, and J. L. Madden, *Arch. Phys. Med. Rehab.* **54**, 96 (1973).
421. K. L. Williams, *J. Radiol. Electr. Med. Nucl.* **48**, 68 (1967).
422. D. E. Strandness, *Rassegna Med. Cult.* **44**, 32 (1967).
423. E. B. Thompson, W. H. Taylor, and N. R. Cohen, *Arch. Int. Pharmacodyn. Ther.* **191**, 49 (1971).
424. R. M. Howell, R. C. Duell, and T. P. Mullaney, *Oral Surg.* **29**, 763 (1970); see also A. V. Newton and J. M. Mumford, *Dent. Pract. Dent. Rec.* **21**, 84 (1970).
425. P.-O. Glantz and S. Friberg, *Odont. Revy* **22**, 341 (1971).
426. E. N. Peterson, G. D. Dixon, and M. A. Levine, *Obstet. Gynecol.* **37**, 468 (1971).
427. T. W. Davison, K. L. Ewing, N. Sayat, N. P. Mulla, and J. L. Fergason, *Obstet. Gynecol.* **42**, 574 (1973).
428. T. W. Davison, K. L. Ewing, J. L. Fergason, M. Chapman, A. Can, and C. C. Voorhis, *Cancer* **29**, 1123 (1972).
429. K. L. Ewing, T. W. Davison, and J. L. Fergason, *Ohio J. Sci.* **73**, 55 (1973).

430. Yu. M. Gerusov, Yu. P. Polosin, and N. N. Gerusova, in Glazkov *et al.* [*368*], p. 312.
431. A. Benjamin, *Image Dyn.* **3**, No. 5, 10 (1968).
432. M. Gautherie, *J. Phys. (Paris)* **30**, C-4, 122 (1969).
433. J. Tricoire, L. Mariel, and J.-P. Amiel, *Nouv. Presse Med.* **2**, 1117 (1973).
434. J. Tricoire, *Nouv. Presse Med.* **2**, 1518 (1973).
435. O. S. Salawry, H. W. Neubauer, H. S. Selawry, and F. S. Hoffmeister, *Amer. J. Surg.* **112**, 537 (1966).
436. O. S. Selawry and J. F. Holland, *Proc. Amer. Ass. Cancer Res.* **7**, 63 (1966).
437. K. D. Panikrotov, Yu. M. Gerusov, G. G. Maidachenko, and V. A. Vorontsov, in Glazkov *et al.* [*368*], p. 309.
438. J. D. Stevens and W. Rogers, *Vasc. Surg.* **5**, 186 (1971).
439. J. B. Kinn and R. A. Tell, *IEEE Trans. Biomed. Eng.* **20**, 387 (1973).
440. R. T. Hall and T. K. Oliver, *J. Amer. Med. Ass.* **218**, 1700 (1971).
441. R. T. Hall and D. K. Barnett, private communication.
442. J. B. Kinn, *Proc. San Diego Biomed. Symp.* **11**, 287 (1972).
443. V. I. Filatov, A. N. Savvina, and M. V. Mukhina, *Vestn. Khir.* **106**, 111 (1971).
444. V. I. Filatov, A. N. Savvina, and M. V. Mukhina, in Zharkova and Lokotko [*382*], p. 243.
445. R. Hoehn and B. Binkert, *Plast. Reconstr. Surg.* **48**, 209 (1971).
446. T. E. Cooper and J. P. Groff, *J. Heat Transfer* **95**, 250 (1973).
447. R. G. Katz and T. E. Cooper, *Proc. Annu. Conf. Eng. Med. Biol., 26th, 1973* p. 257.
448. T. E. Cooper and W. K. Petrovic, *J. Heat Transfer* **96**, 415 (1974).

LIQUID CRYSTALS AND EMULSIONS

Stig Friberg*

The Swedish Institute for Surface Chemistry
Drottning Kristinas väg 45
Stockholm, Sweden

Kåre Larsson

Lipid Chemistry Laboratory
University of Göteborg
Rännvägen, Göteborg, Sweden

I.	Prefatory Remarks	173
II.	Arrangement of Emulsifier Molecules in the Solid State	174
III.	Emulsions with a Crystalline Emulsifier at the Oil/Water Interface	176
IV.	The Structure of Liquid Crystalline Phases of Emulsifiers	176
V.	Phase Behavior of Different Types of Emulsifier–Water Systems	179
VI.	Molecular Arrangement in Surface Films	182
VII.	Water/Emulsifier Interaction	184
VIII.	Emulsifier/Emulsifier Interaction	186
IX.	Hydrocarbon/Emulsifier Interaction	189
X.	Emulsion Stability	190
XI.	Location of the Liquid Crystal	191
XII.	Flocculation and Coalescence	193
XIII.	van der Waals Forces	193
	References	196

I. PREFATORY REMARKS

The formation of ordered structures of associated molecules plays an important role in the physical properties of emulsions, such as rheological behavior and stability. The significance of such structures, most of which

* Present address: Department of Chemistry, University of Missouri–Rolla, Rolla, Missouri 65401.

possess liquid crystalline-order characteristics, will be considered here. The first part of the paper concerns various relevant arrangements of emulsifier molecules and the interaction between the different components in emulsions. On this basis emulsion stability, flocculation, and coalescence are considered.

II. ARRANGEMENT OF EMULSIFIER MOLECULES IN THE SOLID STATE

In addition to the main subject of this paper, the crystalline state will also be discussed briefly for two reasons. In this state of order of emulsifier molecules it is possible to obtain a complete determination of the structure, and this knowledge is useful in the understanding of the molecular packing in less ordered states. Furthermore, an interfacial film of truly crystalline type is sometimes observed in emulsions, particularly those involving lipids, and therefore this state is also directly related to emulsion structure.

The most common types of emulsifier in foods are monoglycerides. Monoglycerides can be prepared in a highly pure form on an industrial scale by molecular distillation. It is even possible using fully hardened fats of animal or vegetable origin to obtain a product more free of other homologs than is desired. There are two types of interfacial films separating the oil and water phases in technical emulsions involving monoglycerides. The first and most important type consists of a liquid crystalline phase; a large variety of other such emulsion structures will be demonstrated in the last part of this paper. The second type is a truly crystalline film of monoglyceride molecules in the interface.

The crystal structure of a monoglyceride [1] is shown in Fig. 1. This is a characteristic structure for all simple amphiphilic molecules that can act as emulsifiers. The structural unit is a molecular bilayer with the nonpolar regions forming a core between the sheets formed by the polar end groups. When the nonpolar region consists of hydrocarbon chains, they are close-packed in an extended zig-zag conformation. The bilayers are held together by strong bonds over the gap between the sheets formed by the polar groups (hydrogen bonds in the structure shown in Fig. 1).

Other emulsifiers differ from the structure shown in Fig. 1 in only a few details. The structure of the polar region depends of course on the groups present and on their configuration, and the only common feature is that the molecules are linked laterally within the bilayer as well as over the gap to the adjacent bilayer. The structure in the hydrocarbon chain region can vary in two ways, and the reason for polymorphism (i.e., the occurrence of alternative crystal forms) is usually due to such differences. The hydrocarbon chains can be close-packed in several different structures, two of which occur most frequently [2]. In one of these every chain plane is parallel, and since the packing can be described by a triclinic unit cell this packing is termed $T_\parallel$. In the other common chain packing every second chain plane is perpendicular to the plane of the others, and this packing is termed $O_\perp$ since the unit cell is orthorhombic. These packing modes will be discussed further in connection with monolayer structures. Another reason for polymorphism is the possibility

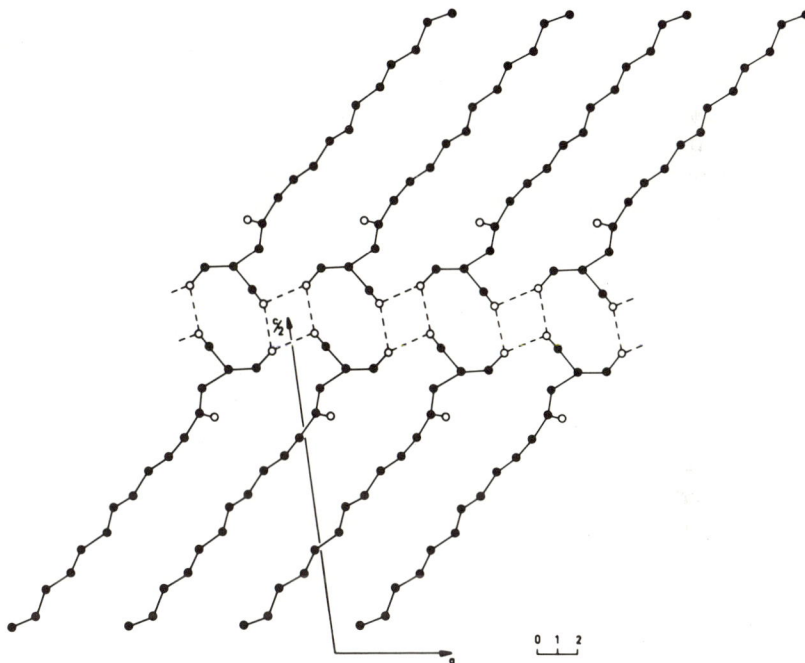

Fig. 1. A projection of the crystal structure of 2-monolaurin along a unit cell axis. All atoms except hydrogens are indicated.

of variation of the angle of tilt of the hydrocarbon chains. By relative displacement of adjacent chains one zig-zag period, the chain packing is not changed, and the angle of tilt can be varied in this way. As an example of polymorphism in simple amphiphilic molecules, n fatty acids can be mentioned [3]. There are three fatty acid crystal forms, one with chain packing $T_\parallel$ and the other two with chain packing $O_\perp$. The two forms with chain packing $O_\perp$ differ from each other only in the angle of tilt of the hydrocarbon chains toward the end group plane.

III. EMULSIONS WITH A CRYSTALLINE EMULSIFIER AT THE OIL/WATER INTERFACE

Crystals of amphiphilic substances form very thin plates parallel to the bilayer plane. These plates are easy to cleave, so that the methyl end group planes form new interfaces, since there are only relatively weak van der Waals forces over the gap formed by these planes. Crystals of amphiphilic molecules have a unique property of exhibiting either hydrophobic or hydrophilic surface properties. If the crystals are grown from an organic solvent or from the melt, the dominating surfaces (parallel to the bilayer units) consist of the methyl end group planes. If, on the other hand, crystallization takes place in an aqueous environment, such as a micellar solution or a dispersion of liquid crystalline particles (described below), the corresponding crystal surfaces will consist of the polar end group planes. These hydrophobic or hydrophilic surface properties, respectively, can easily be demonstrated by the water contact angle.

Oil in water emulsion can thus be formed where the emulsifier crystals expose hydrophobic surfaces toward the oil phase and hydrophilic surfaces toward water. If, for example, monolaurin is dissolved in an oil phase and dispersed mechanically in water while cooling from a temperature above to one below the melting point of monolaurin, an emulsion with this type of crystalline interface can be formed. Much smaller amounts of emulsifier are needed, however, if the chain length of the monoglyceride is increased. The reason for this is that the interfacial film then consists of crystalline monoglyceride layers alternating with water layers. This is a special phase known as the gel phase with rotational disorder of the chains, and it has a tendency to form flexible films. This type of oil/water interface structure occurs in many food emulsions where monoglycerides or phosphatides are used as emulsifiers. The physical properties of emulsions with such interfaces are closely related to those with liquid crystalline interfaces, and these will be treated in detail below.

IV. THE STRUCTURE OF LIQUID CRYSTALLINE PHASES OF EMULSIFIERS

If we consider the type of solid state structure shown in Fig. 1 subjected to heating, it is not surprising that the hydrocarbon chain region, with only weak van der Waals interaction between the chains, can melt although the lamellar arrangement of the bilayers persist, due to the strong forces linking the polar groups together. Such a phase with long-range order and short-range disorder is by definition a thermotropic liquid crystal. Similar phases can also

Liquid Crystals and Emulsions

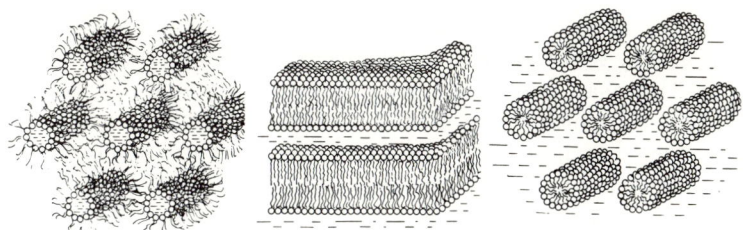

Fig. 2. The three most important structures occurring in aqueous systems of amphiphilic substances.

be formed in aqueous systems of emulsifiers, and they are then termed lyotropic liquid crystals. The pioneering work that revealed the structure of different amphiphile–water phases was done by Luzzati and co-workers [4]. Later studies of numerous systems have shown that there are three important structures (shown in Fig. 2) which occur at different water content.

The most common liquid crystalline phase in emulsifier–water systems is the *lamellar* one shown in Fig. 2. This is related to the crystalline structure with bilayers of emulsifier molecules. The hydrocarbon chains are in a dynamic disordered state which is similar to that of paraffins in the liquid state, and the amphiphile bilayers are separated by water layers. This phase can be unambiguously identified from its X-ray diffraction characteristics. A series of sharp reflections due to the one-dimensional periodicity is thus obtained, from which the repetition unit (the so-called long-spacing) can be calculated. The thickness of the water layer and amphiphile bilayer is directly obtained from this spacing and the volume fraction of the two components. Another feature of the X-ray diffraction pattern common for all the liquid crystalline phases is a diffuse halo at 4.5 Å due to the disordered hydrocarbon chains.

There are two types of hexagonal liquid crystals as shown in Fig. 2. In the one usually termed *hexagonal*, the amphiphile molecules form cylinders arranged in a hexagonal array in a continuous water medium. In the other phase, which usually is termed *inversed hexagonal*, the water and the hydrocarbon chains have changed place. Both these phases give an X-ray diffraction pattern corresponding to a two-dimensional hexagonal lattice, and the spacings can be obtained directly. From the volume fraction of the components the geometrical parameters can be calculated, such as cylinder diameter, as well as distance between the cylinders. If both the hexagonal and the inversed hexagonal structure alternatives are considered in these calculations, it is usually possible to determine the true alternative with regard only to the molecular dimensions (e.g., the radius of an amphiphile cylinder is limited by the length of the extended molecule). There are, however, many differences in properties of the two hexagonal phases, so that identification is no practical problem. The inversed hexagonal phase is observed at very low water content, and it is usually transformed into the lamellar phase with increasing content of water. This phase is observed in amphiphile molecules with large hydrophobic regions, which favors such a curvature (increasing cross-sectional areas from the polar head group toward the hydrophobic part of the molecule). As will be discussed, the inversed hexagonal phase exists only in cases of amphiphiles insoluble in water, whereas the hexagonal one is observed only in cases where the amphiphile can form micellar water solutions. This latter hexagonal phase occurs in amphiphiles with a large polar head group, so that a curvature with decreasing molecular cross sectional areas from the polar group toward the hydrocarbon region is favored.

In addition to the three liquid crystalline phases discussed above there are also phases with *cubic* symmetry. Although these have long-range order in three dimensions they are usually considered as liquid crystals since they have the same type of short-range disorder as the true liquid crystals (liquid hydrocarbon chains separated by a polar group interface from the water). Only one type of these phases is considered to have a definitely established structure. That is the phase formed by anhydrous amphiphiles, and the structure was analyzed in detail in the case of anhydrous strontium myristate [5]. The structure, which is shown in Fig. 3, consists of two interwoven networks formed by the polar groups associated to rods in a hydrocarbon chain matrix. There are other types of cubic amphiphile–water structures, and at least one seems to consist of water aggregates separated by lipid bilayers in a body-centered arrangement of space filling polyhedra [6].

Finally, there is a curved lamellar type of liquid-crystalline structure that is the most important with regard to emulsions. It is formed by water-insoluble amphiphiles when water is added to the lamellar liquid crystal above its limit

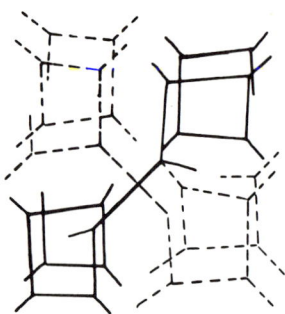

Fig. 3. The rod system (two three-dimensional networks) formed by the polar groups in the cubic phase of anhydrous sodium myristate [5].

of swelling. This limit is in the range 15–20 Å in water layer thickness in all neutral amphiphiles, and from various physical studies it can be concluded that the layer consists of ordered or bound water. When more water is added, particles with a concentric arrangement of water layers alternating with amphiphile bilayers are formed. Such spherical particles are formed by most membrane lipids. They are often used as membrane models under the name *liposomes*. In many papers on phase behavior of amphiphile–water systems this structure is described as a colloidal dispersion of the common lamellar liquid crystal in water. This is not a true description since the lamellar bilayers are always curved when an excess of water is present. Whether the dispersion of liposomes is a two-phase region can perhaps be questioned, since the particle size sometimes corresponds to an isotropic solution [7].

V. PHASE BEHAVIOR OF DIFFERENT TYPES OF EMULSIFIER–WATER SYSTEMS

The general features of the phase diagrams of binary amphiphile–water systems according to our present knowledge on numerous amphiphiles can be classified into two types of phase diagrams.

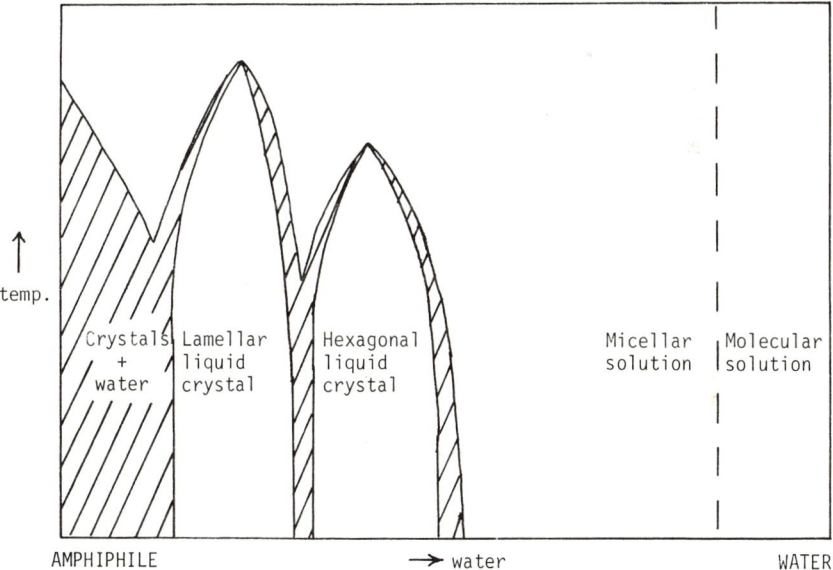

Fig. 4. Characteristic phase diagram of an amphiphile–water system with a strongly polar amphiphile, e.g., soaps, alkyl sulfates, quarternary ammonium salts, and lysolecithins. Two-phase regions are shaded.

Amphiphiles in which the polar groups dominate with relation to the hydrocarbon chains [hydrophilic lipophilic balance (HLB) values larger than about 15] show phase diagrams of the type indicated in Fig. 4. The molecules associate above a certain concentration, the critical micellization concentration (cmc), into micelles, and at higher concentrations a hexagonal liquid crystal is formed. If the water content is reduced further, a lamellar liquid crystal is formed. The temperature must be above the hydrocarbon chain melting point (Krafft temperature) in order to form these aqueous phases. This temperature is often below 0°C in amphiphiles giving this type of phase behavior—soaps, alkyl sulfates, amine and ammonium salts, or phosphatides with only one fatty acid residue. There can be large variations in the chain melting temperature with variation of the chain length in each homologous series. If the size of the hydrocarbon region is increased, a limit will be reached where the amphiphile is insoluble in water at all temperatures. Usually, however, it still gives aqueous phases, and the type of phase diagram obtained is shown in Fig. 5. Alkyl sulfates are an example of the effect of chain length on the water interaction. Sodium dodecyl sulfate has a phase diagram similar to that shown in Fig. 4, whereas sodium docosyl sulfate behaves as the amphiphile type shown in Fig. 5. Emulsifiers of this latter kind give a lamellar

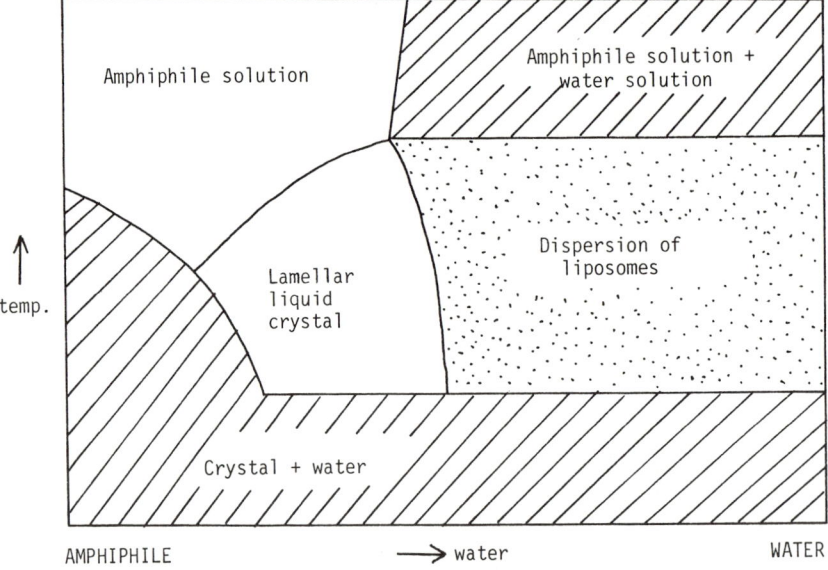

Fig. 5. Characteristic phase diagram of an amphiphile–water system in which the hydrophobic region of the amphiphile is relatively large, e.g., monoglycerides and lecithins. Two-phase regions are shaded.

liquid crystal, and at high water concentrations a dispersion of liposomes is formed. At low water content an inversed hexagonal, and sometimes also a cubic liquid crystal can be formed. Emulsifiers used in foods, such as monoglycerides and phosphatides (HLB values smaller than about 5), often behave according to the diagram in Fig. 5 with a hydrocarbon chain melting point above room temperature. Instead of crystals plus water, as shown in the figure, a gel phase can be obtained by cooling to room temperature.

It should be mentioned that there are a few amphiphiles that do not give aqueous phases of the type shown in Figs. 4 and 5. They form gel phases instead, i.e., the amphiphile bilayers are crystalline. Two types of such amphiphiles have been found to exhibit such phases. If the hydrocarbon chain region is rigid, and thus cannot be disordered as a hydrocarbon chain, gel phases are formed provided that the polar groups have enough affinity for water. Such gel phases have been observed in cholesterol sulfate and cholesterol phosphate [8]. Another type of amphiphile showing the same behavior is illustrated in Fig. 6. The amines are quite unique amphiphiles with regard to their molecular geometry. The polar group is so small that it can be accommodated into the hydrocarbon chain close-packing, and the reason the crystallinity of the bilayer persists when water goes in and forms

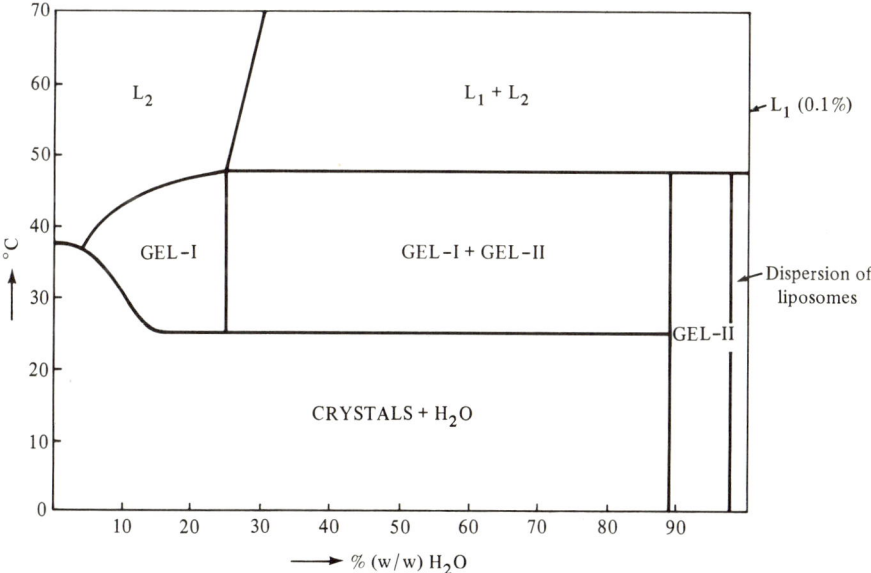

Fig. 6. Phase diagram of the binary system tetradecylamine water.

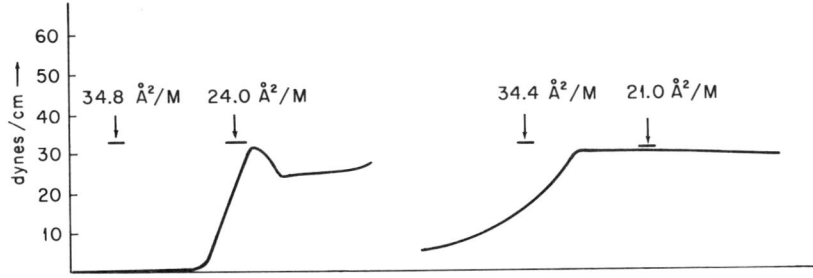

Fig. 7. Pressure area isotherms of stearic acid with a methyl branch in position 17 (to the left) and in position 9 (to the right) at room temperature.

layers between the polar groups is probably related to these molecular dimensions. The two phases gel-I and gel-II shown in Fig. 7 have an identical bilayer structure, and the occurrence of two phases must be interpreted as being due to a forbidden range of water layer thickness [*9*].

VI. MOLECULAR ARRANGEMENT IN SURFACE FILMS

Some aspects of oriented amphiphile layers at the air/water interface that are relevant to the structure of emulsions will be considered here. The surface balance is one of the most sensitive methods for studying the molecular interaction in monolayers, particularly in the hydrophobic region.

A direct photographic recording of the pressure versus molecular area of two methyl branched octadecanoic acids is shown in Fig. 7 to illustrate how minor variations in the hydrocarbon chain structure give drastic differences in monolayer structure. When the methyl branch is located in the 9 position the chains are in a liquid-condensed state, whereas the chains are crystalline in a tilted array when the branch is in the 17 position. If the solid-condensed phases are considered it is obvious from numerous studies that the chains are close-packed as in the solid state [*10, 11*]. Therefore the hydrocarbon chain tilt can only adopt a few values, as the chains can only have a relative displacement of one whole zig-zag period (cf. Section II). If the chains are arranged according to the orthorhombic chain packing $O_\perp$, the chains can be vertical or tilted about 67° or 56°. In the case of the triclinic chain packing $T_\parallel$, the chains cannot be vertical, and the most common angle of tilt is about 72°. When solid-condensed monolayer phases are observed, the calculated angles of tilt are usually in agreement with any of these values. Higher angles can be obtained, however, in the case of bulky polar end groups and branches in the chains. Solid-condensed monolayer phases with very low viscosity are sometimes also observed, and from the cross section per hydrocarbon chain and other properties it can be concluded that the chains are extended and arranged hexagonally with rotational disorder [12].

The most used classification of monolayer phases is that proposed by Harkins [13]. The phases termed CS and L2 by Harkins correspond to close-packed chains in a vertical and tilted structure, respectively, whereas the LS phase corresponds to rotational disorder of vertical chains. The so-called liquid-condensed phase L1 corresponds, on the other hand, to a hydrocarbon chain structure disordered as in lamellar liquid crystals. These correlations between structure in monolayers and in the three-dimensional state is of course important for the structure at the oil/water interphase of emulsions. It should be pointed out that most monolayer studies have been performed at the air/water interface, but all information available from oil/water interface studies indicates that the structural properties discussed here are the same in the two interfaces. The correlations between the two- and three-dimensional states are based on agreement in molecular dimensions (cross section per hydrocarbon chain), critical temperature for polymorphic phase transitions, and phase transition energies (cf. Lundquist [11]).

An interesting surface film phenomenon is the formation of multilayers when amphiphile monolayers are compressed beyond the collapse point. This has recently been discussed in detail elsewhere [14], and only one aspect relevant to emulsions will be considered here. In the case of amphiphile molecules so weakly polar that they do not work as emulsifiers (for example, ethyl stearate), the monolayers grow by monolayers into multilayers. Emulsifier molecules forming monolayers at the air/water interface, however, grow into multilayers by discrete units equal to molecular bilayers, indicating the significance of the bilayer units in emulsions.

The oil phase in many emulsions consists of molecules which orient in contact with water, for example, the triglycerides constituting edible oils. Whether the liquid crystalline interfacial film, which will be discussed below, consists of the emulsifier only or if molecules from the oil are solubilized in this film can be analyzed by the surface balance technique. The monolayer behavior of binary mixtures of the oil and the emulsifiers are then examined. Two extreme cases are shown in Fig. 8, when there are complete mutual molecular solubility and when the two components are completely immiscible. In the first case the mixture behaves as one component, and if the molecular area at collapse is beyond that corresponding to linear additivity, the components possess a condensation effect in relation to one another. This is the situation shown in Fig. 8, and cholesterol, for example, exhibits this effect in monolayers of liquid triglycerides. The opposite effect on triglyceride oil monolayers can be obtained by branched fatty acids, and such expansion effects in emulsifier bilayers should be expected to result in less stable emulsions. When there is no molecular solubility, the component with the smallest

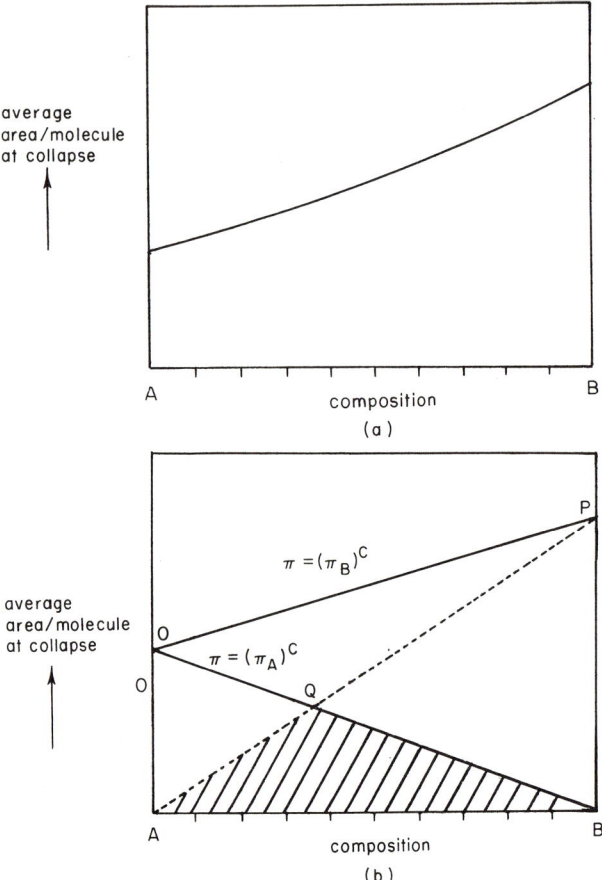

Fig. 8. Alternative collapse behavior of binary monolayer mixtures at the air/water interface (components A and B with collapse pressures π_A and π_B, respectively).

collapse pressure (component B with collapse pressure π_B in Fig. 8) is squeezed out from the monolayer, and at the final collapse of the monolayer (at surface pressure π_A), it contains only component A.

VII. WATER/EMULSIFIER INTERACTION

It is obvious from the earlier discussion in this paper that the interaction between the hydrophilic part of the emulsifier and the water molecules is a decisive factor in the transformation of a crystalline lipid to a liquid crystal.

The presence of water bound to the hydrophilic groups facilitates a structural rearrangement to a less ordered state.

The application in the food industry of this disordering action of water on the emulsifiers as such has been treated in several articles [15, 16]; the use of the liquid crystalline state to obtain a fine dispersion of additives, the function of which depends on this to a high degree, is interesting [17].

In order to understand the role of liquid crystals in emulsions, the adverse of the disordering water/emulsifier interaction, a corresponding ordering effect has to be considered also. This enhanced order has been demonstrated in several systems in which the mixing of a *liquid* emulsifier and water gave rise to a liquid crystalline phase [18]. Investigations on water/aminooctane [19] and on nonionic surfactants [20, 21] have also provided illustrative examples on the ordering effect of water/emulsifier interaction.

The structure of a lamellar liquid crystalline phase contains comparatively large areas of consecutive water/lipid layers, and it appears reasonable to use the structural conditions in the lamellar liquid crystal as a model to better understand the structure the emulsifier attains at the oil/water interface. The first investigation directly aimed at a comparison of these conditions has recently been published [22]. X-ray studies showed that of two emulsifiers, the monoglyceride (Dimodan LS), and the glycerolactapate nitrate (GLP), only the former gave rise to a crystalline phase in combination with water. Surface rheology determinations displayed a difference in the behavior at the oil/water interface. Dimodan LS, which forms the liquid crystal with water gives monomolecular layers, which develop a strong viscoelastic behavior with time, as opposed to GLP which shows no liquid crystals with water and no viscoelasticity of the monolayers. The results indicate an ordering effect of the water/oil interface—an effect which caused a regular packing of the emulsifier at the interface with time for the emulsifier that formed a liquid crystalline phase when combined with pure water. The ordering influence of the water/oil interface has been observed earlier [23]. An emulsion that contained anisotropic layers of sufficient dimensions to be observed in the microscope (Fig. 9). After prolonged standing these layers of micron thickness disappeared and the emulsion spontaneously separated into two isotropic liquids.

Both these cases focus attention on the emulsifier/water interaction as a decisive factor in obtaining the packing in planar layers which is the prerequisite not only in obtaining a liquid crystalline phase but also in forming an interfacial layer of the emulsifier at the interface in an emulsion.

It is also necessary, however, to take into consideration factors other than the emulsifier/water consideration to form a coherent pattern of the importance of different phenomena on the structure of the emulsifier at the interface.

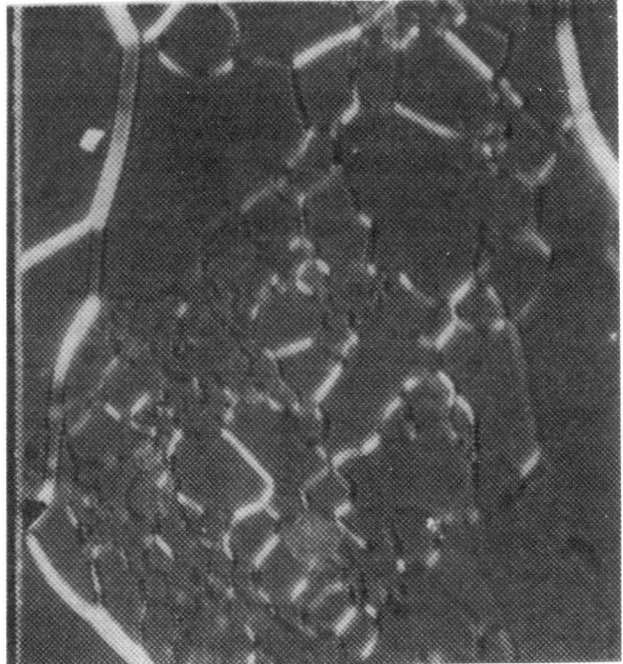

Fig. 9. Thick anisotropic layers may be observed in an emulsion, which at storage separates into two isotropic liquids.

Two of these will be mentioned in the following treatment: the formation of "complexes" between emulsifiers present at the interface and the oil/emulsifier interaction.

VIII. EMULSIFIER/EMULSIFIER INTERACTION

The first factor—the "complex" formation—has been frequently cited since its introduction [24] as the reason behind the optimum emulsion stability that is experienced when two emulsifiers are combined in certain ratios. Even the formation of liquid crystalline phases has recently been referred to as complex formation [25]. The combination of a predominantly hydrophilic emulsifier such as lauryl sulfate with a more hydrophobic amphiphile such as lauryl alcohol gives an increased adsorption and enhanced stability [26]. Contrary to the opinion of earlier authors [25, 27], the results of a recent publication [26] could verify that complexes with a definite composition are not formed at the interface. This result should also be expected; the possibility of the presence of detectable complexes between a long-chain alcohol and an alkyl sulfate at the interface appears very limited, as illustrated by the results from molecular spectroscopy determinations.

Complexes among different long-chain amphiphiles are common in solids [28, 29] and certainly also exist in liquid form [30]. These complexes, e.g., the one between carboxylic acid and the corresponding soap [30] have been found to be completely disintegrated, when water was added in the amount of 40% (W/N). The hydrogen bonds, which gave rise to the soap/acid complex [29], were considerably stronger (NMR; $\Delta\delta \simeq 100$ cps) than the corresponding bond between alcohol and soap (NMR; $\Delta\delta \simeq 10$ cps) [31]. Considering that the spectroscopic evidence for the former disappeared when water was added, the possibility of the presence of entities with any detectable bonds to justify the notation complex between alcohols and ionized surfactants appears extremely small. In order to understand optimal stabilizing action at certain emulsifier ratios, increased adsorption at low emulsifier concentration as pointed out by Vold and Mittal [26] is important, and the energetics of the packing conditions of the interface are of interest. The importance of the packing giving rise to the preferential radius of curvature of the interfacial layer has been indicated for microemulsions by several authors [32, 33]. It is of interest to note that there is an energy difference between the emulsifier layer when it is packed in a regular pattern with a hydrophobic amphiphile and when it is penetrated less regularly by the solvent. One important factor is the anisotropy of the polarizability of the C–C bond, which means that the interaction between similarly oriented chains may increase by a factor of 1.7 compared to nonoriented chains at identical interchain distance [34]. In addition to this factor it appears probable that the mean interchain distance is lower when amphiphile and emulsifier are regularly packed than when solvent molecules are inserted into the emulsifier hydrophobic parts of a surface layer. The distance dependence of van der Waals attraction potential is considerable [35].

These estimations conform perfectly to the results by Vold and Mittal [26] and give indications that the maximum stability of emulsions and the optimal surface characteristics experienced for certain ratios of amphiphiles rather may be explained from enhanced interaction potentials of nonspecific dispersion forces than from agglomeration into specific complexes.

They also serve to illustrate the usefulness of the liquid crystalline structure as a model to obtain information on the structural arrangement of emulsifiers at the interface. The combination of lauryl sulfate and the corresponding alcohol, the behavior of which at an interface has been illustrated recently [26], may be compared to the corresponding behavior of the pair sodium octanoate and octanol [36].

Figure 10 illustrates the formation of normal micelles (A) at low alcohol/soap ratio while the high ratio (B) gives the inverse micelles. When the two amphiphilic compounds are present in excess of a minimum concentration (*L*) and the weight concentration alcohol/soap is ratio approximately in the range 1–2.5, the packing will be in parallel layers forming lamellar liquid crystal. It is easy to recognize the similarity to the conditions at the interface; a proper balance between the predominantly hydrophobic soap facilitates packing conditions in harmony with the conditions at an interface with a large radius of curvative compared to the micelles. Increased length of the emulsifier hydrocarbon chains would be expected to widen the cosurfactant/surfactant

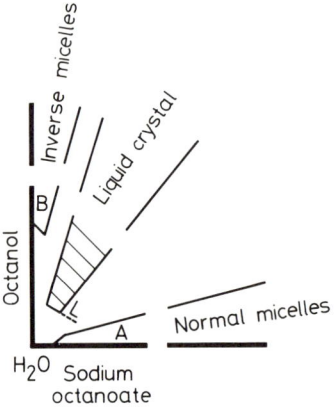

Fig. 10. Mixing ratios of emulsifier combinations may give rise to the spontaneous formation of plane layers in a certain ratio range.

ratio range for obtaining the liquid crystalline phase; the combination decanol/potassium oleate is an example. The weight ratio range 0.33–3 of the two compounds may give rise to a lamellar liquid crystal. This range is considerably larger than the one quoted for the C_8 pair.

It is essential to observe that the similarity is restricted to the actual conditions at the interface, a too-soluble hydrophobic cosurfactant will be dissolved in the oil phase and its concentration at the interface will be too low to result in liquid crystalline packing. At or in excess of the *L* concentration (Fig. 10), knowledge of the structure of the liquid crystalline phase may be valuable in order to draw conclusions about the basic phenomenon giving rise to the structure at the interface. It is important, however, to realize that in addition to the water/emulsifier interaction the emulsifier/hydrocarbon interaction may be an important factor.

IX. HYDROCARBON/EMULSIFIER INTERACTION

The contributions clarifying the phase conditions in systems of water, hydrocarbon, and nonionic emulsifiers for micellar systems [37–43] and systems including liquid crystals [20, 44–50] have brought attention to the importance of the hydrocarbon interaction and its influence on the associated conditions giving rise to ordered structures of liquid crystalline character.

A prominent example is found [49] when water and a nonylphenol adduct are combined with an aliphatic (Fig. 11A) or aromatic (Fig. 11B) hydrocarbon. The composition marked by P in Fig. 11A will form a two-phase system containing one aqueous and one hydrocarbon solution—an emulsion with limited stability in the disperse state. Exchanging the *aliphatic* hydrocarbon hexadecane with the *aromatic* hydrocarbon *p*-xylene, and keeping an identical weight ratio between the constituents, gives rise to a one phase system—a lamellar liquid crystal (Fig. 11B). The influence of the nature of the hydrocarbon on phase conditions in such systems may be deducted using the HLB temperature system [51, 52]; its usefulness in predicting micellar solubilization has been demonstrated in several investigations [37–40].

The examples given here have verified the necessity of including interactions between *all* the constituents in order to give reliable predictions for the behavior of the system. Of the association structures, the different molecular interactions giving rise to the liquid crystalline structure have been demonstrated to have a pronounced influence on emulsion stability and a few facts related to this stability will be treated.

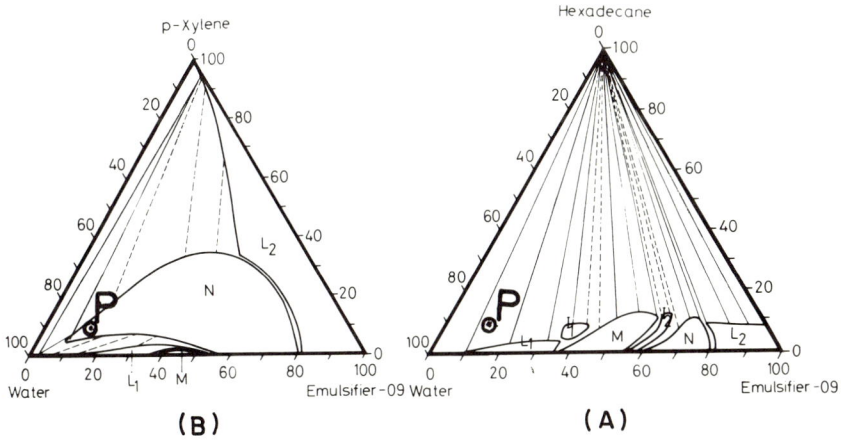

Fig. 11. Changing an aliphatic hydrocarbon (A) in an emulsion to an aromatic hydrocarbon (B) may change it from being an unstable two-liquid emulsion into thermodynamically stable one-phase system, a liquid crystal (composition P).

X. EMULSION STABILITY

Although phase diagrams of systems of water, oil, and emulsifier had been published [37–40] and the conditions in various parts of such systems discussed [48], the first contribution demonstrating the pronounced stabilizing effect of the presence of a liquid crystalline phase in an emulsion was published first in 1969 [53]. It was followed by several publications [54–57] displaying, in some cases, the drastic [56] effect of the liquid crystalline phase.

One case [58] merits further discussion since the results illustrated the distinction between "complex" formation between emulsifiers and collective ordered association such as the formation of liquid crystals. The system [58] contained water and p-xylene and the emulsifiers were 1-aminooctane and 1-octanic acid in varying ratios.

The amine and the acid reacted with each other to form an amine salt. Since this reaction implies a pair formation it may be denoted a complex formation. The complexes in the form of amine salt exist over a wide range of emulsifier ratios [58] and also when the complexes are dissolved in the oil phase [59]. The amine/water interaction gave an ordered association into a liquid crystalline phase with a lamellar structure. This phase absorbed acid molecules to an acid/amine molar ratio of 1.5 and hydrocarbon plus water to an extent that compared with the amine/acid ratio.

The two parameters could therefore be varied in the system to find the separate influence of complex formation and ordered association on emulsion stability. The results [58] gave clear distinction concerning the influence of complexes and of ordered association. The presence of a liquid crystalline phase caused an enhanced stabilization, while two-phase emulsions were unstable regardless of the presence of the amine/acid complex (Fig. 12).

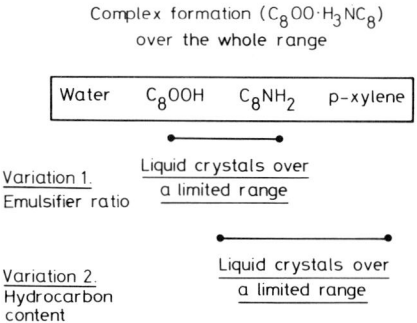

Fig. 12. A system [58] containing water, p-xylene, an amine, and a carboxylic acid has served [56] to distinguish between the influence of molecular complexes and ordered associations of surfactants at the interface.

XI. LOCATION OF THE LIQUID CRYSTAL

In order to understand the reason for the stabilizing action of the liquid crystal, knowledge about its location in the dispersion is one necessary prerequisite. The easiest method is by microscopy.

Figure 13 demonstrates the optically anisotropic, wide regions observed in emulsions containing liquid crystals. Caution is necessary, however, when interpreting pictures such as the one in Fig. 13. Figure 9 shows an example of a

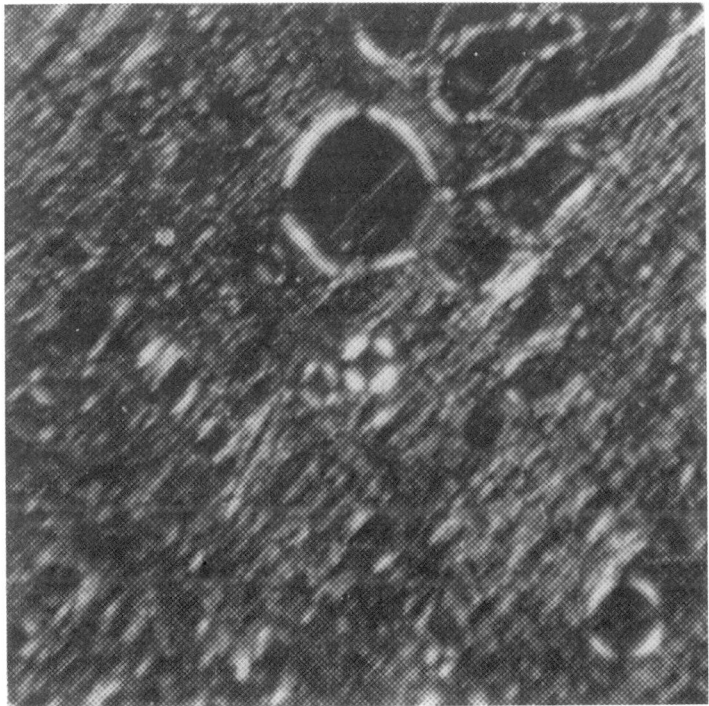

Fig. 13. The liquid crystal phases are easily observable in an emulsion from their radiance in polarized light.

disperse system with optically anisotropic layers, but which spontaneously separates into two isotropic liquids on storage. Electron microscopy, on the other hand [60], is useful since it provides detailed information about the layer structure close to the interface. Figure 14 shows an example of a lamellar phase between two flocculated droplets. The arrow points to the film between

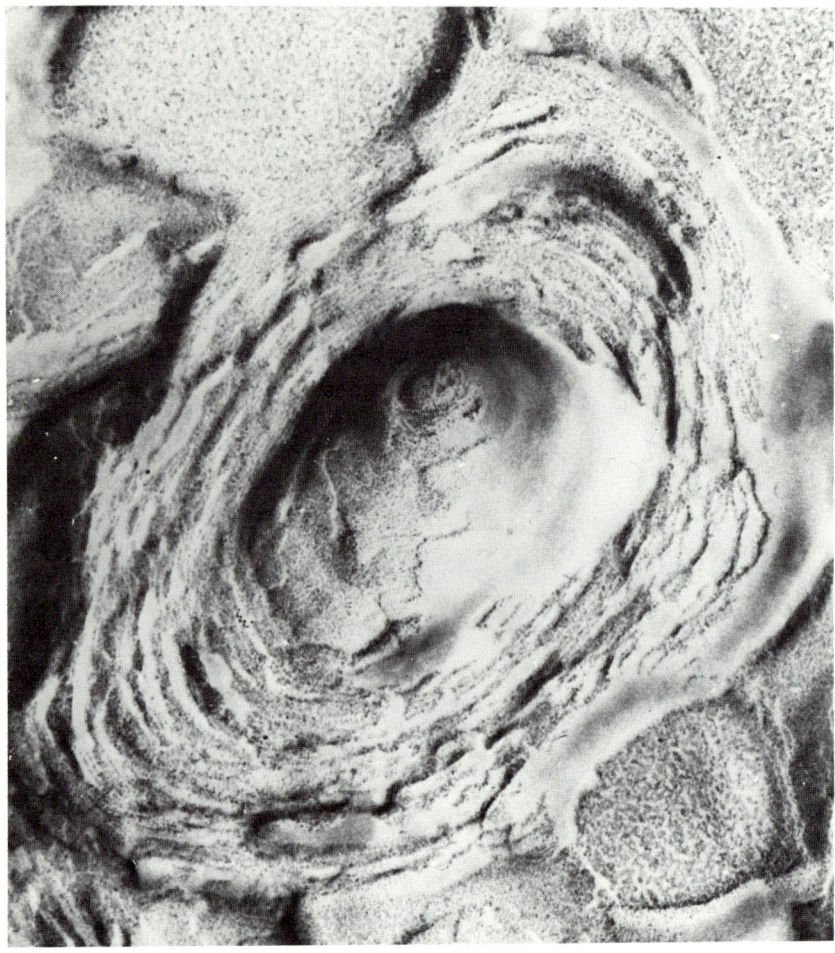

Fig. 14. Electron microscopy enables observation of the arrangement of layers in a liquid crystal phase between two droplets.

them; the ordered arrangement of the layers between the droplets interfaces is easily recognizable.

These pictures indicate that an energetically favorable arrangement of the dispersion is with the liquid crystal at least partly localized at the oil/water interface. An evaluation of the stabilizing mechanism of the liquid crystalline phase, when located at the interface, would accordingly be meaningful.

XII. FLOCCULATION AND COALESCENCE

Emulsified droplets describe an irregular path in the liquid and may, if sufficiently small, serve to illustrate Brownian movement. The van der Waals attraction forces, arising from the London dispersion forces between atoms, will, however, become dominant at sufficiently short distances and the droplets will be forced toward each other and flocculate (Fig. 15). The flocculation is followed by coalescence; the time for this process to occur depends on the stability of the thin film between the droplets in the flocculated state.

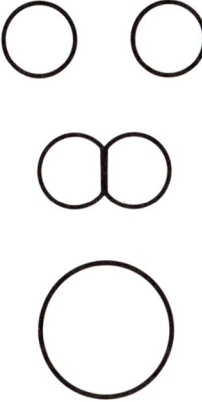

Fig. 15. Emulsifier droplets flocculate first and coalesce later.

The effect of the presence of the liquid crystal at the interface depends on two pertinent factors for flocculation and coalescence. The first is the modification of the van der Waals attraction potential due to the presence of liquid crystalline layers, and the second is the retardation effect of the viscous liquid crystalline phase on the thinning of the film in the coalescence process.

XIII. VAN DER WAALS FORCES

An attraction potential is always experienced between identical particulate matter. Hamaker [61] treated the phenomenon in a first approximation as being additive, independent of interaction with other neighbors. From the attraction potential V_A between two atoms at a distance r

$$V_A = -\beta/r^6 \tag{1}$$

the attraction between macroscopic bodies may be found by intergration. For two spheres

$$V_A = -A/6 \cdot \{2a^2/(R^2-4a^2)+2a^2/R^2+1r(R^2-4a^2)/R^2\} \quad (2)$$

which expression at short distances simplifies to

$$V^A = -Aa/12d \quad (3)$$

where A is the Hamaker constant, a the radius of the droplet, and R the center to center distance.

More complex expressions have been given by Lifshitz [62] directly using the dielectric conditions in the space between macroscopic matter. Lifschitz's expressions have been modified by Parsegian [63] to more easily tractable calculation equations. The following treatment involves the kinetics of drop

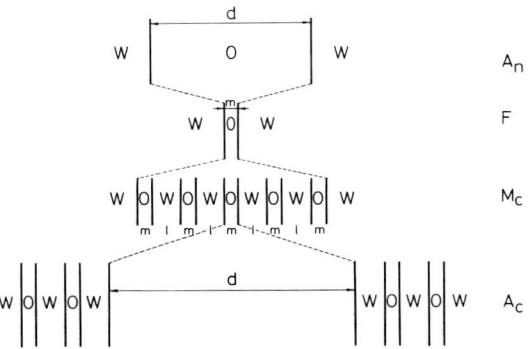

Fig. 16. When a droplet covered with a liquid crystal phase coalesces, this has to take place over several layers of the liquid crystalline phase (see text).

encounter at short distances and Eq. (3) will in principle serve for an evaluation of the conditions. The comparison may be limited to the attraction potential if the hydrodynamic conditions at drop encounter are postulated not to be influenced by the presence of the liquid crystalline phase. At close approach the droplets will flatten at the contact point, and it is reasonable to use the potential for plane parallel layers, which at short distances may be approximated

$$V = -A/12\Pi d^2 \quad (3)$$

in which d equals the distance between the planes. Using this equation and summarizing for the different layers we may, according to the model (Fig. 16), obtain a first approximation expression for the ratio of attraction potentials

Liquid Crystals and Emulsions

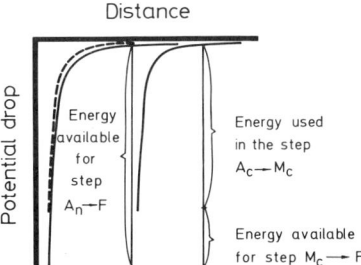

Fig. 17. The energy changes during flocculation and coalescence are completely different for an emulsion covered with a liquid crystalline phase from an uncovered one. (———) Distance measured between the drop surfaces. (----) Distance measured between outer layers of the droplet covered with a liquid crystal.

between covered and uncovered droplets as they approach each other

$$R = d^2 \left[\sum_{q=0}^{n} \sum_{p=0}^{n} \frac{1}{[d+(p+q)(1+m)]^2} + \sum_{q=0}^{n-1} \sum_{p=0}^{n-1} \frac{1}{[d+21+(p+q)(1+m)]^2} - \sum_{q=0}^{n} \sum_{p=0}^{n-1} \frac{2}{[d+1+(p+q)(1+m)]^2} \right]$$

The influence of the presence of a liquid crystal is illustrated by Fig. 17. It is obvious from the figure that the difference in potential drop in the flocculation process (A_n to F and A_c to M_c in Fig. 16) is not pronounced between the two cases; the presence of the liquid crystal, however, reduces the potential drop. This means that the drops covered with a liquid crystalline layer absorb almost all of the total potential drop available to reach the common state F (Fig. 16) in the preliminary flocculation step $A_c \to M_c$. For the coalescence process over the central layered structure ($M_c \to F$; Fig. 16) of thickness $2n(1+m)$ only a lesser part ($\approx 25\%$) of the energy is left.

There is also a comparison between the potential at equal distances between the innermost surface of the emulsion drops; the noncovered drops have in the example 99.7% of the potential energy left for the continued thinning of a less viscous film compared to 25% for the covered droplets in state M_c. The viscosity of the liquid crystal is higher than for the oil—a factor that further enhances the stabilization. Its influence may be approximately evaluated, using the Reynolds equation [64] for two parallel plates with viscous medium in between. The thinning rate depends on the pressure Δp and the radius according to Reynolds equation

$$d(1/h^2)/dt = 4\Delta p/3\Pi r_0^3 \eta; \qquad (4)$$

or

$$dh/dt = -4h^3 \Delta p/9\Pi r_0^3 \eta$$

Using the attraction potential to obtain Δp

$$\Delta p = -k \cdot dV/dh = k_1/h^3$$

it is found that

$$dh/dt \sim 1/\eta.$$

Since the viscosity of the liquid crystal is of the magnitude of 100 times the viscosity of the liquids in emulsions, the combination of reduced attraction potential and increased viscosity should give considerable enhancement of the emulsion stability.

These estimations are considered as a first approximation; a more complete treatment needs to account for the complex rheological behavior of a liquid crystalline phase between two walls and to improve information about the location of the liquid crystalline phase. These estimations, however, serve to illustrate the importance of the presence of multilayers at the interface for emulsion stability.

REFERENCES

1. K. Larsson, *Ark. Kemi* **23**, 35 (1964).
2. K. Larsson, *J. Amer. Oil Chem. Soc.* **43**, 559 (1966).
3. E. von Sydow, *Ark. Kemi* **9**, 231 (1956).
4. V. Luzzati, H. Mustacchi, A. Skoulios, and F. Husson, *Acta Crystallogr.* **13**, 660 (1960).
5. V. Luzzati, A. Tardieu, and T. Gulik-Krzywicki, *Nature (London)* **217**, 1028 (1968).
6. K. Larsson, *Chem. Phys. Lipids* **9**, 181 (1972).
7. N. Krog and K. Larsson, *Chem. Phys. Lipids* **2**, 143 (1968).
8. S. Abrahamsson, K. Larsson, and I. Pascher, to be published.
9. K. Larsson and A. Al-Mamun, *Chem. Phys. Lipids* **12**, 176 (1974).
10. T. Burch, K. Larsson, and M. Lundquist, *Chem. Phys. Lipids* **2**, 129 (1968).
11. M. Lundquist, *Chem. Scr.* **1**, 5 (1971).
12. K. Larsson, *Nature (London)* **213**, 383 (1967).
13. W. D. Harkins, "The Physical Chemistry of Films," 2nd ed. Reinhold, New York, 1954.
14. K. Larsson, *Surface Colloid Sci.* **6**, 261 (1973).
15. E. A. Flack and N. Krog, *Food Trade Rev.* **40**, 27 (1970).
16. N. Krog and A. P. Borup, *J. Sci. Food Agr.* **24**, 691 (1973).
17. N. Krog, *in* "Food Emulsions" (S. Friberg, ed.), pp. 67–139. Dekker, New York, 1976.
18. P. Winsor, "Solvent Properties of Amphiphilic Compounds." Butterworth, London, 1954.
19. R. Collison and A. C. S. Lawrence, *Trans. Faraday Soc.* **25**, 662 (1958).
20. K. Shinoda and H. Saito, *J. Colloid Interface Sci.* **26**, 70 (1968).
21. K. Shinoda, *J. Colloid Interface Sci.* **34**, 278 (1970).
22. J. V. Boyd, N. Krog, and P. Shennan, *Soc. Chem. Ind. London Monogs.* p. 99 (1974).
23. S. Friberg and L. Rydhag, *Kolloid Z. Z. Polym.* **244**, 233 (1971).
24. J. H. Schulman and E. G. Cockbain, *Trans. Faraday Soc.* **36**, 661 (1940).

25. P. A. Saunders, *J. Soc. Cosmet. Chem.* **21**, 377 (1970).
26. R. D. Vold and K. L. Mittal, *J. Colloid Interface Sci.* **38**, 451 (1962).
27. J. T. Davies and G. R. A. Mayers, *Trans. Faraday Soc.* **56**, 691 (1960).
28. H. C. Kung and E. D. Goddard, *J. Phys. Chem.* **68**, 3465 (1964).
29. S. Friberg, L. Mandell, and P. Ekwall, *Acta Chem. Scand.* **20**, 2632 (1964).
30. S. Friberg, L. Mandell, and P. Ekwall, *Kolloid Z. Z. Polym.* **233**, 955 (1969).
31. G. Gillberg and P. E. Ekwall, *Acta Chem. Scand.* **21**, 1630 (1967).
32. L. M. Prince, *J. Colloid Interface Sci.* **23**, 165 (1967); **29**, 216 (1969).
33. M. Robbins, *Amer. Chem. Soc. Nat. Coll. Symp., 48th, 1974, Preprints* p. 174.
34. F. Fowkes, *J. Colloid Interface Sci.* **28**, 493 (1968).
35. D. Langbein, *Phys. Kondens. Mater.* **15**, 61 (1972).
36. P. Ekwall, L. Mandell, and K. Fontell, *Mol. Cryst. Liquid Cryst.* **8**, 157 (1969).
37. K. Shinoda, T. Nakagawa, B. Tamamushi, and T. Isemura, "Colloidal Surfactants." Academic Press, New York, 1963.
38. T. Nakagawa, K. Kyriama, and H. Inove, *J. Colloid Sci.* **15**, 168 (1960).
39. K. Shinoda and T. Ogawa, *J. Colloid Interface Sci.* **24**, 56 (1956).
40. H. Saito and K. Shinoda, *J. Colloid Interface Sci.* **24**, 10 (1967).
41. K. Kan-no and A. Kitahara, *J. Colloid Interface Sci.* **37**, 469 (1971).
42. A. Kitahara, T. Ishikawa, and S. Tanimori, *J. Colloid Interface Sci.* **23**, 243 (1967).
43. K. Shinoda and H. Kunieda, *J. Colloid Interface Sci.* **42**, 381 (1973).
44. R. Salisbury, E. E. Leuallen, and L. T. Chawkin, *J. Amer. Pharm. Ass. Sci. Ed.* **43**, 117 (1954).
45. B. W. Burt, *J. Soc. Cosmet. Chem.* **16**, 465 (1965).
46. F. Lachampt and R. M. Vila, *Amer. Perfum. Cosmet.* **82**, 29 (1967).
47. J. Swarbrick, *J. Soc. Cosmet. Chem.* **19**, 187 (1968).
48. T. Mitsui and Y. Machida, *J. Soc. Cosmet. Chem.* **20**, 199 (1969).
49. S. Friberg and L. Mandell, *J. Amer. Oil Chem. Soc.* **47**, 149 (1970).
50. J. S. Marland and B. A. Mulley, *J. Pharm. Pharmacol.* **23**, 561 (1971).
51. K. Shinoda and H. Arai, *J. Phys. Chem.* **68**, 3485 (1964).
52. K. Shinoda, *Int. Congr. Surface Active Substances, 5th, 1968.* Vol. II, p. 275.
53. S. Friberg, L. Mandell, and M. Larsson, *J. Colloid Interface Sci.* **29**, 155 (1969).
54. S. Friberg and L. Mandell, *J. Pharm. Sci.* **59**, 1001 (1970).
55. S. Friberg and L. Mandell, *J. Amer. Oil Chem. Soc.* **47**, 149 (1970).
56. S. Friberg and L. Rydhag, *Kolloid Z. Z. Polym.* **244**, 233 (1971).
57. S. Friberg, *J. Colloid Interface Sci.* **37**, 291 (1971).
58. S. Friberg and G. Söderlund, *Kolloid Z. Z. Polym.* **243**, 56 (1971).
59. S. Friberg and B. Sarthz, *Int. Congr. Surface Active Substances, 6th, 1972.* Vol. II, p. 825.
60. S. Friberg, P. O. Jansson, and E. Cedergren, *J. Colloid Interface Sci.* (in press).
61. H. C. Hamaker, *Physica* **4**, 1058 (1937).
62. J. E. E. Dzyaloskinski, E. M. Lifshitz, and L. P. Pitaevski, *Advan. Phys.* **10**, 165 (1959).
63. V. A. Parsegian and B. W. Ninham, *Nature (London)* **224**, 1197 (1969).
64. O. Reynolds, *Phil. Trans.* **177**, 157 (1886).

VIBRATIONAL SPECTROSCOPY OF LIQUID CRYSTALS

Bernard J. Bulkin*

Department of Chemistry
Hunter College of the City University of New York
New York, New York

I. Introduction.	199
II. The Vibrational Spectrum as a Source of Information.	200
III. Internal Vibrations in Nematic Phases	201
IV. Intermolecular Modes in Nematics and Nematogenic Crystals	213
V. Vibrational Spectra of Smectic Phases	222
VI. Vibrational Spectra of Cholesteric Phases.	225
VII. Vibrational Spectra of Lyotropic Mesophases.	227
References	229

I. INTRODUCTION

Vibrational spectroscopy (infrared and Raman spectroscopy) has been used for many years to elucidate molecular structure and to study strong intermolecular interactions such as hydrogen bonding. It is natural that many research groups should have thought to apply these techniques to the study of liquid crystals.

Although studies of the vibrational spectra of liquid crystals began many years ago, there has been a flurry of activity in the years since 1968. Some of the early work has already been the subject of reviews [1, 2] and will not be exhaustively discussed again in this review. Instead, we will focus on the work done since 1968, mentioning earlier work only in passing. No attempt is being made to be exhaustive in the review although most of the relevant papers are at least cited, if not discussed.

* Present address: Department of Chemistry, Polytechnic Institute of New York, Brooklyn, New York 11201.

As with many spectroscopic techniques, the application of vibrational spectroscopy to mesophases has proceeded in stages. In the early work to be discussed, and in some of the work still being published today, the main contribution has been the observation of phenomena in the spectra-intensity changes, frequency shifts, etc.—which occur at or near the various phase transitions. Further, as is characteristic of the application of spectroscopic studies to liquid crystals, a certain amount of attention has been focused on measurement of order parameters.

There now seems to be sufficient background of this sort available, however, to begin to make some generalizations about the sort of expectations one has for the vibrational spectra of liquid crystals. Further, we can begin to make some quantitative assignment of the phenomena observed to changes at the molecular level. Several such papers have appeared during 1974 and 1975 and will be discussed extensively.

This review is organized as follows. First, the type of information which one can potentially obtain from vibrational spectroscopic studies will be discussed briefly. This is necessary because the application is a fairly unique one and calls for some novel approaches to the data. Second, the studies in the literature will be reviewed according to the types of mesophase being discussed, with some inevitable intermingling in the case of compounds with multiple liquid crystalline phases. Thermotropic materials will be discussed first, followed by lyotropic mesophases. Most of the work to date has been on the former.

II. THE VIBRATIONAL SPECTRUM AS A SOURCE OF INFORMATION

The study of vibrational spectroscopy is conveniently subdivided in two ways. The first is a technique used to measure the spectrum, i.e., the mechanism by which radiation is coupled to the molecular vibrations. In this paper we discuss primarily the techniques of infrared and Raman spectroscopy, but there exist other methods, such as inelastic scattering of slow neutrons, ultrasonic spectroscopy, and dielectric relaxation, which can also probe vibrational spectra. Some work on mesophases using these techniques will be mentioned briefly.

The second subdivision comes in the spectral region under examination, constituting a separation between low and high energy vibrations. This is of the utmost importance for the study of mesophases, because it distinguishes between intramolecular and intermolecular vibrations. For organic materials the dividing line is approximately 150 cm^{-1}; it is unlikely that lattice modes or any intermolecular vibrations for a mesophase or mesogenic material will be observed at higher frequency. It is possible that internal vibrations will be observed below 150 cm^{-1}, however, and hence the distinction is not an absolute one in any sense. Furthermore, it is posible for the internal and external modes to be coupled in the potential energy; some examples of this have been observed in nematogenic crystals.

Three pieces of information are obtained in any particular region of the vibrational spectrum of a molecule, namely, frequency maximum, intensity, and band shape. The first two are the most commonly exploited, the last has only been utilized in a few studies of anisotropic phases, but is potentially quite interesting. In addition, with aligned samples the polarization of a band, either in the infrared or Raman spectrum, can yield valuable information about molecular order. Raman depolarization ratios can also be measured in isotropic phases. More detailed discussions of these techniques can be found in any number of textbooks.

From the point of view of data analysis, aside from qualitative conclusions, one may attempt calculations of the vibrational spectrum, either intramolecular or intermolecular. In the case of mesogenic molecules, which are always complex organic systems, such a calculation is on the very edge of feasibility. However, a few have been attempted, one a rather extensive normal coordinate calculation for a Schiff base, a calculation of intermolecular vibrational frequencies for a nematogenic crystal, and a model calculation for the vibrational modes of a nematic phase. All are discussed in more detail below.

Considerable caution is required in observing and interpreting changes in infrared and Raman spectra of liquid crystals. At the phase transitions, with application of fields, and with other alignment procedures, these materials undergo changes in refractive index and optical constants. Such changes may appear as relative intensity changes, frequency shifts, or band shape changes, which are not to be interpreted at the molecular but rather at the macroscopic level. Examples of problems of interpretation of this sort are discussed below.

III. INTERNAL VIBRATIONS IN NEMATIC PHASES

Perhaps the most common analysis performed on the vibrational spectra of organic molecules is the assignment of the various absorption (infrared) or scattering (Raman) maxima to group frequencies, e.g., $C=O$ stretch, CH_2 deformation, etc. These assignments, for large molecules, are very approximate, since there is considerable mixing of the modes in the potential energy distribution. The extent of this mixing depends on symmetry and mechanical coupling forces, as well as on other interactions within the molecules. In particular, if there is a spectral region in which a molecular vibration appears isolated from all others, then we may presume this to be a fairly pure vibrational mode. In liquid crystalline molecules, such cases do occur, but rarely. The number of variations is large, and the potential for proximity of two or more modes is great. The CN stretching vibration in molecules such as the cyanobiphenyls or CN containing Schiff bases is often cited as an example of a pure, isolated vibrational mode. These modes generally occur near 2200 cm^{-1}, and there are no other fundamentals in such molecules in this region. Even in this case, however, one must be aware of the possibility of interaction between overtones or combination bands and the fundamental (Fermi resonance), which can complicate the band position and shape.

What is the reason for studying assignments of the vibrational spectrum? In fact, little or no information about ordering or intermolecular interactions is gleaned from the assignments themselves. Since most liquid crystalline molecules have well-established molecular structures with little or no symmetry, structural questions are not being answered. The main use of such studies arises if one observes different temperature dependencies for different fundamental modes as a phase transition is being traversed. We shall see, as well, that the order parameter can be measured from infrared spectra. This measurement depends on knowing something about the band assignments. If one wishes to draw conclusions about molecular rotations from a study of band shapes, one should know the assignment of the mode under consideration. Finally, there have been several studies relating Raman spectroscopy to chain melting in end chains of nematogenic and smectogenic molecules, and for such studies one must be sure of the assignment of the modes to the end chain.

Study of the internal vibrations of nematic materials began in the 1940's with the Raman study of *p*-azoxyanisole (PAA) by Freyman and Servant [3] and continued sporadically through the 1960's with more work on PAA, anisaldazine [4] and the alkoxybenzoic acids [5]. Maier and Englert [6] managed to assign most of the vibrational modes observed in the infrared spectrum of the alkoxy azoxy benzenes in a reasonable way using an extensive series of model compounds, all of which were substituted benzenes or azobenzenes. This approach is useful, though approximate, because it gives a general idea of which types of molecular modes are expected in each region. At the time the work was carried out, the study of group frequencies in organic chemical spectra was still being developed. Now, it is common practice to be able to analyze a spectrum at the level of sophistication in their paper, as many spectra are well known for a wide range of model compounds. Similar group frequency analyses of the spectrum of *p*-butyl-*N*-(*p*-methoxybenzylidene) aniline (MBBA) in the infrared [7] and Raman [8] have been done recently. In addition, the work on the alkoxy azoxy benzenes was repeated [9] using Raman spectroscopy with essentially the same modes being observed. This is not surprising, since the molecules have no symmetry, hence all modes are both infrared and Raman active. Bulkin *et al.* [10] used Raman spectroscopy to show that PAA was nonplanar in solution, in agreement with X-ray data [11].

A much more extensive analysis of the vibrational spectrum of MBBA has been carried out by Vergoten [12, 13]. In this work, benzylidene aniline was used as a model for the internal vibrations of all but the tails of the MBBA molecule. We will discuss the tail vibrations further as they are of some interest. The spectra of MBBA and BA are compared in Table I. It is clear that reasonable assignments can be made, though the normal coordinate computation on which they are at least partially based is a very approximate one.

TABLE I

OBSERVED FREQUENCIES, RELATIVE INTENSITIES AND ASSIGNMENTS OF MBBA AND OBSERVED FREQUENCIES OF BA[a]

Frequency (cm^{-1})	MBBA[b] Relative intensity	Assignments (Wilson's notation)	BA[b] Observed frequency
180	6		
315	1		250
340	2		
410	4	16a	405
			540
630	5	6b	620
			650
720	2		
760	2		750
775	4		770
790	4		
825	3		825
886	3	δ (C—H)[a]	875
934	1	17b	920
975	7	δ 5	970
			1000
1014	1	18a (I and II)	1025
1105	4		
1164	70	9a (I and II)	1165
1182	37	ν (ϕ—N)	1190
1245	5	ν (ϕ—C)	1240
1305	5		1320
1370	4	δ (C—H)[a]	1370
1422	12	19b (I and II)	1450
1503	9	19a (I and II)	1480
1575	53	8a (II)	1575
1596	100	8a (I)	1590
1626	41	ν (C=N)	1630

[a] Reprinted from Vergoten [12], by permission of Heyden and Son, Ltd.
[b] MBBA, p-butyl-N-(p-methoxybenzylidene)aniline; BA, benzylidene aniline.

It seems clear that if more detailed information about the vibrational assignments were of interest, spectra of deuterated species would have to be obtained.

Bulkin et al. [7] have shown, in an infrared spectroscopic study of MBBA, that assignments can be aided by the examination of spectra of homogeneous and homeotropically aligned samples. In such samples, modes having a dipole moment change along the long axis of the nematic molecule can be separated from those in the plane perpendicular to the long axis. With the usual sample configuration of an infrared transmission measurement, these data can be obtained without the use of a polarizer. Potentially, with the use of a polarizer, modes in the plane perpendicular to the long axis could also be separated, but this is generally not observed because of rapid molecular rotation, which averages the intensities, or because the distribution of orientations about the long axis is completely random. Bulkin et al. [7] have attempted such an observation with MBBA but always found complete absence of any polarization. It may be that low frequency modes associated with the hydrocarbon chains are good candidates for such an observation. More importantly, such an observation should be attempted with some of the more rigid smectic phases, where questions of orientation and rotation about the long axis are of interest.

Several workers have examined the temperature dependence of the spectra of nematic molecules as the transitions from crystal (c) to nematic (n) to isotropic (l) liquid are traversed. One may summarize the observations as follows: (1) At the crystal-to-nematic phase transition several bands in the midinfrared spectrum usually disappear. This seems to be a change in integrated intensity rather than a broadening. (2) At the c–n transition, several bands appear to change the position of the absorption maximum by a few wavenumbers. (3) In the 200–400 cm^{-1} region, anomalous splitting and broadening of a band seems to occur near the c–n transition. This is assigned to the accordion mode of the end chain. (4) Very little change occurs in intensity or frequency at the n–l transition. (5) With aligned samples, changes in infrared dichroism or Raman depolarization ratios are readily observed at the n–l transition. In some cases, it has been claimed that these show some anomalies in the vicinity of the transition.

An example of the disappearing bands at the c–n transition is shown in Fig. 1. Bulkin et al. [14] have discussed the possible reasons for such disappearing bands, concluding that the most probable reason is that these bands arise from sum and difference modes between lattice vibrations and internal vibrations. Nikolov and Simova [15] have offered a slightly different explanation, discussing the changes in terms of induced dipoles relating to the intermolecular interactions. While definitive work has not yet been carried out, it seems that the disappearing bands are an indication of the coupling between molecules,

Vibrational Spectroscopy of Liquid Crystals 205

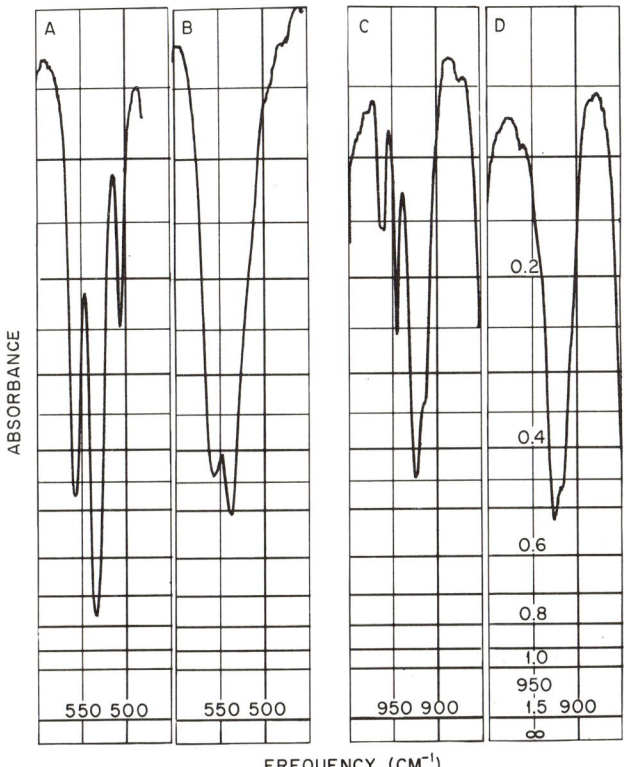

Fig. 1. Typical changes in the infrared spectra of 4,4′-bis(alkoxy)azoxybenzenes at the crystal–nematic transition. A and B are crystalline and nematic 4,4′-bis(methoxy)azoxybenzene, respectively. C and D are crystalline and nematic 4,4′-bis(ethoxy)azoxybenzene, respectively. The crystal spectra are at 35°C, the temperature of the unheated sample in the infrared beam. Nematic melt spectra are at ~2°C above the c–n transition temperature. Reprinted from Bulkin [30], with permission of the American Institute of Physics.

and hence of importance in understanding ordering forces in the crystal as compared to those in the liquid crystal. Bulkin *et al.* [14] have also claimed that there are pretransition effects in the crystalline phase, i.e., that the intensity of the bands which will disappear begins to change several degrees below that temperature at which actual melting occurs. Such pretransition effects have been observed by several groups, though not by all, as shall be discussed further. However, it should be pointed out that the effects observed by Bulkin *et al.* [14] are not corrected for either density or Boltzmann factors. Recently, such corrections have been made in an infrared spectroscopic study of the same compounds by Kirov and Simova [16], and the pretransition effects are still observed in that case.

Kapustin et al. [17] studied the infrared spectra of PAA and 4,4'-azoxydiphenetole in the 700–1800 cm^{-1} region with results that are similar to those reported by other workers. Spectra were presented in all stable phases. They claimed on the basis of group frequency arguments that bands associated with the vibrations of the end groups and NNO linkage changed more than ring vibrations did.

The frequency changes have been reported at the c–n transition for nematics by several workers [9, 18–20]. Again, there is some indication that these changes are not discontinuous, but may occur over a range of 2°–3°. The pre- or posttransition effects have not been clearly proved in this case. The reason for the frequency shifts has been the subject of considerable speculation. It is tempting to follow the lead of some workers and attribute these effects to changes in either molecular conformation [9] or intermolecular forces [18, 20]. In the former case, it should be noted that many low rotational barriers exist within the nematogenic molecules, so some variation in conformer population is possible. If there are changes in intermolecular forces resulting in a difference in mean field near a vibrating molecular group, this could affect the position of the frequency maximum. However, it should be noted that the volume changes (hence the changes in interatomic distances) are small at the phase transition, leading one to expect little change in frequency of absorption modes which experience only secondary perturbation from such interactions in any case. It seems reasonable that the shifts observed (2–4 cm^{-1}) could simply be due to changes in the relative intensities of overlapping bands that are unresolved under the conditions of the measurements. This explanation would lead one to conclude that the frequency shifts and disappearing bands are two different ways of observing the exact same phenomena, except that in one case the disappearing band is never clearly resolved.

One interesting observation along these lines is that reported by Schwartz and Wang [21] in the spectrum of p-(p-ethoxyphenylazo)-phenylundecylenate (PPU) and in the spectrum of p-(p-ethoxyphenyl)-heptonoate. In the nematic phase, striking changes in the relative intensity of several Raman bands were observed as a function of applied electric field strength. In this case, certain bands increased in integrated intensity, while others showed the opposite effect. By a careful analysis of spectra in the isotropic phase, as well as from comparative infrared spectroscopic data, it was concluded that the changes were due to a change in the relative populations of cis and trans isomers about the azo linkage, induced by the electric field. An explanation of this unexpected shift in equilibrium was given in terms the collective stabilization due to the large ensemble of molecules aligned by the field.

The question of chain fluidity in nematic (and more recently in smectic, see Section V) phases has been the subject of several Raman investigations by Schnur [22, 23]. He has studied the alkoxy azoxy benzenes in all spectral

Vibrational Spectroscopy of Liquid Crystals

regions, but with particular emphasis on the region from 200 to 400 cm^{-1}, where it is asserted that an accordionlike vibration of a short (3-8 carbon) all trans hydrocarbon chain attached to a benzene ring should occur. An approximate calculation [22] was carried out to confirm the assignment, although again one must assume that there is some mixing with ring modes and other modes of the chain.

The result of these studies indicate that near the c–n transition, and to some extent in the nematic phase as well, there are conformational changes taking place in the end chains. An illustration of the changes observed for the C_6

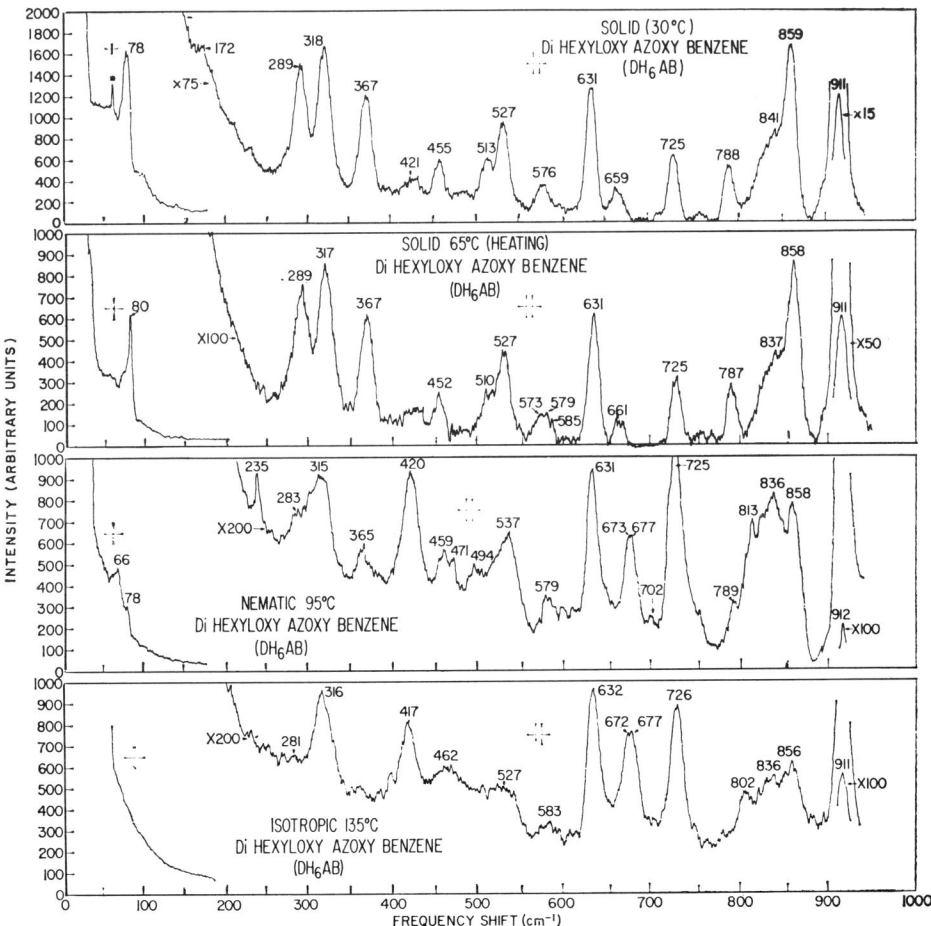

Fig. 2. Tracing of observed Raman spectra of dihexyloxyazoxybenzene (C_6). Reprinted from Schnur [23] with permission of Gordon and Breach Science Publishers.

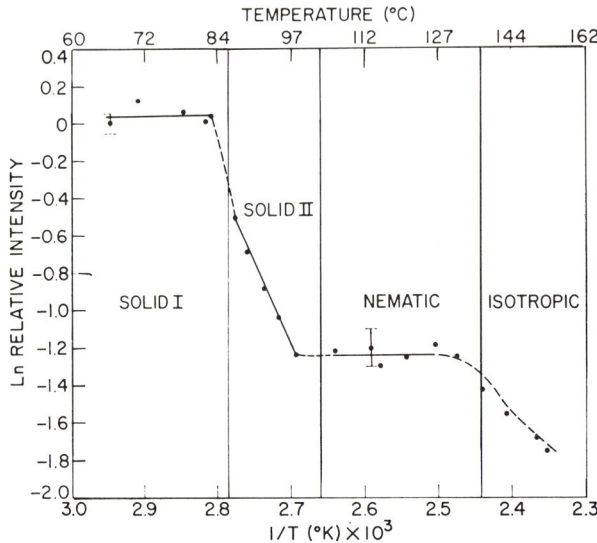

Fig. 3. The relative integrated intensity of logarithmic units of the accordion band in C_4 versus reciprocal temperature. The integrated intensity of the 340 cm^{-1} band was calculated relative to the adjacent 320 cm^{-1} band which appeared to remain reasonably constant. Reprinted from Schnur [23] with permission of Gordon and Breach Science Publishers.

homolog is shown in Fig. 2. Of course, these compounds also show other spectral changes, as discussed above, but the changes observed by Schnur seem to have a distinct phase and temperature dependence. In some of the compounds, solid phase polymorphism exists, and this seems to drastically affect the intensity of the accordion mode. An example is shown for the C_4 homolog in Fig. 3. Note that this might be interpretable as a pretransition effect in the crystal rather than a distinct solid phase.

Additional evidence regarding chain fluidity comes from a paper by Bulkin et al. [24], who showed that in the nematic phase of MBBA the CH stretching vibrations show an increase in bandwidth, so that the entire contour of the CH stretching modes appears to lose resolution. As temperature is increased until just below the nematic–isotropic phase transition, the bands sharpen with a slight shift in the position of the frequency maxima. This behavior was interpreted as arising from a reorganization of butyl chain conformations on the time scale of the infrared measurement. Since the two time scales (internal rotation and CH stretching) overlap for a short temperature range, broadening is observed. Such phenomena are common in nuclear magnetic resonance. Similar reorganization of conformers in MBBA was deduced from ultrasonic measurement of Jahnig [25] and is consistent with these data.

Lugomer and Lavrencic [26] have recently reexamined the Raman intensity data for a wide variety of nematic materials with different length carbon end chains in an attempt to interpret these data in terms of the elastic constants. They have presented evidence that the Raman intensities follow the so-called universal curve of Gruler [27] for the bend/splay elastic constant ratio, K_{33}/K_{11}, which shows a very strange temperature dependence in the nematic phase.

Janik and co-workers [28] have proposed, on the basis of infrared spectra, that two solid forms of MBBA exist, one of which is called the "metastable phase," prepared by rapid cooling from the nematic. This phase can be converted to the stable solid by annealing at about 0°. It is interesting to note that both forms have infrared bands of comparable width. If the metastable form were a glassy liquid crystal, as has been suggested by others [29] broader bands would be expected.

The biggest differences between the infrared spectra of the two solid forms of MBBA shown by Janik are in the 100–400 cm^{-1} region. This region principally contains modes associated with the end chains, leading one to conclusions similar to those of Schnur regarding the role of these end chains in bringing on the c–n transition. Janik has also presented evidence for pretransition effects at the c–n transition for MBBA.

At the nematic to isotropic transition, there are few, if any, real changes observed in the spectra. One likely candidate is the change in intensity of the accordion mode observed by Schnur [23]. There has been confusion on this transition from Raman spectroscopic observations in the past. This confusion arises as follows: In the Raman spectrum of a low symmetry, nematogenic molecule, the Raman bands are all polarized, but may have widely varying depolarization ratios. In an unaligned, or imperfectly aligned sample, the Raman exciting radiation and the scattered light have their electric vectors rotated due to refractive index discontinuities. Such an effect is well known in the Raman spectrum of powders [30]. When the n–l transition occurs, this scrambling of polarization disappears, and the result is a different set of relative intensities. This has been documented in a more quantitative fashion by Bulkin et al. [31]. Thus any changes in relative intensity of bands (and changes in apparent frequency maxima which may occur from overlapping bands changing in relative intensity) must be carefully checked to correct for the band polarization. The effects can also be checked by study of the infrared spectra in most cases.

Finally, we turn to the question of making good polarization measurements in the nematic phase. This can be done with well-aligned samples. The problem is relatively simple in the infrared region, indeed Maier and Englert [6] made such measurements, using thin films aligned by rubbing NaCl plates. It is also possible to use the technique of Guyon et al. [32] to coat KBr plates with MgF_2

in order to obtain spectra of well-aligned homogeneous nematics. Homeotropic alignment in the infrared may be achieved by a lecithin rubbing applied to the KBr plates [7]. The absorption due to lecithin is either negligible or readily accounted for. There have been few attempts to use external electric fields to align samples for infrared work, but it should be noted that if one uses untreated alkali halide plates for infrared spectra, the effective electric field at the surface may be extremely high, leading to alignment.

Because thicker samples are required, in general, the question of measuring polarization in the Raman spectrum has been a more difficult one. Some initial attempts were unsuccessful, with authors observing a depolarization ratio of 1.0 in the nematic phase. This is indicative of complete scrambling of the radiation, as discussed above. If changes from a ratio of 1.0 are observed at the n–l transition, as has been reported [33], then this is not surprising. Lavrencic and Lugomer [34] obtained spectra of 4,4'bis(heptyloxy)azoxy benzene in a magnetic field of 1000 G. They assert that at the phase transitions ($\pm 2°$ on either side) the depolarization ratio shows a minimum. Several bands were examined, and all showed this effect. The origin is not clear, but one wonders whether it might be similar to the effects observed in transparency curves of liquid crystals shown by Poziomek et al. [35] for several cases.

The most thorough treatment of oriented nematics in the Raman has been done by Priestly et al. [36, 37] and Jen et al. [38]. Important details are also available in the thesis of Jen [39]. They have shown that it is, in principle, possible to measure both the first ($\cos^2 \theta$) and second ($\cos^4 \theta$) terms in the order parameter or distribution function from the Raman spectrum. Experimentally, they have used samples of MBBA in which a probe molecule that also forms a nematic phase N-p-butoxybenzylidene-p'-cyanoaniline (BBCA) is dissolved. The sample is aligned by rubbing (or a related techinque) two plates, and 180° Raman scattering is observed. The BBCA molecule shows a strong CN stretching vibration, which is approximately along the long axis and is isolated from other intramolecular vibrations. From these measurements, the terms in the order parameter can be deduced. Comparison with existing mean field theories of the nematic order parameter has been made from these measurements. It is asserted that the fit is better with the Humphries–James–Luckhurst [40] approach than with the Maier–Saupe [41, 42] theory for the P_2 term, but that neither does particularly well with the P_4 observations. The authors have discussed possible reasons for the strange P_4 results, including those which may arise from the measurement technique itself.

Boyd and Wang [43] have applied the Raman technique to the study of the effect of pressure on nematic materials. In this case, the Raman spectrum simply becomes the analytical probe by which the phase transition is detected. It was shown that the c–n and n–l transitions are effectively first order (follow the Clausius–Clapeyron equation) when measured in this way for one case (PPU).

This technique is useful in that it also allows one to determine the change in volume at the phase transition; in both cases (c–n and n–l) the higher temperature phase had the lower density.

Ohnishi [44] has applied the principles of infrared spectra of aligned nematics to study molecular orientation in the dynamic scattering mode of MBBA. The results allow one to follow the change in orientation of C≡N and C—O moieties in the molecule with applied voltage. Because the changes were small, an infrared difference technique was used.

There have been a few attempts to obtain information about the degree of motional freedom, or the barriers to motion of molecules, from the study of the shapes of bands in infrared and Raman spectra of nematic materials. Simova and Kirov [45–48] in a series of papers, have extensively studied the temperature dependencies of the bandwidths for the homologous 4,4'dialkoxyazoxy benzenes, MBBA, p-butyl-N-(p-ethoxybenzylidene)aniline (EBBA), p-aminophenylacetate (APAPA), p-pentoxybenzoic acid, and dibenzene p,p'-diaminobiphenyl, examining several bands in the infrared spectrum of each molecule. They use the approach due to Rakov [49] which essentially relates the half-bandwidth to a "preorientation time." An Arrhenius-type temperature dependence for the preorientation time is predicted, with a consequent activation barrier U_{or}.

In their work the values of τ and U_{or} were determined by Kirov and Simova in the crystal, nematic, and isotropic phases. The experimental data on bandwidths for several bands of PAA are shown in Fig. 4. Table II summarizes the

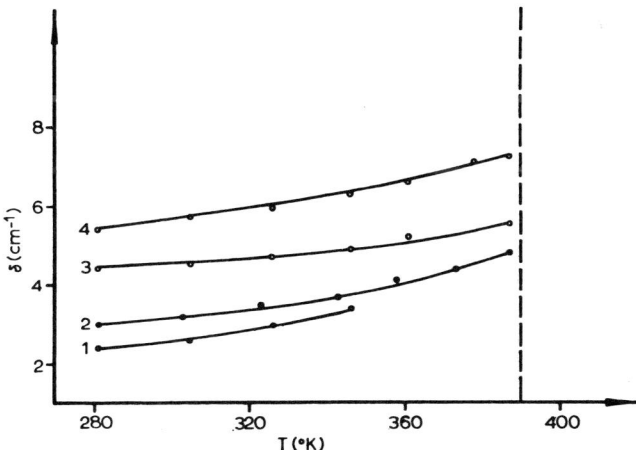

Fig. 4. p-Azoxyanisole (PAA) (crystal phase). Temperature dependence of the half-bandwidth δ (cm^{-1}) vs. the absolute temperature T(°K). Curve 1: 724 cm^{-1}; curve 2: 1299 cm^{-1}; curve 3: 638 cm^{-1}; curve 4: 668 cm^{-1}. Reprinted from Simova and Kirov [47], by permission of Elsevier Scientific Publishing Company.

TABLE II

Values of U (kcal/mole) of PAA Molecules in Crystal, Nematic, and Isotropic Phases Determined by Different Methods[a]

Phase of PAA	Ir U_{OR}	NMR	Neutron scattering U_D	Viscosity U_F
Crystal phase	4.2	5.6		
Nematic phase	9.2	9.1		2.2
Isotropic phase	6.3	5.7	8	5.7

[a] Reprinted with permission of Elsevier Scientific Publishing Company from Simova and Kirov [47].

values of U_{or} obtained from these data. The authors have also discussed the values for this quantity obtained from NMR, neutron scattering, and viscosity data, suggesting reasons for the discrepancies that exist. While the interpretation of all the data presented in these papers is not yet clear, and probably awaits more extensive studies of rotational or translational barriers by other techniques, the data themselves are extremely valuable for future workers to use.

Lugomer [50] has examined the bandshape for an infrared band in the liquid crystalline phases of 4,4'-diheptyloxyazoxybenzene. While his work has been limited to this one band, in smectic, nematic, and isotropic phases, the approach has been a different one from that of Simova and Kirov. In this case, a rotational correlation function (essentially a normalized Fourier transform) of the band profile has been calculated. These correlation functions (see Rothschild [51] for a discussion of the method) can lead to rotational correlation times, with the results for this case being shown in Table III. Because these times are rather rapid compared to those calculated from an analogous NMR approach of Samulski [52], it is concluded that only the rotation of a portion of the molecule—namely, the benzene rings—is being observed. The different times

TABLE III

Rotational Correlation Function Data for 4,4'-Bis-(Heptyloxy)azoxybenzene

Phase	Time (sec)	Average jump angle (degrees)
Smectic	8.2×10^{-12}	30
Nematic	5.3×10^{-12}	60
Isotropic	4.1×10^{-12}	63

observed in smectic, nematic, and isotropic phases are then concluded to be due to 30°, 60°, and greater than 60° jumps of the benzene rings in the three phases, respectively. The main difficulty with this approach at the moment is that the theoretical development which underlies it has been carried out primarily for isotropic phases of high symmetry molecules. To transfer this theory over to the anisotropic molecules of low symmetry in anisotropic fluid phases may require more of a theoretical adjustment. Thus we can say that the computed correlation functions are most certainly different, but just what corrections should be made to them, and what conclusions deduced, remains a bit speculative.

IV. INTERMOLECULAR MODES IN NEMATICS AND NEMATOGENIC CRYSTALS

One of the areas in which vibrational spectroscopy yields quite unique information is the study of intermolecular or lattice vibrations. These are seen close to the exciting line in the Raman spectrum, hence a good rejection of stray light and a relatively low Rayleigh scattering level are needed for their observation. In the infrared spectrum, the lattice modes occur in the far infrared region, generally below 150 cm^{-1} for organic crystals. They tend to absorb very strongly, particularly in molecules such as nematogenic crystals, which usually have dipolar groups present. While in the Raman spectrum measurements with single crystals are readily made, and yield definitive assignment of the lattice modes to a particular symmetry species, in the far infrared single crystal measurements are difficult. This further complicates the observations, because the crystallite size may be comparable to the wavelengths of the radiation, leading to spectral distortions.

In the nematic phase we again have the possibility of observing "pseudo-lattice" vibrations, i.e., intermolecular motions characteristic of the long range order in this phase. It should be noted that the intermolecular potential in such a phase with organic crystals, probably is proportional to an intermolecular separation term of $1/r^3$ or higher power of $1/r^n$, so the force constants, which depend on the second derivative of $1/r^n$ fall off rather rapidly with distance. However, in cases where there is long range order, it may be possible to propagate phonons even in liquid crystals. As we shall see, there is not much evidence for this in the spectra of nematics, but there is some in the case of the more highly ordered smectic phases.

Although a substantial number of papers on nematics have appeared, almost all the attention has been concentrated on two materials, PAA and MBBA. We will discuss these two cases in some detail, referring relatively briefly to other work that has been carried out.

The Raman spectrum of PAA in the lattice vibration region was first published by Bulkin and Grunbaum [53], and shortly thereafter appeared in a paper of Amer, Shen, and Rosen [54]. While the latter paper did not show many of the bands seen in the former, a subsequent full paper of Amer and Shen did reproduce the data of Bulkin in the crystal. Schnur has also reproduced these data [55] as have other workers [56]; however, a recent report again showed low resolution and high background scattering with a lack of fine details in the Raman spectrum [57]. Bulkin and Prochaska [58] examined the Raman spectrum of a single crystal of PAA, at $-90°C$, and were thus able to assign the lattice vibrations to particular symmetry species. Their spectrum, taken in two views to show the A_g and B_g modes clearly, is shown in Fig. 5.

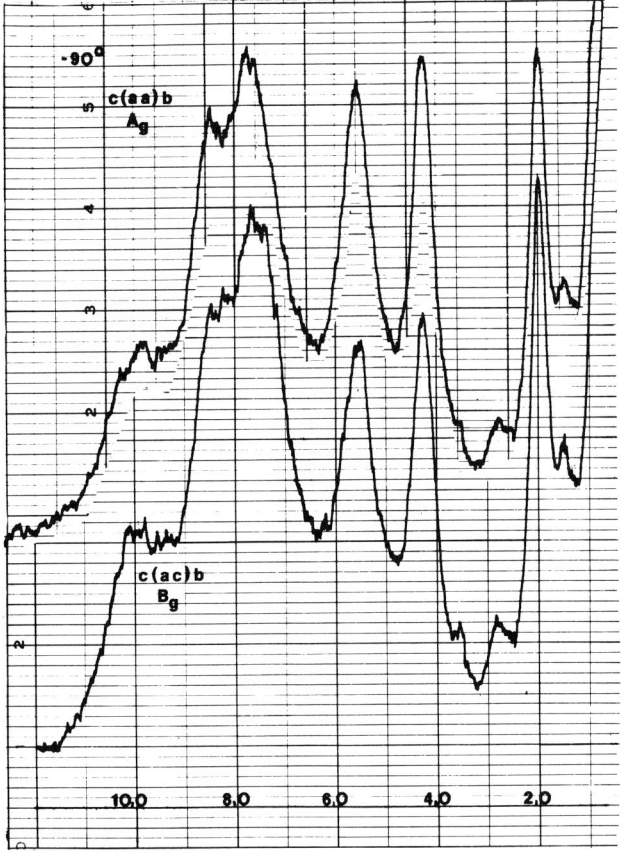

Fig. 5. Raman spectrum of PAA (single crystal, $-90°$) in the lattice vibration region. Reprinted from Bulkin and Prochaska [58], by permission of the American Institute of Physics.

As the c–n transition is approached, Bulkin and Prochaska [58] observed pretransition effects which they interpreted as being indicative of a soft mode, such as is commonly observed near ferroelectric phase transitions. They asserted that this soft-mode-like behavior occurs with a mode or modes which initially (i.e., 5° below the phase transition) is in the vicinity of 70 cm^{-1}. It appears to shift rapidly toward zero frequency in the 2°–3° below the phase transition. Amer and Shen [9] did not find evidence for such pretransition effects and attributed this observation to problems with temperature control in the experiments of Bulkin and Prochaska. Sakamoto et al. [57] also failed to observe this effect, but in their case the lattice vibration spectrum even 8° below the phase transition shows only two broad bands, instead of the six distinct maxima observed by other workers.

There is other evidence for the pretransition effects described above. (1) It should be pointed out that to a certain extent there is a matter of definition of a pretransition effect. Some workers have seen the transition occurring over a 2°–3° range and stated that it was abrupt, yet this is the range over which it was asserted that the soft mode behavior occurred. (2) Riste and Pinn [59], using neutron scattering techniques, have observed identical pretransitional behavior at the PAA phase transition. Their results on integrated Bragg intensity vs. temperature are shown in Fig. 6. Haller and Cox [60] have also observed such behavior in thermodynamic measurements. As we shall see below, there are pretransition effects observed in the Raman spectra of other liquid crystalline materials at the c–n phase transition. The soft mode behavior was predicted in theoretical treatments of Kobayashi [61] and Ford [62].

The far infrared spectrum of crystalline PAA has been reported by Bulkin and Lok [63]. The important points to be noted from their data are as follows:

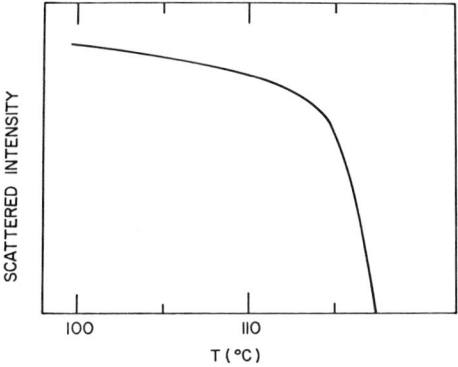

Fig. 6. Variation of integrated Bragg intensity with temperature observed in neutron scattering from PAA near the c–n transition. Reprinted from Riste and Pynn [59], with permission of Pergamon Press.

TABLE IV

OBSERVED AND CALCULATED LATTICE VIBRATION FREQUENCIES OF PAA[a]

Symmetry species	Translatory			Rotatory		
Ag						
Observed	30	52	70	16	74	91
Calculated	28	55	62	20	69	95
Bg						
Observed	30, 37	52	70	16	74, 90	95
Calculated	37	60	69	25	82	101
Au						
Observed	50	70	—	135, 150	84	50?
Calculated	51	69	0	130	89	55
Bu						
Observed	—	—	70?	115	84?	50?
Calculated	0	0	67	119	78	52

[a] Reprinted with permission of the American Chemical Society from Bulkin and Grunbaum [64].

(1) There exist lattice modes in the far infrared spectra at higher frequencies than any observed in the Raman. (2) The differences between far infrared and Raman spectra in the external mode region show that here unit cell symmetry (C_{2h}^5) governs the selection rules, rather than molecular or site symmetry (both C_1).

Bulkin and Grunbaum [64] have carried out a calculation of the lattice vibrations of PAA. In Table IV the results of their calculation are shown, compared with experimental data from Bulkin and Prochaska [58] and Bulkin and Lok [63]. The calculation is based on a model potential function for the unit cell, which uses atom–atom interactions of the form

$$V_n = -a/r^6 + be^{-cr}$$

where r is the nth atom–atom distance and a, b, and c are parameters. In the calculation, the parameters are transferred from the literature, without any refinement on the experimental data. The potential is differentiated twice with respect to all motional coordinates to give the force constants, from which the frequencies are then calculated.

The results in Table IV are grouped according to motions about the same type of coordinate. Thus, when the frequencies in a particular column are close to one another, this is indicative of weak coupling between the molecules in the unit cell with respect to motion along that coordinate. A large splitting reflects a strong coupling. As expected for anisotropic molecules, both cases occur. It is interesting to note that the lowest and highest frequency bands observed in the spectrum result from very strong coupling, giving rise to a large splitting. Likewise, the cases of weak coupling lead to bands in the 70 cm^{-1} region, which is where the putative soft mode was seen. This is as expected for such a mode. This calculation seems possible now for all nematogenic or smectogenic crystals for which crystal structure data are available. Raman spectra of the higher homologs of PAA have been studied by several groups in the lattice vibration region. Interestingly, none shows as well-defined a spectrum as does PAA. Bulkin [14] had previously commented that is seemed, on the basis of infrared data, to be more difficult to obtain highly ordered crystals of these materials. Schnur [22, 23] has pointed out that there is considerable solid polymorphism in these materials, and in macroscopic samples one wonders whether this might be affecting the resolution in the lattice vibration spectrum.

The spectra of various members of the series have been published by Schnur [22, 23], Amer and Shen [9], and Sakamoto et al. [19, 20, 57]. Although Schnur shows spectra for all the compounds with alkyl chains from C_1 to C_7 (except C_3) it is not always possible to measure the frequencies from his spectra because of a highly compressed scale of presentation. In a few cases the frequencies are given. For the C_2 compound, reported by all three groups, there seems to be general agreement that only one band, with a maximum at about 54 cm^{-1}, is observed. All workers had difficulty resolving this line from the Rayleigh background. Sakamoto et al. [20] report that there appears to be another weak band near 90 cm^{-1}.

Amer and Shen report only one band at 37 cm^{-1} in the butyl homolog, while Schnur has presented evidence that there are two solid forms of this compound, one showing lattice vibration frequencies at 47, 80, and 93 cm^{-1}, the other (higher temperature solid) showing 41 and 78 cm^{-1} bands. Solid polymorphism is quite common in organic crystals, and so it is not surprising that it should be observed here. Indeed, for PAA different solid forms have been documented by Schnur previously. It is possible that disagreements between various groups as to the lattice vibration region spectra observed for these materials can almost always be resolved on the basis of solid polymorphism. In this regard, it would be helpful, when Raman spectra are taken, to make some X-ray measurement, even if it is only a powder diagram, on the sample. This would assure some standardization of the crystalline form.

For the C_5 homolog, Sakamoto et al. have reported two bands, one at 40 cm^{-1}, the other at 60 cm^{-1}. Schnur also reports two bands, but the frequencies appear to be slightly different. For the hexyl compound, Amer and Shen did not observe any bands, while Schnur has seen a band in the 78–80 cm^{-1} region. From his spectra, there are clearly some lower frequency modes as well which are not resolved from the background. The spectra of the heptyl compound will be discussed later in our treatment of smectic materials.

Thus far we have concentrated on discussing the observations in the crystalline phase of the alkoxyazoxybenzenes. What of the spectra in the nematic phase? In all papers, the authors seem to agree that there is Raman scattering in the region below 100 cm^{-1}, which appears in the nematic phase as distinct frequency maxima on the Rayleigh wing and which subsequently disappears in the isotropic phase.

Not surprisingly, there are differences in the positions of the quoted frequency maxima. These are due to the difficulty of accurately measuring the positions of broad bands on an intense background. It would seem desirable to carry out some computer deconvolution of the Rayleigh background in an attempt to find these bands more accurately. This information will be valuable in any attempt to model the nematic intermolecular vibrations. Also of some controversy is the question of whether the observed intermolecular Raman bands in the nematic phase disappear continuously or discontinuously at the n–l phase transition. Again, more work is needed to establish this point. Of the alkoxyazoxy benzenes only PAA has been studied in the far infrared region [63]. Here one observes a broad, intense band near 100 cm^{-1}, which seems to have a low frequency shoulder near 50 cm^{-1}. In this case, however, the band does not disappear in the isotropic liquid, only in dilute solution in CCl_4 (and probably other solvents). This is similar to the behavior seen in the far infrared for many polar liquids of organic compounds. The high frequency maximum observed for such a mode (100 cm^{-1}) as compared to values of 50–75 cm^{-1} for lighter molecules is indicative of a strong intermolecular coupling.

The other major class of compounds that has been studied in the low frequency region is the Schiff base nematics. MBBA has been studied in the most detail. The Raman spectrum of MBBA in the lattice vibration region was first reported by Billard et al. [65], and subsequently by Borer et al. [66] and Vergoten et al. [13]. The spectrum of Billard et al. is shown in Fig. 7 at two temperatures in the solid state. At low temperatures, even in the polycrystalline sample, MBBA shows a very well-defined vibration spectrum. Nine distinct frequency maxima are observed between 38 and 180 cm^{-1} at $-173°$, but the two highest modes, at 179 and 181, are explained as internal vibrations. It seems possible that the mode at 142 cm^{-1} may also have a contribution from internal vibrations; this could be checked by study of solution spectra. Borer et al. are in essential agreement with these results.

At temperatures close to the c–n phase transition, Billard et al. have noted that the spectrum becomes quite broad and most of the fine detail is lost. Thus in the spectrum at 15°, about 5–7° below the c–n phase transition, the bands are already quite broad. A spectrum given by Borer et al. at 2° also shows considerable broadening. This seems to be indicative of similar pretransition effects to those described earlier, although no soft mode was observed.

The far infrared spectrum of solid MBBA has been reported by several

Vibrational Spectroscopy of Liquid Crystals

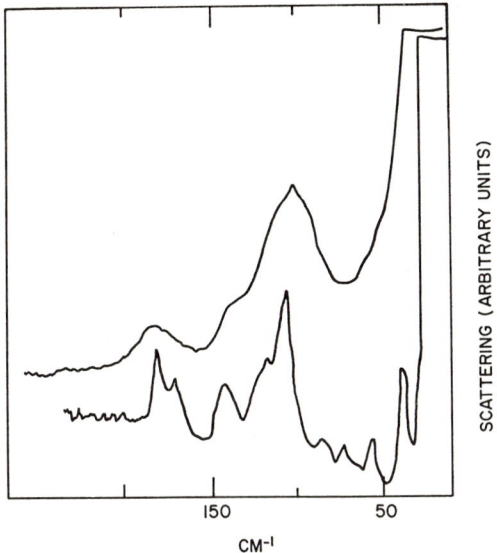

Fig. 7. Raman spectrum of solid MBBA in the lattice vibration region. Upper curve: 15°; lower curve: −100°. Reprinted from Billard et al. [65], by permission of the French Academy of Science.

groups. The most accurate spectrum of the polycrystalline sample of MBBA seems to be that of Evans et al. [67]. These data are shown in Fig. 8. Sciesinska et al. [68] have made observations as low as 80 cm^{-1}, while Bulkin and Lok [63] have mainly discussed the spectrum in the nematic phase. Again, one notes that the most intense absorption in the far infrared comes at higher

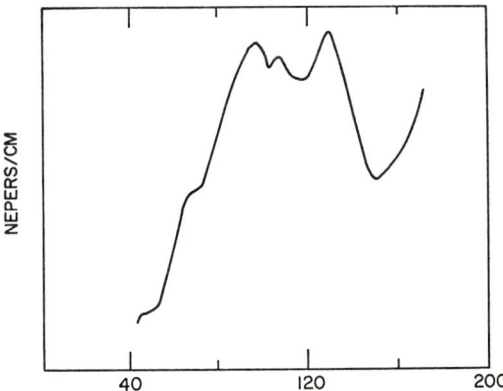

Fig. 8. Far infrared absorption of solid MBBA at 10°. Reprinted from Evans et al. [67], by permission of the Chemical Society.

frequencies than does the Raman scattering in this region. This seems to be explainable, as in the case of PAA, by the nature of the intermolecular interactions, which couple and split the lattice modes. There has been no discussion of the temperature dependence of the far infrared spectrum of solid MBBA.

Sciesinska et al. [68] have questioned the attribution of the absorption below 170 cm^{-1} to intermolecular modes, contending that the torsional vibrations and other low frequency internal modes of the MBBA molecule account for this absorption. They cite as evidence changes in the spectra which they associate with different phases of MBBA solid and the persistence of the absorption in solution. The experiments of Evans et al. [67], and Bulkin and Lok [63] indicate that this is not the case, however. Of primary importance is the fact that in dilute solution, in longer pathlength cells (so that the number of MBBA molecules being observed is constant), the absorption below 170 cm^{-1} disappears, while that above 170 cm^{-1} is virtually unchanged. This is shown clearly in Fig. 9, in which the far infrared spectrum is presented as a function of temperature in a 10.2 w/w% cyclooctane solution.

The data on MBBA in the nematic and isotropic phases in the low frequency region are consistent with those for PAA. Again, some broad scattering is observed in the nematic phase Raman spectrum. Vergoten [12] has placed this at 35 cm^{-1} in the nematic phase at 25°. He has asserted that this band shifts to 23 cm^{-1} at the n–l transition and disappears at higher temperatures in the isotropic liquid. It seems possible that with the lower temperatures of

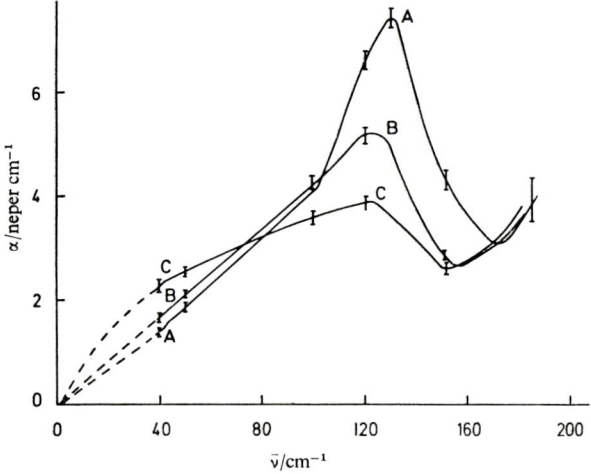

Fig. 9. Far infrared spectrum of MBBA as 10.2% solution in cyclooctane at 24°, 1.5 mm path (curve A); 51°, 1.75 mm path (curve B); 65°, 1.5 mm path (curve C). Reprinted from Evans et al. [67], by permission of the Chemical Society.

observation here as compared to the alkoxyazoxybenzenes, background Rayleigh scatter may be lower and observations in the low frequency region simplified. In addition, the measurements of Vergoten show a very low scattered light background, probably due to the instrumentation being used.

In the far infrared, one again sees a strong absorption band, which at 23° shows its maximum frequency at 130 cm^{-1}. This seems to be higher than that observed for PAA. In the isotropic phase the band shifts slightly to lower frequency (~ 123 cm^{-1}). As mentioned above, the band does disappear in solution, but only on either considerable dilution [63] or by raising the temperature somewhat at moderate dilutions [67]. Thus, while the absorption is of intermolecular origin, the near neighbor interactions involved are very strong. This is also reflected in the frequency maximum coming so much higher than that for other polar liquids.

Evans et al. [67] have carried out some model calculations to explain the observations in the far infrared spectrum of MBBA. In these calculations an idealized packing is assumed which reproduces the experimentally observed density at reasonable internuclear distances. The model used is a modification (Hoffman and Larkin [69]) of that originally due to Brot [70], which has been of some utility in explaining the far infrared absorption of other polar liquids. To reproduce the observed spectrum using this model, it is found that the potential well experienced by an MBBA molecule in the field of its neighbors needs to be considerably narrower and steeper than that of other polar liquids. While the model does not correctly reproduce the observed integral intensities, it does give insight into the nature of the potential well.

Evans et al. have also examined the rotational barrier predicted by a potential of the form already discussed for PAA

$$V = a \exp(-br) - c/r^6$$

in which the only intermolecular interactions considered are those between the benzene rings packed in an idealized geometry. Relative energy of static configurations, in which the central molecule of a hexagonally packed array is rotated, were calculated. To improve the calculations, some cooperative rotation of the other molecules was also allowed in some configurations. Actually, it is not the height of the barrier but the shape of the well that is more important in the far infrared absorption, and this may be sensitive to other factors including the end chain interactions.

The Raman spectra of a number of other Schiff base compounds have been observed by Sakamoto et al. [57]. In the compounds generally known as 20-3, 50-6, and 10-3, they did not see any low frequency modes in the nematic phase. They attribute the absence of such intermolecular scattering to weaker interactions between molecules than in the alkoxyazoxybenzenes. However, this seems unlikely in view of the results already discussed. While Sakamoto *et al.* [*19, 20, 57*] explain this in terms of the absence of the azoxy dipole, it has already been noted by Bulkin and Grunbaum [*64*] that this dipole makes a relatively small contribution to the total intermolecular forces.

In one case, the orthohydroxy substituted 10-3, Sakamoto *et al.* have observed a well-defined Raman band at 55 cm^{-1} (40°), which, with increasing temperature, gradually broadens, so that by 82° it is no longer visible. They have explained this observation as a soft mode associated with intermolecular hydrogen bonding. Spectra of deuterated samples would be of some interest in checking this assignment.

While infrared and Raman measurements have not been able to observe intermolecular modes below about 20 cm^{-1} in the nematic phase thus far, Dimic and Osredkar [*71*] have observed a mode at 15 cm^{-1} by neutron scattering. This is seen in the inelastic scattering of cold neutrons from PAA, anisaldazine, the heptyl- and hexyl-oxyazoxybenzenes, and anhydrous sodium stearate. Just what this could be is somewhat mysterious, but the explanation of a damped torisonal oscillation of some sort given by the authors is certainly plausible. It would appear that for the forseeable future neutron scattering may be the most promising technique for observing the lowest frequency intermolecular oscillations.

V. VIBRATIONAL SPECTRA OF SMECTIC PHASES

To date, relatively little work has been done on the spectra of thermotropic smectic phases. In retrospect, one can see that it would have probably been more fruitful to concentrate on these phases rather than on the nematics, and this will very likely be true in the future. The reason is simply that the smectic phases, particularly such phases as smectics B and H, are much closer to crystalline phases than are nematics. As a result one can probably learn more about the long-range intermolecular order from vibrational spectra than was true with nematics. Thus while there are few changes observed at the n–1 transition in the vibrational spectrum, at such transitions as smectic B–C (smB–smC) there are substantial changes. One might expect that such changes would be more interpretable as well, since one might be able to use a group theoretical approach to the changes in selection rules at the crystal–smectic phase transition more readily than at the crystal–nematic phase transition. To date, group theory has not been exploited in this area.

Early studies on the longer alkyl chain alkoxybenzoic acid dimers [*5*] included some work on internal vibrations of smectic phases with chain lengths of C_7, C_9, and C_{16}. These papers concentrate on infrared dichroism measurements on the internal vibrations of the materials as a function of temperature.

Schnur [22] has investigated the alkoxyazoxybenzenes, but the only smectogenic compound he studied was the heptyloxyazoxybenzene. In this case he found that the rather well-defined spectrum in the lattice vibration region (as contrasted with less well-defined spectra for the homologs from C_2-C_6) contained distinct frequency maxima at 34, 39, and 76 cm^{-1}. In the smectic phase, only a sharp band at 81 cm^{-1} was seen, which shifts to 78 cm^{-1} in the nematic phase and disappeared completely in the isotropic liquid. This is somewhat puzzling, as no such distinct mode is observed in the other members of this series in the nematic phase spectrum. Again, as in the other members of the series, the accordion mode of the chains, assigned to a 271 cm^{-1} band in the crystal, decreases greatly in intensity in the smectic phase and cannot really be seen above the noise in the higher temperature phases.

Amer and Shen [9] studied the Raman spectrum of diethylazoxybenzoate, and diethylazoxycinnamate, which have smectic A phases. These compounds show a single low frequency Raman mode in the crystalline phase at 22 and 26 cm^{-1}, respectively. In the smectic phase, this mode appears to shift, with the maximum at about 14 cm^{-1}. It vanishes abruptly at the smA–I transition. The mode is strongly overlapped by the Rayleigh line in the smectic phase, but it does seem to be of considerably lower intensity than in the crystal. One wonders if this could possibly be the same mode as that seen by neutron scattering discussed above. As usual, the modes are explained as being some sort of rotatory oscillation, but in this case one which is primarily associated with strongly interacting functional groups, such as C=O, within the molecule. This would account for the seemingly small mass dependence in the smectic phase. Probably such an explanation is not necessary, since a relatively small frequency shift in such a low frequency mode would be sufficient to account for the mass difference. The authors have also studied the internal vibrations of this compound, as have Zhdanova et al. [4]. No frequency shifts or relative intensity changes with temperature were observed such as have been with nematic materials. However, it is interesting that the authors noted a broadening of the bands associated with the c–smA and smA–I transitions. This indicated that band shapes may be of use in interpreting rotatory oscillations in the smectic phase. Since the question of molecular rotation in the smectic phases is now of some interest, vibrational spectroscopy may prove to be an important technique.

Sakamoto et al. [57] have examined the Raman spectra of the compound ethyl-(p-methoxybenzylidene amino) cinnamate, which has both a smectic A and a nematic phase. The spectra in all phases were approximately the same, in both the middle and low frequency regions. The authors studied the relative intensity (relative to what?) of a band at 1598 cm^{-1} as a function of temperature and found that it changed discontinuously with temperature at the phase transitions. When the sample was previously oriented between glass plates by a rubbing technique, however, the intensity of this band appeared to change gradually over a 9° range at the sm–n transition. While the authors attribute this to a second-order phase transtion occurring in this sample, it should be recognized that a continuously varying order parameter could cause such a change, and this does not in itself imply a second-order phase transition.

Finally, the interesting compound terpthal-bis-butyl aniline (TBBA) has been studied. This material has seven known fluid phases, viz., isotropic, nematic, three stable smectics (B or H, C, and A), and two monotropic liquid crystalline phases not yet characterized.

Schnur and Fontana [72] studied all of the stable phases in the low frequency and higher frequency Raman spectra. They have two important sets of observations. In the low frequency region, the lattice vibration spectrum, which contains a number of bands, appears in their spectra to be almost the same in the smectic B phase as it is in the crystal. At the smB–smC transition, most of this structure disappears, and the spectra in the higher temperature phase all show only the broad scattering characteristic of these phases. Note that a band is observed by these workers in the 22 cm^{-1} region of the crystal, which shifts to 19 cm^{-1} in the smB phase, and disappears in the higher temperature phases. If this were the same mode occurring in this region referred to earlier, we would expect to see it persist in the higher temperature phases.

A curious observation is seen in the 1550–1650 cm^{-1} region. In the solid and smB phases, one band is seen in this region at 1590 cm^{-1}. This is also true in the smB phase, but in the smC phase two new bands appear, at about 1560 and 1620 cm^{-1}. This has not been observed in any other spectra of liquid crystals. The authors do not expound on the possible explanations of this phenomenon. Nonetheless, it is straightforward to enumerate mechanisms by which new bands could appear in the spectrum. In particular, the most usual are the appeareance of new conformations (in this case perhaps due to the multiple minimum potential which likely exists about the Schiff base linkages), or a breaking of symmetry, resulting in a lowering of selection rules. In this regard, it should be noted that the smectic B phase is expected to have a higher symmetry than the smC phase. In addition, TBBA is one of the few liquid crystalline molecules which has a fairly high molecular symmetry; it probably possesses an approximate or exact center of symmetry. The intersection of the molecular symmetry and smB symmetry could result in selection rules which would cause modes to be forbidden in smB but allowed in lower symmetry phases. This point should be explored further, as other molecular modes should show the same behavior if this is the case.

Dvorjetski et al. [73] have also investigated TBBA, concentrating their attention on the low frequency region Raman spectrum in the lower temperature phases (smB and the two monotropic phases, called VI and VII). In the solid phase their observations are in agreement with those of Schnur and

Fontana, but this is not true in the smB phase. Here, at 130°C (same temperature as Schnur and Fontana used), Dvorjetski et al. only observe broad, ill-defined scattering. In phase VI, at 80°C, they observe some additional structure, while in phase VII, at 67°C, the original crystalline spectrum is recovered intact. There is no apparent explanation of the difference in observations between these two groups, and resolution must await further experimental work.

Dvorjetski et al. have studied the temperature dependence of the 19 cm^{-1} band in the solid state, approaching the c–smB transition. After correcting for background and Boltzmann factor, they find that this mode shows a small pretransition effect in its frequency, shifting toward zero frequency over a temperature range of about 4°C. They thus feel that this mode behaves as a soft mode and interpret their results in terms of the types of molecular motion possible in the various phases. They conclude that the data are consistent with free molecular rotation about the long axis in the smectic B phase. This rotation freezes out progressively as the molecules are cooled to phases VI and VII. Further work is needed to verify these conclusions in other regions of the spectrum.

VI. VIBRATIONAL SPECTRA OF CHOLESTERIC PHASES

Cholesteric phases have not been actively investigated from the viewpoint of mesophase structure elucidation via the vibrational spectrum as have the other phases. However, as discussed below, there have been applications of the cholesteric medium in the study of infrared optical activity, as well as in the interface between infrared spectrometers and the gas chromatograph.

L'vova and Sushchinskii [74] had, in the early 1960's, reported the infrared spectrum of cholesteryl propionate. More recently, Bulkin and Krishnan [75] discussed the Raman spectra of a number of cholesteryl esters in crystal, cholesteric, and isotropic phases, as well as the spectra of mixtures of cholesterics. In the middle frequency region (200–2000 cm^{-1}), no significant changes in the spectra were observed with the phase transitions. While some small changes in relative intensity and frequency may occur, it is reasonable to summarize the spectra by saying that the changes at the c–ch phase transition are much less than those at the c–n phase transition. Thus, while it is often useful to recognize the similarities between the cholesteric and nematic phases, from the point of view of intermolecular forces in the crystal they may be quite different.

Bulkin and Krishnan have observed changes in the spectra in the C–H stretching region of the spectrum (280–3100 cm^{-1}) which they associate with changes in the environment of angular methyl groups on the steroid framework.

These groups, which protrude from the cholesteric layer, may be a good probe of order in the neighboring layer. An interesting case is the mixture of cholesteryl nanoate–cholesteryl chloride (60/40 w/w) which shows changes at the ch–l phase transition. The spectrum of this mixture in the cholesteric phase is different from that of a weighted avarage of the spectra of the two components, but in the isotropic liquid the construct spectrum and actual spectrum are identical. This lends weight to the idea that the mixture is a unique phase with different chemical environments from those of the individual cholesteric phases.

Myasnikova [76] has examined the infrared spectra of six cholesteryl esters between 25° and 200°C. It was found that in componds with long CH_2 chains, e.g., palmitate, stearate, etc., bands in the 1200–1330 cm^{-1} region, associated by the author with a crystalline chain, disappear at the c–ch or ch–l phase transitions. These bands are not observed for compounds such as cholesteryl formate, acetate, and propionate. They are explained as a combination of an internal mode at 1175 cm^{-1} and lattice vibrations, which is the same explanation as that given for disappearing nematic bands. Consistent with the results reported by Bulkin and Krishnan, changes were observed for vibrations of CH_3 groups. In addition, some changes were observed for C=O vibrations.

In the low frequency region, Chang [77] reported that several cholesteryl esters showed two sharp bands at 18 and 24 cm^{-1} in the Raman spectrum of the crystalline phase. He has discussed the assignment of these bands to lattice vibrations. It is strange that the modes do not appear to be mass sensitive, as one would expect for a true lattice vibration. While this might indicate the presence of an instrumental artifact, it is also possible that only a portion of the molecule is librating at these frequencies.

More recently, Fleury and Vergoten [78] presented spectra of a number of cholesteric compounds in the low frequency region. Their results in the crystalline phase indicate the presence of a number of low frequency lattice vibrations, which are summarized in Table V. They indicate that while Chang's results are not quite accurate as to frequency, bands in the regions he discussed

TABLE V

Lattice Vibration Frequencies of Cholesteryl Esters

Compound	Observed frequency maximum (cm^{-1})
Cholesteryl chloride	10.7, 16.8, 28, 40
Cholesteryl benzoate	7, 14.5, 23
Cholesteryl palmitate	6, 17, 26, 40
Cholesteryl stearate	5.2, 16.8, 25.6, 38

do occur, with frequencies varying by 1–3 cm^{-1} in each compound. It will be interesting to attempt a dynamical model which will allow one to understand these modes.

There have been several reports of infrared optical activity induced by the chirality of the cholesteric medium [79, 80] as well as one report for a twisted nematic phase [24] which yields the same result. Most of the cholesteric phases used in the experiments have been produced by dissolving an optically active solute in MBBA or some other nematic liquid crystal. The theory behind the effect has been discussed [81] as has the instrumentation necessary for making the measurements [82].

Lephardt and Bulkin [83] have employed the cholesteric phase as a medium for fractionating gas chromatograph effluents from carrier gas in order to obtain infrared spectra of these effluents "on-the-fly." In these measurements the cholesteric phase simply serves as an inert solvent, which can also indicate by means of its birefringence the presence of a compound to be studied spectroscopically.

VII. VIBRATIONAL SPECTRA OF LYOTROPIC MESOPHASES

Infrared spectroscopy has been used for many years in the study of soaps, lipids, and related lyotropic phases. Some of this work has been reviewed in the literature previously by Chapman [84]. Particularly noteworthy are numerous papers by Friberg and co-workers [85, 86] which explored the nature of hydrogen bonding in such systems.

More recent efforts have concentrated on the spectra of hydrocarbon chains and chain melting processes in lyotropic phase transitions. Chapman [84] has previously reported that information about chain packing can be obtained from the infrared spectra in the 700 cm^{-1} region, where (1) a sharp band is seen at 720 cm^{-1} for all hydrocarbon chains parallel, (2) two bands for those arrangements in which every second chain is plane perpendicular to the rest, and (3) a broad band is seen for disordered chains.

Larsson [87] has investigated the Raman spectra of a number of long chain lipids and hydrocarbons at various temperatures and in various crystalline and liquid crystalline states. His work has concentrated on the C–H stretching region, where a number of changes are observed that are sensitive to such arrangements. While these materials are not lyotropic mesophases as such, they are discussed in this section because the work forms a background to the spectroscopic studies of phospholipid–water gels discussed below. Some of the changes reported by Larsson are analogous to those seen by Bulkin and Krishnan [75] for cholesteryl esters. Some work on spectra of deuterated lipids is included. Heintze [88] has investigated the spectrum of tripalmitin in some detail, concentrating on CH_2 group vibrations.

Chapman [89] investigated the temperature dependence of the infrared spectrum of phospholipids, noting that at high temperatures the spectra broaden and lose many details characteristic of infrared spectra of crystalline materials. While this work was carried out on pure solid phospholipids, Bulkin and Krishnamachari [90] later reported that when a crystalline phospholipid is mixed with 20% water to form a gel phase, the sharp bands of the crystalline solid are again transformed, so that a spectrum nearly analogous to Chapman's high temperature solid spectrum is obtained. The exception is a sharp band in the infrared spectrum at 1470 cm^{-1}, which is as sharp in the spectrum of the gel as it is in the crystalline solid. When this gel is heated, however, it undergoes a well-known endothermic phase transition, and during this transition the band at 1470 cm^{-1} broadens considerably. A plot of this broadening (either by monitoring half-bandwidth or peak intensity) vs. temperature is a sensitive way to monitor the phase transition. In a subsequent study, this was shown for several lipid–water gels [91] and in a recent report [92] the spectroscopic changes for a mixed lipid system, phosphatidyl choline–phosphatidyl serine, were used to construct the phase diagram at constant water content.

Raman spectra of the phospholipid–water systems have also been investigated by several groups. Lippert and Peticolas [93] studied the spectrum below 2000 cm^{-1}, concentrating particularly on the 1000–1200 cm^{-1} region, where C–C stretching modes are seen. Bulkin and Krishnamachari [91] have also studied this region, as well as the C–H stretching region. A recent report by Brown et al. [94] also focuses attention on the C–H stretching region, with results similar to those of Bulkin and Krishnamachari. All of these data can be summarized by stating that the Raman spectra indicate clearly that the phase transition is a chain melting process, in which an all-trans extended chain is converted to a liquidlike state with many conformations. While it would be interesting to attempt observation of the accordion mode in such compounds, as discussed previously for nematics by Schnur [23], experiments along these lines have not been particularly successful. Brown et al. attribute a mode at 125 cm^{-1} in dipalmitoyl phosphatidyl choline, and 161 cm^{-1} in dipalmitoyl phosphatidylethanolamine, to this vibration. The latter frequency is the same as that seen for pure palmitic acid.

There has been considerable discussion as to the assignments of the vibrations of the polar head groups of the phospholipids. Lippert and Peticolas [93, 95] have proposed some assignments in their early papers, modified slightly in the paper of Brown et al. [94]. A recent paper by Spiker and Levin [96] discusses the assignments in the 2800–3000, 1000–1150, and 700–800 cm^{-1} regions in some detail, in an attempt to understand which vibrations arise from the polar head groups of diplamitoyl phosphatidyl choline. Model compounds are used in this approach.

Colbow and Clayman [97] have investigated the far infrared absorption of the water–lecithin system. A strong absorption is seen in this system. Above 120 cm^{-1} there is virtually no transmission in a 0.1 mm path length. Below 120 cm^{-1}, transmission increases until about 35 cm^{-1}, then begins to fall off again rapidly. The authors noted that with thermal cycling of the sample (heating to 60°C and recooling to 24°), the transmission of the sample increases. An explanation for this observation is presented in terms of the reorientation of the polar head groups above the phase transition. The absorption is attributed to a vibration of the positively and negatively charged groups against one another. A hysteresis effect is presumed to be the explanation for why the spectrum at 60°C and that at 24°C after cycling are the same and show greater transmission than before cycling.

Finally, the Raman measurements have been applied to a study of the spectra of hemoglobin-free erythrocyte ghosts. A preliminary report of such a spectrum was published by Bulkin [98], and more recently Lippert et al. [99] and Wallach [100] reported higher quality spectra, including both lipid and protein vibrations and pointing out possible problems in the previously reported data. It seems clear that such spectra allow one to draw conclusions as to the degree of fluidity of lipid chains in membrane vesicles and other model membrane systems. One can expect this to be an active field in the future.

REFERENCES

1. S. Chandrasekhar and N. V. Madhusudana, *Appl. Spectrosc. Rev.* **6**, 189 (1972).
2. A. Saupe, *Mol. Cryst. Liquid Cryst.* **16**, 87 (1972).
3. R. Freyman and R. Servant, *Ann. Phys.* **20**, 131 (1945).
4. A. S. Zhdanova, L. F. Merozova, G. Peregudov, and M. Sushichinskii, *Opt. Spektrosk.* **26**, 209 (1969).
5. S. Khadzaeva, *Izv. Akad. Nauk SSSR, Ser. Khim.* **11**, 2409 (1969).
6. W. Maier and G. Englert, *Z. Elektrochem.* **62**, 1020 (1958).
7. B. J. Bulkin, T. Kennelly, and W. B. Lok, *in* "Liquid Crystals and Ordered Fluids" (R. Porter and J. Johnson, eds.), Vol. 2, p. 85. Plenum, New York, 1974.
8. J. P. Heger and R. Mercier, *Helv. Phys. Acta* **45**, 886 (1972).
9. N. M. Amer and Y. R. Shen, *Solid State Commun.* **12**, 263 (1972).
10. B. J. Bulkin, F. T. Prochaska, and D. L. Beveridge, *J. Chem. Phys.* **55**, 5828 (1971).
11. W. R. Krigbaum, V. Chatani, and P. G. Barber, *Acta Crystallogr., Sect. B* **26**, 97 (1970).
12. G. Vergoten, *Advan. Raman Spectrosc.* **1**, 219 (1972).
13. G. Vergoten, R. Demal, and G. Fleury, *Trav. Soc. Pharm. Montpellier* **33**, 321 (1973).
14. B. J. Bulkin, D. Grunbaum, and A. Santoro, *J. Chem. Phys.* **51**, 1602 (1969).
15. N. K. Nikolov and P. D. Simova, *Proc. Bulg. Acad. Sci.* **26**, 1013 (1973).

16. N. Kirov and D. Simova, *Mol. Cryst. Liquid Cryst.* **30**, 59 (1975).
17. A. P. Kapustin and O. A. Kapustina, *Sb. Dokl. Vses. Nauch. Konf. Zhidk. Krist. Simp. Ikhor, 1973*, pp. 202–205 (1973).
18. H. Itoh, H. Nakatsuka, and M. Matsuoka, *J. Phys. Soc. Jap.* **34**, 891 (1973).
19. A. Sakamoto, K. Yoshino, U. Kubo, and Y. Inuishi, *Jap. J. Appl. Phys.* **13**, 359 (1974).
20. A. Sakamoto, K. Yoshino, U. Kubo, and Y. Inuishi, *Jap. J. Appl. Phys.* **13**, 1285 (1974).
21. M. Schwartz and L. H. Wang, private communication (1973).
22. J. M. Schnur, *Phys. Rev. Lett.* **29**, 1141 (1972).
23. J. M. Schnur, *Mol. Cryst. Liquid Cryst.* **23**, 155 (1973).
24. B. J. Bulkin, D. Grunbaum, T. Kennelly, and W. B. Lok, *Proc. Int. Conf. Liquid Cryst., 1973; Pramana Suppl.* p. 155 (1975).
25. F. Jahnig, *Chem. Phys. Lett.* **23**, 262 (1973).
26. S. Lugomer and B. Lavrencic, *Solid State Commun.* **15**, 177 (1974).
27. H. Gruler, *Z. Naturforsch. A* **28**, 474 (1972).
28. J. M. Janik, J. A. Janik, and W. Witko, *Acta Phys. Pol. A* **44**, 483 (1973).
29. M. Sorai, *Proc. Int. Conf. Liquid Cryst. 1975*, in press; *Pramana Suppl.* p. 503 (1975).
30. B. J. Bulkin, *J. Opt. Soc. Amer.* **59**, 1387 (1969).
31. B. J. Bulkin, J. O. Lephardt, and K. Krishnan, *Mol. Cryst. Liquid Cryst.* **19**, 295 (1973).
32. E. Guyon, J. E. Proust, and L. Ter-Minassian-Saraga, *Solid State Commun.* **11**, 1227 (1972).
33. C. H. Wang, *J. Amer. Chem. Soc.* **94**, 8605 (1972).
34. B. Lavrencic and S. Lugomer, *Advan. Raman Spectrosc.* **1**, 215 (1972).
35. E. J. Poziomek, T. J. Novak, and R. A. Mackay, *Mol. Cryst. Liquid Cryst.* **15**, 283 (1972).
36. E. B. Priestly, P. S. Pershan, R. B. Meyer, and P. H. Dolphin, *Raman Mem. Vol.* **14**, 93 (1971).
37. E. B. Priestly and P. S. Pershan, *Mol. Cryst. Liquid Cryst.* **23**, 369 (1973).
38. S. Jen, N. A. Clark, P. S. Pershan, and E. B. Priestly, *Phys. Rev. Lett.* **31**, 1552 (1973).
39. S. Jen, Ph.D. Thesis, Harvard University (1975).
40. R. L. Humphries, P. G. James, and G. R. Luckhurst, *J. Chem. Soc., Faraday Trans.* **68**, 1031 (1972).
41. W. Maier and A. Saupe, *Z. Naturforsch. A* **13**, 564 (1958).
42. W. Maier and A. Saupe, *Z. Naturforsch. A* **15**, 287 (1960).
43. J. D. Boyd and C. H. Wang, *J. Chem. Phys.* **60** 1185 (1974).
44. Y. Ohnishi, *Jap. J. Appl. Phys.* **12**, 1079 (1973).
45. P. Simova and N. Kirov, *Phys. Lett. A* **37**, 51 (1971).
46. P. Simova and N. Kirov, *Spectrochim. Acta A* **29**, 55 (1973).
47. P. Simova and N. Kirov, *Advan. Mol. Relaxation Processes* **5**, 235 (1973).
48. P. Simova and N. Kirov, *Acta Phys. Acad. Sci. Hung.* **35**, 183 (1974).
49. A. V. Rakov, *Tr. Fiz. Inst. Akda. Nauk. SSSR* **22**, 117 (1964).
50. S. Lugomer, *Mol. Cryst. Liquid Cryst.* **29**, 141 (1974).
51. W. Rothschild, *J. Chem. Phys.* **53**, 990 (1971).
52. E. T. Samulski, C. R. Dybowski, and C. G. Wade, *Phys. Rev. Lett.* **29**, 340 (1972).
53. B. J. Bulkin and D. Grunbaum, in "Liquid Crystals and Ordered Fluids" (R. Porter and J. Johnson, eds.), Vol. 1, pp. 303–310. Plenum, New York, 1970.
54. N. M. Amer, Y. R. Shen, and H. Rosen, *Phys. Rev. Lett.* **27**, 718 (1970).
55. J. M. Schnur, M. Hass, and W. Adair, *Phys. Lett. A* **41**, 326 (1972).
56. A. Gruger, N. LeCalve, and F. Romain, *J. Mol. Struct.* **21**, 97 (1974).

57. A. Sakamoto, K. Yoshino, U. Kubo, and Y. Inuishi, *Jap. J. Appl. Phys.* **13**, 1691 (1974).
58. B. J. Bulkin and F. T. Prochaska, *J. Chem. Phys.* **54**, 635 (1971).
59. T. Riste and R. Pynn, *Solid State Commun.* **12**, 407 (1973).
60. J. Haller and R. Cox, in "Liquid Crystals and Ordered Fluids" (R. Porter and J. Johnson, eds.), Vol. 1, p. 395. Plenum, New York, 1970.
61. K. K. Kobayashi, *Mol. Cryst. Liquid Cryst.* **13**, 137 (1971).
62. W. G. F. Ford, *J. Chem. Phys.* **56**, 6270 (1972).
63. B. J. Bulkin and W. B. Lok, *J. Phys. Chem.* **77**, 326 (1973).
64. B. J. Bulkin and D. Grunbaum, *J. Phys. Chem.* **79**, 821 (1975).
65. M. J. Billard, M. Delhaye, J. C. Merlin, and G. Vergoten, *C. R. Acad. Sci., Ser. B* **273**, 1105 (1971).
66. W. J. Borer, S. S. Mitra, and C. W. Brown, *Phys. Rev. Lett.* **27**, 379 (1971).
67. M. Evans, M. Davies, and I. Larkin, *J. Chem. Soc., Faraday 2 Trans.* **69**, 1011 (1973).
68. E. Sciesinska, J. Sciesinski, J. Twardowski, and J. D. Janik, *Inst. Nucl. Phys.* 847/PS (1973).
69. R. Hoffman and T. W. Larkin, *J. Chem. Soc., Faraday Trans. 2* **68**, 1729 (1972).
70. C. J. Brot, *Physique* **28**, 789 (1967).
71. V. Dimic and M. Osredkar, *Mol. Cryst. Liquid Cryst.* **19**, 189 (1973).
72. J. M. Schnur and M. Fontana, *J. Phys. (Paris)* **35**, L53 (1974).
73. D. Dvorjetski, V. Volterra, and E. Wierner-Avnear, *Phys. Rev. A*, **12**, 681 (1975).
74. A. S. L'vova and M. M. Sushchinskii, *Opt. Spektrosk. (Akad. Nauk. SSSR, Otd. Fiz-Mat. Nauk)* **2**, 266 (1963).
75. B. J. Bulkin and K. Krishnan, *J. Amer. Chem. Soc.* **93**, 5998 (1971).
76. T. P. Myasnikova, *Sb. Dokl. Vses. Nauch. Konf. Zhidk. Krist. Simp. Ikhor 1973*, p. 206.
77. R. Chang, *Mol. Cryst. Liquid Cryst.* **12**, 105 (1971).
78. G. Fleury and G. Vergoten, *Mol. Cryst. Liquid Cryst.* **30**, 223 (1975).
79. I. Chabay, *Chem. Phys. Lett.* **17**, 283 (1972).
80. E. H. Korte and A. Schrader, *Angew Chem., Int. Engl. Ed.* **11**, 226 (1972).
81. G. Holzwarth and N. A. W. Holzwarth, *J. Opt. Soc. Amer.* **63**, 324 (1973).
82. E. H. Korte and A. Schrader, *Messtechnik* **12**, 371 (1973).
83. J. O. Lephardt and B. J. Bulkin, *Anal. Chem.* **45**, 706 (1973).
84. D. Chapman, "The Structure of Lipids." Methuen, London, 1964.
85. S. Friberg, *J. Coll. Interface Sci.* **37**, 291 (1971).
86. S. Friberg and G. Soderlund, *Kolloid Z. Z. Polym.* **243**, 56 (1971).
87. K. Larsson, *Chem. Phys. Lipids* **10**, 165 (1973).
88. W. Heintze, *Kiel. Milchwirtsch. Forschungsber.* **26**, 3 (1974).
89. D. Chapman, *Ann. N. Y. Acad. Sci.* **137**, 745 (1966).
90. B. J. Bulkin and N. Krishnamachari, *Biochim. Biophys. Acta* **211**, 592 (1970).
91. B. J. Bulkin and N. Krishnamachari, *J. Amer. Chem. Soc.* **94**, 1109 (1972).
92. B. J. Bulkin and N. Krishnamachari, *Mol. Cryst. Liquid Cryst.* **24**, 53 (1973).
93. J. L. Lippert and W. L. Peticolas, *Proc. Nat. Acad. Sci. U.S.* **68**, 1752 (1971).
94. K. G. Brown, W. L. Peticolas, and E. Brown, *Biochem. Biophys. Res. Commun.* **54**, 358 (1973).
95. J. L. Lippert and W. L. Peticolas, *Biochim. Biophys. Acta* **8**, 282 (1972).
96. R. C. Spiker, Jr. and I. W. Levin, *Biochim. Biophys. Acta* **388**, 361 (1975).
97. K. Colbow and B. P. Clayman, *Biochim. Biophys. Acta* **323**, 1 (1973).
98. B. J. Bulkin, *Biochim. Biophys. Acta* **274**, 649 (1972).
99. J. L. Lippert, L. E. Gorczyca, and G. Meiklejohn, *Biochim. Biophys. Acta* **382**, 51 (1975).
100. D. F. H. Wallach and S. P. Verma, *Biochim. Biophys. Acta* **382**, 542 (1975).

EQUILIBRIUM THEORY OF LIQUID CRYSTALS

J. L. Ericksen

The Johns Hopkins University
Baltimore, Maryland

I. Introduction	233
II. Energetics	236
A. Nematics	237
B. Cholesterics	246
C. Smectics	249
III. Equilibrium Equations	257
A. Nematics and Cholesterics	257
B. Smectics	262
IV. Linear Theory	267
V. Nonlinear Problems	270
Ordinary Differential Equations	270
VI. Linear Problems	289
References	293

I. INTRODUCTION

Our purpose is to cover the continuum theory of liquid crystals as it applies to static equilibrium. We try to elucidate concepts and techniques which seem useful and to indicate agreements and disagreements with experiment. We have not found it easy to tie together the diverse researches, but have made some attempt to do so. To keep the length within reasonable bounds, we have omitted much of the analytical detail which is available in research papers.

We assume that the reader has some familiarity with the elements of liquid crystal research. Since we cover only a narrow part, the reader interested in a broader view should consult other expositions. Among those of a broader nature are the reviews of Brown and Shaw [1], Brown et al. [2], Chatelaine [3], Chistyakov [4], Durand and Litster [5], Friedel [6], Friedel and Friedel

[7], Mauguin [8], Saupe [9, 10], and the books of de Gennes [11] and Gray [12]. Of these, de Gennes' book contains the most information on continuum theory. One review, by Stephen and Straley [13], came to our attention only after we had essentially exhausted the time available to complete this article. Covering the physics of liquid crystals, it includes discussions of much of the work on continuum theory. Various types of practical applications are discussed by Fergason et al. [14] and Heilmeier [15], among others. An historical article has been written by Kelker [16].

There are various surveys which concentrate on continuum theory. Of these, the most comprehensive is the Russian article of Aero and Bulygin [17], covering both statics and dynamics. The well-known paper by Frank [18] still serves well as a brief introduction to our subject, the equilibrium theories.

There are various mixtures whose macroscopic behavior is so like that of pure materials that they are well-described by the same type of continuum theory, with the obvious understanding that moduli may vary with concentration. Generally, our discussion refers to materials which are, in this sense, essentially pure.

It seems pertinent to discuss some points which have caused confusion and delayed progress. For this, we restrict our attention to the nematics, where the situation is clearest. The situation seems not to be very different for the cholesterics. The continuum theory now in common usage is, in essence, the creation of Oseen. After a series of preliminary studies, Oseen [19] proposed equations slightly more general than those now commonly used, deducing them from a molecular model. For years, most workers in the field were antagonistic toward this type of theory, Zocher [20–22] being a notable exception. Largely, the objection was that it seemed to presume perfect alignment of molecules. Then, as now, it was believed that alignment is far from perfect. In fact, Oseen's equations are better than their derivation. Frank [18] provided a straightforward macroscopic derivation which helped clear the air. To better understand the situation, we recall that the molecules in such nematics are, generally and roughly, rigid rods with one end indistinguishable from the other. A configuration of one such can then be described by a unit vector $\mathbf{l}$,

$$\mathbf{l} \cdot \mathbf{l} = 1, \tag{1}$$

with $\mathbf{l}$ physically indistinguishable from $-\mathbf{l}$. In terms of any sensible distribution function, we would then expect that, in terms of thermal averages, the odd order moments should vanish, i.e.,

$$\langle l_i \rangle = \langle l_i l_j l_k \rangle = \cdots = 0. \tag{2}$$

For this reason, the discussion of Ericksen [23] is unsound. Optically, these are uniaxial materials and the magnetic susceptibility is also of the uniaxial form, at least in cases where such information is available. Taking into account Eq. (1), it is then plausible that

$$\langle l_i l_j \rangle = S n_i n_j + \tfrac{1}{3}(1-S)\delta_{ij}, \tag{3}$$

where **n**, called the *director*, is a unit vector parallel to the optic axis,

$$\mathbf{n} \cdot \mathbf{n} = 1, \tag{4}$$

and is physically indistinguishable from $-\mathbf{n}$. Here S, the degree of orientation, is a scalar satisfying

$$2S = 3\langle (\mathbf{l} \cdot \mathbf{n})^2 \rangle - 1, \tag{5}$$

$S = 0$ corresponding to random alignment, $S = 1$ to perfect alignment. In principle, it could be negative. As is discussed by Brown *et al.* [2], experimental values for S vary from material to material and with temperature, but commonly lie in the range

$$0.3 \leq S \leq 0.7.$$

Thus alignment is far from perfect. It now seems clear that the fluctuations here indicated are the primary cause of the strong light scattering occurring in these phases. Empirically, the optic axis often varies from point to point, generally in a rather smooth fashion. Oseen's equations are best interpreted as applying to it, rather than the molecular vector **l**. Plausibly, one might ignore other such structure variables if, at fixed temperature, molecular directions are distributed isotropically about the mean and if this dispersion is insensitive to external effects. Apparently, this is more or less true in situations encountered in practice, judging from the fact that the elementary theory is reasonably successful. It is conceivable Eq. (3) would hold, but that higher moments would not have corresponding transversely isotropic forms. It is to be expected that at least fourth-order moments will influence mechanical response so, to some degree, the success of the mechanical theory argues against this. One might expect departures from the simplistic view to become more serious very near orientation defects or phase transitions. For use in the isotropic phase, near the nematic–isotropic liquid transition (the clearing point), de Gennes [*11*, Chapter 5; *24*, *25*] has proposed and discussed the validity of equations which seem appropriate. Rather similar theories have been proposed by Fan and Stephen [*26*], Fisher [*27*], and Sullivan [*28*]. In all of these, **n** and S are replaced by a second-order tensor simply related to $\langle l_i l_j \rangle$. Fisher and Sullivan espouse an old view—namely, that the effects of variable fractional orientation, covered by such theory, are also important in the nematic phase. A similar theory was studied earlier by Hand [*29*] in connection with dilute polymer solutions, wherein the role of the aforementioned fourth-order moments is described. Additional discussion of the molecular view of liquid crystal theory is given by Lubensky [30] and de Gennes [11].

Turning to another point, some workers have found it difficult to accept the idea that, kinematically, orientation and macroscopic motion should be regarded as independently variable. It is true that changing orientation will bring into play forces which may cause motion and vice versa. However, there is no simple relation between the two. In the classical viscometric experiments of Miesowicz [31, 32], samples are subjected to oscillatory shearing motion while **n** is held essentially fixed, in several different directions, by applying strong magnetic fields. Other observations wherein orientation is quickly changed, without producing perceptible motion, are mentioned by de Gennes [11, Chapter 5]. The following presumptions underlie the successful interpretation of static experiments on nematics and cholesterics. Begin with two identical samples. Now stir one, however you like, to induce large shearing strains in one, not experienced by the other. Now form the two into samples of identical size and shape, with identical orientations. This is feasible, though it does require some expertise. Having done this, the two samples will be physically indistinguishable. Some more intimate connection between orientation and deformation might occur in some of the smectic liquid crystals, some of which appear to resemble more closely conventional elastic solids; knowledge concerning this point is still rather scanty. In brief, the better-studied liquid crystals do exhibit a kind of elasticity but not elasticity of the conventional type. There are in the literature proposed theories presuming too strong a link between orientation and deformation. One is the nonlinear theory of Coleman [33], who does express serious doubts as to the applicability of his theory. Wang [34] discusses a rather similar theory. A linear theory of this kind was proposed, though rather quickly abandoned by Martin et al. [35]. The theory of nematics proposed by Lee and Eringen [36] seems to ascribe some independence to orientation and motion, but they are subtly linked in their treatment of material symmetry. After correcting an error in their analysis of a viscometric flow, Shahinpoor [37] finds that the predictions are physically unacceptable.

II. ENERGETICS

Normally, theories of mechanical equilibrium can be based on the notion that equilibrium configurations correspond to extremals of the energy of an isolated system; such configurations are judged stable if the energy is a minimum, unstable otherwise. General experience with this type of approach indicates that it is rather reliable except for one point. When the energy criterion indicates stability, there may in fact be instabilities associated with nonequilibrium phenomena; we later mention cases where this occurs. Within some limits, it is feasible to show, using dynamical theory, that the energy criterion provides a necessary condition for stability of equilibrium. Analyses of this kind can be based on the dynamical theories proposed by Leslie [38, 39]. Reasoning involved in obtaining results of this general kind is discussed by Gurtin [40], for example, though he does not treat liquid crystal theory specifically. For this, it seems necessary to assume that the entropy flux is the heat flux, divided by temperature. Leslie's theory permits departures from this relation for cholesterics. Jenkins [41] has explored an implication of the possible difference, but pertinent experimentation is lacking. Apart from this, the basic idea requires rather conventional adjustments to accommodate idealistic situations where the energy associated with infinite regions is infinite, etc.; we shall not dwell on such matters. With these words of caution, we freely employ the traditional energy approach. In thermodynamic terms, the energy is best interpreted as Helmholtz free energy.

We begin with a discussion of commonly considered contributions to the energy for the three main types of liquid crystals. Equations are first given in direct or cartesian tensor form, which lends itself to physical interpretations. We then indicate variations which are more convenient for problem solving.

A. Nematics

1. Bulk Energies

The Oseen–Frank theory is based on the premise that, with a nematic idealized as being incompressible, the mechanical bulk energy, E_1, is expressible in the form

$$E_1 = \int \mathscr{W}(\mathbf{n}, \nabla \mathbf{n}) \, dv, \tag{6}$$

where the density $\mathscr{W}$ is invariant under arbitrary orthogonal transformations,

$$\mathscr{W}(\mathbf{Q}\mathbf{n}, \mathbf{Q}\nabla\mathbf{n}\mathbf{Q}^T) = \mathscr{W}(\mathbf{n}, \nabla \mathbf{n}), \tag{7}$$

for all $\mathbf{Q}$ with

$$\mathbf{Q}^{-1} = \mathbf{Q}^T. \tag{8}$$

Here $\mathbf{n}$ is the director discussed earlier, indicating the direction of the optic axis. In part, this covers Galilean invariance; rigidly rotating a sample does not change its energy. Invariance with respect to reflections is here considered appropriate since the molecules are symmetric with respect to reflection. Frank [18], but not Oseen [19], argues that $\mathbf{n}$ and $-\mathbf{n}$ are physically indistinguishable, leading to the additional restriction

$$\mathscr{W}(-\mathbf{n}, -\nabla \mathbf{n}) = \mathscr{W}(\mathbf{n}, \nabla \mathbf{n}), \tag{9}$$

a restriction now accepted by most workers. Nehring and Saupe [*42*] discuss forms appropriate when this restriction is not imposed and argue, rather plausibly, that it is appropriate to include some dependence on second derivatives of **n**.

Another assumption, compatible with experience, is that $\mathscr{W}$ is a minimum when **n** is constant. Then, as is discussed by Frank [*18*], the special form

$$2\mathscr{W} = K_1(\nabla\cdot\mathbf{n})^2 + K_2(\mathbf{n}\cdot\operatorname{curl}\mathbf{n})^2 + K_3\|\mathbf{n}\wedge\operatorname{curl}\mathbf{n}\|^2 \\ + (K_2+K_4)[\operatorname{tr}(\nabla\mathbf{n})^2 - (\nabla\cdot\mathbf{n})^2], \tag{10}$$

should represent a good approximation for $\nabla\mathbf{n}$ sufficiently small, the K's being constants or, more realistically, functions of temperature. These moduli are relabeled in various ways in the literature, a common variant being the k_{11}, k_{22}, etc., introduced by Frank [*18*], with

$$K_1 = k_{11}, \qquad K_2 = k_{22}, \qquad K_3 = k_{33}, \qquad K_4 = k_{24}. \tag{11}$$

We employ this form freely, though some analyses extend easily to more general forms of $\mathscr{W}$. Typically, the K's are of the order of 10^{-6} dynes. Data from various sources are tabulated by de Gennes [*11*, Chapter 3]. Roughly, it is the smallness of these moduli which makes it easy to alter orientation. As will be discussed later, some of the moduli can be much larger near smectic–nematic phase transitions. As is discussed by Ericksen [*43*], the K's are subject to the inequalities

$$K_1 \geqslant 0, \qquad K_2 \geqslant |K_4|, \qquad K_3 \geqslant 0, \qquad 2K_1 - K_2 - K_4 \geqslant 0. \tag{12}$$

Oseen [*19*] noted that

$$(\nabla\cdot\mathbf{n})^2 - \operatorname{tr}(\nabla\mathbf{n})^2 = \nabla\cdot(\nabla\cdot\mathbf{n}\mathbf{n} - \mathbf{n}\cdot\nabla\mathbf{n}), \tag{13}$$

so, by the divergence theorem, the last term in Eq. (10), integrated over the volume, can be transformed to a surface integral. This means that it will not influence interior equilibrium equations. Apparently, Frank [*18*] overlooked this result and it was forgotten until it was rediscovered by Ericksen [*44*]. For this reason, some workers concede that the influence of this term is not observable and discard it or use it to rearrange Eq. (10). A commonly occurring form occurs if we set $K_2 = -K_4$ or

$$2\mathscr{W} = K_1(\nabla\cdot\mathbf{n})^2 + K_2(\mathbf{n}\cdot\operatorname{curl}\mathbf{n})^2 + K_3\|\mathbf{n}\wedge\operatorname{curl}\mathbf{n}\|^2. \tag{14}$$

For some types of boundary value problems, Eqs. (10) and (14) can yield different results, as is discussed by Warren [*45*], for example. Differences could also arise in stability analyses, since the value of K_4 influences the energy, in general. Thus some workers prefer to retain the term in question. From molecular theory, Nehring and Saupe [*42*] deduce the relation

$$2K_4 = K_1 - K_2, \tag{15}$$

which, by Eq. (12), requires that $3K_1 \geq K_2$. Data which we have seen are consistent with the latter inequality, except near nematic–smectic transitions, and we have found no alternative method for estimating the numerical value of K_4, K_1 and K_2 being measurable. As is discussed by Frank [18], the moduli K_1, K_2, K_3 measure resistance to three simple types of distortion, called splay, twist, and bending, respectively.

Because of its greater tractability, the special case

$$2\mathcal{W} = K \operatorname{tr}(\nabla \mathbf{n})(\nabla \mathbf{n})^T \tag{16}$$

is used occasionally for qualitative analysis, following Oseen [46]. A calculation, employing identities which hold when

$$\mathbf{n} \cdot \mathbf{n} = 1, \qquad \mathbf{n} \cdot \nabla \mathbf{n} = 0, \tag{17}$$

shows that this form results when

$$K_1 = K_2 = K_3, \qquad K_4 = 0. \tag{18}$$

For some special types of vector fields, fewer restrictions are required for equivalence. In all plane problems and, more generally, whenever $\mathbf{n}$ is normal to a family of surfaces,

$$\mathbf{n} \cdot \operatorname{curl} \mathbf{n} = 0, \tag{19}$$

so the term multiplied by K_2 contributes nothing to $\mathcal{W}$ or to the equilibrium equations. This special form is degenerate in the sense that it is invariant under independent rotations of $\mathbf{n}$ and the coordinates. For this reason, it is sometimes called the *isotropic form*.

Another volume energy encountered is that associated with a static magnetic field $\mathbf{H}$. Generally, a naive estimate suffices, essentially that introduced by Zocher [20]. That is, an applied field is given, satisfying the vacuum equations

$$\nabla \cdot \mathbf{H} = \operatorname{curl} \mathbf{H} = 0, \tag{20}$$

in the region to be occupied by the liquid crystal. It gives rise to a magnetization

$$\mathbf{M} = \chi_\perp \mathbf{H} + \chi_a \mathbf{H} \cdot \mathbf{nn}, \qquad \chi_a = \chi_\parallel - \chi_\perp, \tag{21}$$

where $\chi_\parallel$ and $\chi_\perp$ are constant susceptibilities for the field parallel and perpendicular to $\mathbf{n}$, respectively. The corresponding elementary estimate of field energy is then

$$E_2 = \int \mathcal{W}_M \, dV = -\tfrac{1}{2} \int \mathbf{M} \cdot \mathbf{H} \, dv. \tag{22}$$

For common nematics, $\chi_a > 0$, which means that this energy is minimized with $\mathbf{n}$ parallel to $\mathbf{H}$, and χ_a is of the order of 10^{-7} cgs units; we have found no one worrying about corrections for the induced field. There is the usual dielectric analog

$$\mathbf{D} = \varepsilon_\perp \mathbf{E} + \varepsilon_a \mathbf{E} \cdot \mathbf{nn}, \tag{23}$$

with corresponding energy

$$E_3 = \int \mathcal{W}_E \, dv = -\frac{1}{8\pi} \int \mathbf{D} \cdot \mathbf{E} \, dv, \tag{24}$$

but with a complication. In some practice, as illustrated in the works of Deuling and Helfrich [47] or Gruler and Meier [48], for example, it may not be reasonable to take $\mathbf{E}$ as a fixed vacuum field, independent of $\mathbf{n}$. Rather, one estimates the field induced in the liquid crystal, which does depend on $\mathbf{n}$. There are, of course, situations in which the dependence on $\mathbf{n}$ is slight. In deducing equilibrium equations, the assumption that $\mathbf{E}$ is a fixed function of $\mathbf{x}$, produces equations agreeing with those commonly used. As is mentioned by Schadt [49], ε_a is positive for some nematics, negative for others. In practice, this theory is less reliable than the theory of magnetism. Apart from but not unrelated to this, the nonequilibrium effects of small conductivities, charge injection, etc., can be significant with dc fields. Switching to ac fields suppresses some such problems, but one can then get anomalous alignment, as is discussed by Carr [50], for example. Some instabilities of a more or less static nature are not predictable from equilibrium theory. Electrohydrodynamic theory, outside the range of our discussion, can cope with such effects. Even within the confines of equilibrium theory, there are complications. This is an effect vaguely similar to piezoelectricity, first discussed by Meyer [51]. For it, de Gennes suggests the name "flexo-electric effect," and this name is being used by various workers. That is, if $\mathbf{n}$ somehow becomes nonuniform, it may produce a polarization, hence a space charge. Here the energy does not neatly split into a sum of mechanical and electrical parts, because of this interaction. For the combined energy $\mathcal{W}_c$, the proposal is that

$$2\mathcal{W}_c = K_1 [\nabla \cdot \mathbf{n} - (e_{11}/K_1)\mathbf{E} \cdot \mathbf{n}]^2 + K_2 (\mathbf{n} \cdot \text{curl } \mathbf{n})^2$$
$$+ K_3 \|\mathbf{n} \cdot \nabla \mathbf{n} - (e_{33}/K_3)(\mathbf{E} - \mathbf{nE} \cdot \mathbf{n})\|^2, \tag{25}$$

here omitting the usual term associated with K_4. Helfrich [52] suggests that this underlies the effect observed by Haas et al. [53] and proposes a simple experiment. Corresponding experiments by Schmidt et al. [54] conform to theory and give for the modulus e_{33} a positive value of the order of 10^{-5} dyne$^{1/2}$ for MBBA (p-methoxybenzylidene-p'-n-butylaniline). Meyer [51] mentions several phenomena likely to be associated with this effect. Deuling [55] proposes a method for measuring e_{11} and e_{33}, employing both electric and magnetic fields. Helfrich [56] discusses an associated electrooptical effect. The flexoelectric effect can produce periodic orientation patterns somewhat like those induced by electrohydrodynamic effects, as is discussed, e.g., by Fan [57]. Derzhanski and Petrov [58] discuss a model calculation of the flexoelectric moduli. These and other electrical phenomena are discussed in more detail by Helfrich [59]. Since **H** is an axial vector while **E** is polar, a precise magnetic analog of the flexoelectric effect is not to be expected, but some similar effect might occur. Possibly, some such effect has been seen by Cheung *et al.* [60]. Henceforth, we restrict our attention to effects of magnetic fields or to effects of electric fields governed by the simple dielectric analogy.

Optical theory underlying the interpretation of most observations involves using a constitutive equation of the type given by Eq. (23) for **D**, with the usual assumption that the magnetization produced by radiation is negligible. Of course, one must venture into fluctuation theory to treat scattering effects. There are optical effects such as the Faraday rotation observed by Schenz *et al.* [61] which are outside the range of this theory. However, for most purposes, this theory or the implied geometrical optics counterpart seems adequate. References to work on optics will be mentioned later, in the discussion of cholesterics. When **n** = const., the optical theory is of the simple "textbook" variety. In most cases of interest, **n** is not uniform and *a priori*, may not even be known. Even if it is known, the corresponding physical optics calculations can be formidable. Consequently, the optics used in practice incorporates some errors not inherent in the basic optical theory.

Effects of a gravitational field can be included with the usual estimate

$$E_4 = \int \rho \phi \, dv, \tag{26}$$

where ϕ is the gravitational potential, a given function of position, and ρ is the (constant) mass density. With the small samples commonly used, this is neglected in most analyses. There are exceptions, e.g., in the work of de Gennes [62], who studies instabilities involved when one fluid layer floats on another.

Adding up contributions of the type indicated, we see that the total volume energy E tends to be of the form

$$E = \int U(\mathbf{n}, \nabla \mathbf{n}, \mathbf{x}) \, dv. \tag{27}$$

2. Interfacial Effects

It is now well known that another medium, in contact with a liquid crystal, tends to produce a characteristic orientation in the liquid crystal at the interface. For example, Grandjean [63] reports observations of several cases of contact of liquid crystals with solid single crystals, one case having been observed earlier by Mauguin [64]. For cases where the interface is a well-defined crystallographic plane, the crystal imposes an orientation at the interface, bearing some simple relation to the lattice vectors of the solid. In some cases there are two or more preferred or "easy" directions but, as one might expect, the possibilities are discrete. On a free surface of MBBA, **n** makes an angle of about 15° with the normal to the interface, according to the observations of Bouchiat and Langevin Cruchon [65]. With a free surface of PAA (*p*-azoxyanisole), **n** lies tangent to the interface, according to Naggiar [66]. In either of the latter cases, there are clearly infinitely many easy directions. When PAA is in contact with glass, prior treatment of the glass has an influence. Sometimes Chatelaine [67] is credited with having discovered that sufficient rubbing of the glass produces an orientation in the plane of the interface, parallel to the direction of rubbing. Much the same effect on orientation of dyes had been noted earlier by Zocher and Coper [68], and Zocher [21] clearly describes using the technique on nematics. As will be discussed later, there is some theoretical reason to think that this is at least partly associated with undulations or scratches produced by the rubbing, a factor which might not first occur to one. Zocher [22] mentions observations which support this view. The more recent observations of Berreman [69] and Wolff *et al.* [70] provide more support.

That mechanisms more of a physicochemical nature can also play an important role is rather clearly established by Proust *et al.* [71]. At present, it seems that the mechanical effects can dominate in some cases, physicochemical mechanisms in others. A similar orientation or one tilted relative to the surface can be obtained by evaporating thin metallic films on glass, according to Janning [72] and Guyon *et al.* [73]. According to Dixon *et al.* [74], nonuniformity of the film helps produce such effects. As is emphasized by Kahn [75], oblique orientations, in conjunction with fields with $\varepsilon_a < 0$ or $\chi_a < 0$, can produce uniform orientations, a fact of some importance for certain devices. To obtain an orientation normal to the interface, one can treat the glass with suitable acids or detergents, as is discussed by de Gennes [11, Chapter 3]. Friedel [6, p. 346] suggested that acids might sometimes roughen the glass. From theory, it is not trivial to predict what orientation should result for a surface so roughened. Possibly, it would promote orientation roughly parallel to the mean normal to the wall. Our understanding of why such interfacial effects occur does leave something to be desired. Some of their physical chemistry is studied by Uchida *et al.* [76], and Rapini [77] discusses some possible mechanisms. In a study aimed at checking earlier speculations, Haller [78] finds no easy correlation between alignment properties of a nematic–solid interface and its wetting properties.

One possibility is to use such empirical information to set boundary values for **n**, presuming that other influences do not alter it; we follow de Gennes

Equilibrium Theory of Liquid Crystals   243

[*11*, Chapter 3] in calling this the condition of *strong anchoring*. This simple assumption is very commonly made and seems to yield satisfactory predictions in many cases. With it, the distinction between Eqs. (10) and (14) disappears. Schmidt *et al.* [*54*] note that the anchoring of MBBA to a lecithin-coated glass is quite weak. According to Haller [*78*], the strong anchoring condition may fail at a glass–MBBA interface. Zocher [*21*] reports that observable departures from strong anchoring can be induced, using strong magnetic fields, when the orientation is induced by rubbing. Thus some caution should be exercised. An alternative obtains from introducing an interfacial energy, for which one needs some sort of constitutive equation. For cases where the adjoining medium is an isotropic fluid or a relatively rigid isotropic solid, the trend has been to follow the lead of older writers such as Friedel [*6*] and Oseen [*46, 79*], introducing a surface energy of the form

$$\mathscr{E} = \int f[(\mathbf{n}\cdot\mathbf{v})^2]\, dS, \tag{28}$$

where dS is the usual element of area and $\mathbf{v}$ the unit normal to the interface. When it is a fluid at constant pressure p_0, one should, in principle, account for work which it does, as is commonly done in ordinary hydrostatics. Usually, it is equivalent to interpret stress on the liquid crystal as the actual stress, less the stress associated with this pressure, so, effectively, we set $p_0 = 0$. Exceptions can occur when only part of the boundary involves contact of this type. In most researches on liquid crystals, this is of no import, so, with these words of caution, we gloss over this point. From observations such as are mentioned above, we can estimate for what angle between $\mathbf{n}$ and $\mathbf{v}$ $\mathscr{E}$ is a minimum, in special cases. This is enough to permit some qualitative understanding of some of the observed droplet shapes, as is discussed by Chandrasekhar [*80*], if one is willing to accept oversimplified assumptions concerning interior orientations. Actually, interior orientation patterns can be rather complex, so such simple reasoning can mislead. Press and Arrott [*81*] discuss numerical calculation of the orientation of MBBA droplets floating in water. Obviously, some information or hypothesis concerning f is needed for quantitative analysis. If the argument changes but little from its minimizing value, expanding f about this gives, as a first approximation,

$$f = A_1(\mathbf{n}\cdot\mathbf{v})^2 + A_2, \tag{29}$$

where A_1 and A_2 are constants. Some workers employ this, others leaving the form of f unspecified. Empirical determinations of f seem not to exist. In cases where the strong anchoring assumption works well, it would seem that f must increase substantially when we attempt to move the argument from its minimizing value. Conversely, when f is nearly constant, use of the strong anchoring assumption seems very dubious. As is discussed by Gray [*12*, Chapter VI], the rather different measurements of surface tension of PAA by Ferguson and Kennedy [*82*] and Naggiar [*83*] yield values which are not markedly different, but with different and contrary temperature dependencies; the data of Schwarz and Moseley [*84*] are in better accord with those of Naggiar. Gray suggests that the difference may stem from differences in orientation, which is neither controlled nor observed in these experiments. Perhaps this indicates that one should be cautious in using the strong anchoring condition at a free surface of PAA. On the other hand, the better-defined observations of Naggiar [*66*] seem not to be inconsistent with strong anchoring. In brief, we lack reliable experimental information concerning the form of f.

Using optical methods, Langevin and Bouchiat [*85*] study the interface between isotropic and nematic phases of MBBA. They find that **n** lies in the interface, that the interfacial energy is nearly isotropic and they measure values of the interfacial tension.

There are cases where Eq. (28) cannot be correct, though the strong anchoring condition may then suffice. From the previous discussion of contact with single crystals, it is clear that the interfacial energy should then change if we hold the normal component of **n** fixed, but vary the tangential component. Generalizations of this type are discussed by Rapini [*77*] and Ericksen [*86*].

It is certainly conceivable that gradients of **n** might influence the interfacial energy density, as they do the bulk energy; Dubois-Violette and Parodi [*87*] propose a theory of this type. However, it seems to need revision. One contribution, the normal derivative of the surface normal, seems to us meaningless. Also, if one examines the variation of their energy, it is not clear what could balance terms arising from their assumed dependence on the normal derivative of **n**, unless one introduces into the bulk energy dependence on higher derivatives of **n**, as suggested by Nehring and Saupe [*42*]. Revisions of theory to eliminate such terms would yield equations much like those employed in modern theories of thin elastic shells, such as are discussed by Naghdi [*88*] or the theories of membranes by Jenkins [*89*]. They also resemble the theory of thin films of smectic C liquid crystals proposed by Meyer and Pershan [*90*], who suggest that unusual polarity effects may be important at interfaces. Yun [*92*] also notes the possibility of polarity effects. It can be shown that the surface integral deriving from Eq. (13) depends only on tangential derivatives of **n**. Helfrich [*91*] discusses types of membrane theory of interest for describing lipid bilayers.

At interfaces between two different liquid crystals, or at a singular surface in one, the two orientations could differ, suggesting yet another modification of theory. Rault [*93*] briefly discusses a case of this type. If, for whatever reason, there is an abrupt change of orientation across some curve lying on the interface, it would seem pertinent to introduce corresponding edge energies. Here, we generally restrict our attention to cases where either the strong anchoring condition or Eq. (28) applies.

3. IMPERFECTIONS

Nematic liquid crystals derive their name from the fact that, commonly, their orientations are undefined on curves or small tubular regions, now called *disclinations*. Numerous observations of these, including rules for their attraction, repulsion, dissociation, or recombination, are discussed by Friedel [6]. Oseen [46] gave the first analysis of these, Frank [18] providing sketches of some of the predicted orientation patterns resembling those observed. As estimated using Eq. (10) or Eq. (14), the energy associated with the orientation singularity is infinite. Oseen [46] attempted to remedy this by modifying his theory to allow for compressibility effects; a rough equivalent is to consider the core hollow. For this, it would seem more reasonable to introduce as a variable some measure of the orientation dispersion such as the degree of orientation S [Eq. (5)]. Said differently, here is a place where theories allowing for fractional orientation might be superior. As is discussed by Saupe [94], a molecular theory suggests that the moduli in Eq. (10) are of the form

$$K_\alpha = C_\alpha V^{-7/3} S^2, \tag{30}$$

where V is the molar volume and the C's are certain temperature independent constants. Further, Saupe and Nehring and Saupe [95] note that one implication, the temperature independence of ratios of these moduli, is in rough accord with experiment. This is certainly not true near nematic–smectic transitions, and the calculated ratios are not very close to those measured. Also χ_a or ε_a is expected to be proportional to S; Luckhurst and Smith [96] discuss the observed variation of $K_2/\chi_a S$ with temperature. The natural assumption that $S = 0$ on the disclination curve can then help remove the objectionable singularity, at the expense of complicating the theory. Fan [57] discusses the theory allowing S to vary. A simpler alternative is to consider the core as having a small but finite radius, as is commonly done in the elastic dislocation theory for solids, using Eq. (10) or Eq. (14) only outside this. This approach is rather commonly used. Physically, the core might be conceived as having transformed to the isotropic phase, a possibility explored by Ericksen [97]. Mathematically, it is assumed that whenever $\mathscr{W}$ exceeds a critical value $\mathscr{W}_0$, the material undergoes phase transformation. A likely alternative, mentioned by Lehmann [98, pp. 47–48] is that it is filled with impurities. Cladis and Kléman [99] discuss a possibility which we shall later discuss further, of a nematic core, within which **n** is, roughly, tangent to the disclination curve. There might, but need not, be a discontinuity in **n** in this case. If **n** has a finite discontinuity on a surface near the core, it would seem appropriate to introduce a corresponding surface energy, as is discussed by Rault [*93*]. He discusses how such textures might relate to irregularities observed in some disclination patterns. In brief, any mathematical description of the core is speculative, and it is becoming increasingly clear that not all cores are alike. One consequence is that we can expect to find different solutions of the equilibrium equations which, nominally, could represent more or less the same physical situation. Indeed, imperfection theory is now encountering some such puzzles. Simpler solutions of this type are of some interest for illustrative purposes and for making some order of magnitude estimates, as is covered in the discussion of de Gennes [*11*, Chapter 4]. Examination of the discussion of Friedel [*6*] indicates that his empirical rules involve only the topology of the vector field **n**. Topological considerations can be of real help in gaining an understanding of disclinations, as is discussed by Kléman and Friedel [*100*], Friedel and Kléman [*101*], Bouligand and Kléman [*102*], Rault [*103*] and the Orsay Group [*104*], for example. They might well be regarded as the prerequisite for setting up problems to be solved using the continuum equations. However, in themselves, they involve little of the continuum theory. For this reason, we omit discussion of them.

Point defects also occur. According to de Gennes [*11*, Chapter 4], this was recognized by Friedel, but they are largely ignored in earlier theoretical work. Again, there are vagaries concerning the description of their core, but, as is discussed by de Gennes [*11*, Chapter 4] and Saupe [*105*], recent work provides some understanding of them; they can occur in capillaries, with normal orientation at the wall, for example. As is discussed by Meyer [*106*], they can easily occur on the surface of drops.

In summary, it does not seem worthwhile to record any specific proposals for core energies, though one should bear in mind the possibility of their occurrence, where they will be counted as strong singularities by the elementary theory. Later, we will say a bit more about the analyses of imperfections. We find it difficult to add much to de Gennes' [*11*, Chapters 4, 6, 7] summary of the numerous researches on imperfections in nematics, cholesterics, and smectics, combined with what is in Saupe's [*105*] exposition of defects. Accordingly, our coverage is far from complete.

B. Cholesterics

An abbreviated discussion of cholesterics will suffice, since their theory is quite similar to that of nematics. Again, the bulk mechanical energy is assumed to be given by a constitutive equation covered by Eq. (6), but with two important differences. Here, molecules are not symmetric with respect to reflections, so we require $\mathscr{W}$ to be invariant only with respect to proper orthogonal transformations, replacing Eq. (8) by

$$\mathbf{Q}^{-1} = \mathbf{Q}^T, \quad \det \mathbf{Q} = 1. \tag{31}$$

Second, $\mathcal{W}$ attains a minimum, not for **n** constant, but for **n** of the form

$$\mathbf{n} = (\cos \tau x_3, \sin \tau x_3, 0), \tag{32}$$

where τ is a constant depending on the material and temperature. Values of τ of the order of 10^5 cm^{-1} are common, though much lower values, approaching zero, can be obtained by mixing left- and right-hand varieties. Low values of τ can also be obtained by mixing a little chiral matter with a nematic to make what is, effectively, a cholesteric. The preferred axis, here taken as the x_3-axis, is variously referred to as the helical axis, twist axis, or chiral axis.

Often, this twist is recognizable in variously distorted configurations. As is discussed by Kléman [107], it is both possible and useful to define surfaces to which **n** is tangent, their normals indicating the chiral axes. Then **n** and this axis determine a mutually perpendicular unit vector, providing an orthonormal basis neatly related to the texture. As is also emphasized by Kléman, similar constructs can be useful in analyzing defects occurring in nematics or cholesterics.

It is a curious coincidence that this and $\mathbf{n} = \text{const.}$ are the only configurations devoid of singularities which can satisfy the equilibrium equations for arbitrary forms of $\mathcal{W}$, according to Ericksen [108]. Further, for either, the value of $\mathcal{W}$ is independent of position. It thus seems quite reasonable that minimum energy configurations should be limited to one of these two types. For a cholesteric, we might have $\tau = 0$ at a particular temperature, or a nematic might have its minimum energy in the twisted form. In the latter case, the energy should remain a minimum if τ is replaced by $-\tau$, as is discussed by de Gennes [11, Chapter 1]. Saupe [10] discusses evidence of an apparently different, optically isotropic texture which can occur at higher temperatures, perhaps requiring some different theory. Durand and Litster [5] discuss pretransition effects near the cholesteric–isotropic fluid transition, presenting some evidence that theory appropriate for this is much the same as for nematics.

The theory is not quite as well developed or tested as is the theory of nematics; vestiges of the twisted structure remain to complicate analysis and interpretation of most experiments. Concerning the detailed form of $\mathcal{W}$, there have been rather different proposals. Again more or less in accord with Oseen, Frank [18] proposes

$$2\mathcal{W} = K_1 (\nabla \cdot \mathbf{n})^2 + K_2 (\mathbf{n} \cdot \text{curl } \mathbf{n} + \tau)^2 + K_3 \|\mathbf{n} \wedge \text{curl } \mathbf{n}\|^2$$
$$+ (K_2 + K_4) [\text{tr}(\nabla \mathbf{n})^2 - (\nabla \cdot \mathbf{n})^2], \tag{33}$$

this being a suitably invariant function which is quadratic in the gradients of **n**. It is this which has been used for essentially all analyses, so we only record it. There is one misleading feature of the direct notations used in

Eq. (33). The algebraic signs of **n**·curl **n** and τ get reversed by an improper coordinate transformation. A common practice is to employ right-handed coordinate systems. For simplicity, we assume $\tau \geq 0$ in the few places where the sign matters. Jenkins [109] notes that, if this form is to have a minimum for **n** given by Eq. (32), the moduli should satisfy

$$K_2 + K_4 = 0,$$

$$K_1 \geq 0, \quad K_2 \geq 0, \quad K_3 \geq 0. \tag{34}$$

Also, he suggests that it seems more reasonable to develop a form valid for small perturbations from the twisted state, rather than merely assuming $\mathscr{W}$ quadratic in the gradients. He produces such a form. Rough order of magnitude estimates given by de Gennes [11, Chapter 6] suggest that the difference is generally negligible, but this does involve educated guess work concerning moduli. Nehring and Saupe [42] express the view that, since **n** is not constant in the ground state, higher derivatives of **n** have a greater chance of influencing the energy and, occasionally, their proposal is employed. Thus far, predictions of Eq. (33) seem to be in reasonable accord with experiment. For qualitative work, an analog of Eq. (16) is available, viz.

$$2\mathscr{W} = K \text{ tr } \nabla\mathbf{n}\,\nabla\mathbf{n}^T + M(\mathbf{n}\cdot\text{curl }\mathbf{n}+\tau)^2. \tag{35}$$

For magnetic fields, Eqs. (21) and (22) are commonly used. Here χ_a is commonly negative, the field energy being minimized when **n** is perpendicular to **H**. Also, χ_a tends to be much smaller than for nematics, of the order of 10^{-9} cgs units, while the K's are comparable to those for nematics. If $\varepsilon_a > 0$, one can, with some caution, produce alignment with electric fields. If ε_a and χ_a are both negative, one can employ orthogonal electric and magnetic fields, as is discussed by Carr et al. [110]. Larger and positive values of χ_a can be obtained by the aforementioned mixing of nematics with chiral matter. In principle, with this symmetry, an electric field could produce magnetization, or a magnetic field could polarize, but such effects seem not to have been observed. We note that, again, Eq. (27) covers commonly used forms of the bulk energy.

For optics, Eq. (23) is again used, if not a geometrical optics counterpart. Commonly, for both nematics and cholesterics, **n** varies with position. The corresponding variation in the dielectric tensor gives rise to the striking optical phenomena which are so characteristic of liquid crystals and so often exploited in devices. There is a rather large literature on this topic, both theoretical and experimental, much of it dealing with the case where **n** twists, as indicated by Eq. (32), in a slab of effectively infinite extent, or with small perturbations therefrom. In terms of the equations of physical optics, this is perhaps the simplest case where **n** is not uniform. As is to be expected, it is the twist which gives the cholesterics their enormous rotatory power, and the periodicity does give rise to Bragg reflection. Circular dichroism and selective reflection occur, and transmission involves significant angular dependencies. Agreement between theory and experiment has been checked in some detail and is, on the whole, quite good. A summary of work of this kind is given by de Gennes [*11*, Chapter 6]. Other work is briefly summarized by Prasad and Madhava [*111*]. There are various papers not cited by these writers, e.g., Chou et al. [*112*], Coates and Gray [*113*], Dreher and Meier [*114*], Ennulat [*115*], Fergason [*116*], Teucher et al. [*117*], or van Doorn and Heldens [*118*]. The reader interested in this topic should not find it difficult to extend the list. These matters are somewhat peripheral to the equilibrium theories of concern here. It is important to have at hand reliable optical theory and to know how consistent is the interpretation of **n** as the optic axis, since this is involved in the interpretation of experiment. There are, of course, cases where naive interpretation, or unduly crude analysis of the optics can mislead. Often, threshold effects are used to estimate moduli. It goes without saying that some monitors are more sensitive than others to changes occurring near the threshold. As is analyzed by van Doorn [*119*], there are cases where the nominal optical threshold differs significantly from the actual, which is, in the case he treats, more influential on capacitance measurements. This tendency of "optical thresholds" to differ from the actual is mentioned by various other writers. With splay, bend or other untwisted distortions commonly occurring in nematics, it has not been customary to explore their physical optics in great depth. The optics of individual patterns is, of necessity, treated, and in a manner deemed adequate for the task at hand.

Interfacial effects, imperfections, etc., do exist, as for nematics, but have a more complicated appearance. All indications are that the basic theory for them should be essentially the same as for nematics. Typical defects and textures are covered in the discussions of de Gennes [*11*, Chapter 6] and Saupe [*105*], both of whom present photographs.

C. Smectics

Generally, theories of smectics are even less well tested, being just in the initial stages of development. In this situation, it seems worthwhile to review some of the older thinking, which seems relevant at least to the classical smectics, or smectics A in the classification of Sackmann and Demus [*120*]. In fact, if we are interested only in the equilibrium behavior of smectics A, little else is needed. Here, smectics are viewed as lower temperature phases, relative to the nematic or cholesteric mesophases, wherein molecules form identical coherent layers, which easily glide over each other. The top of a layer is no different from the bottom. Within a layer, molecules can move freely, as in a liquid, and, on the average, orient themselves normal to the surface. The layers tend to be of a certain fixed thickness which is, by macroscopic standards, infinitesimal. Such a fluid is most relaxed when the surfaces are (parallel) planes. They are commonly seen in textures where the layers are curved, involving imperfections. Textures occurring in the various types of liquid crystals are discussed at length by Bouligand [*121–123*]. Such imperfections are restricted to points or curves, not spread out on surfaces. At least roughly, these are thoughts which seem to underlie attempts to form theory, at least up to the time it was discussed by Frank [*18*]. More recently, it has been found that there are smectics with less fluidity and with the molecular axis tilted away from the layer normal. Such modifications encountered in smectics B and C clearly do not fit the older scheme and are ignored for the moment. Generally speaking, the smectic phases are not optically uniaxial, as is discussed by de Gennes [*11*, Chapter 6], the only likely exceptions being the smectics A.

In the continuum limit, we replace the discrete layers by a generally smooth family of surfaces. The assumption that the layer thickness is constant translates into the condition that these are parallel surfaces. Such a family is determined by any surface in the family; one marks off equal distances on its normals to get the other surfaces. Clearly, this is an unusual type of long-range interaction. Unless they are parallel planes, the family will, if it covers enough space, necessarily exhibit singularities, places where curvatures of these surfaces become infinite. Only in exceptional cases are the loci of these singularities points or curves. When they are curves, the surfaces are Dupin cyclides, corresponding to the often-mentioned focal conic textures. The cholesterics have a pseudo-layering, associated with the periodicity of their preferred configuration and also exhibit these focal conic singularities. Thus it is not surprising that, in older work, there was a tendency to lump them with the smectics. Friedel [*6*] as well as Friedel and Friedel [*7*] forcefully argue that the resemblance is only superficial, and it is now generally accepted that the cholesterics are much more like twisted nematics. The Dupin cyclides are characterized by the fact that their principal curvatures are constant along the corresponding lines of principal curvature. Relevant differential geometry is covered in books on the subject, e.g., Eisenhart [*124*] or Hlavatý [*125*]. Combining knowledge of such geometry with results of observation, older writers such as Friedel and Grandjean [*126*], Friedel [*6*], and Bragg [*127, 128*] argued that the parallel surfaces should be either parallel planes or these Dupin cyclides. In a much more recent work, Bouligand [*121*] discusses at length imperfections in smectics and cholesterics, pointing out that more refined observations do not quite accord with the focal conic description. He makes it plausible that dislocations occur to help accommodate distortions not compatible with the picture of parallel surfaces. There are, then, some flaws in these classical views.

To the extent that these classical views apply, the theory is largely a matter of geometry and, as is noted by Frank [*18*], a theory of their mechanics is not really required. However, partly to correct some of Oseen's views, he does present a revised and sounder version of Oseen's theory. The premise is that Eq. (10) applies, but with **n** now constrained to be the unit normal to a family of parallel surfaces. Formally, this requires that

$$\mathbf{n} \cdot \operatorname{curl} \mathbf{n} = \mathbf{n} \wedge \operatorname{curl} \mathbf{n} = 0 \Rightarrow \operatorname{curl} \mathbf{n} = 0. \tag{36}$$

Quite reasonably, Frank [*18*] interprets this as meaning that two moduli, K_2 and K_3 grow very large, idealistically becoming infinite, so that violations of Eq. (36) would create huge energies. If we are to avoid additional constraints, K_4 should also become infinite in such a way that

$$K_2 + K_4 \to \overline{K} \tag{37}$$

where $\overline{K}$ is finite, while K_1 should remain finite. One thus obtains

$$2\mathscr{W} = K_1(\nabla \cdot \mathbf{n})^2 + \overline{K}[\operatorname{tr}(\nabla \mathbf{n})^2 - (\nabla \cdot \mathbf{n})^2], \tag{38}$$

where, from Eq. (12),

$$K_1 \geqslant 0, \qquad \overline{K} \leqslant 2K_1 \tag{39}$$

Interestingly, Frank here comes close to anticipating some recent researches. From molecular theory, Kobayashi [*129*] and McMillan [*130*] concluded that nematic–smectic A phase transitions would, in certain cases, be of second order. Pursuing this, de Gennes [*131*] deduced that, if the temperature T is slightly above the transition temperature T_0, then, as $T \to T_0$, K_2 and K_3 should grow like $(T-T_0)^{-\nu}$, $\nu \approx \frac{2}{3}$. On the other hand, K_1 should depend only weakly on T. Like many, he discards the term involving K_4. As is discussed by Cladis [*132*], impurities can shift the measured value of ν noticeably, as can errors encountered in data reduction. For cholesterics, τ decreases rapidly near a smectic transition. Alben [*133*] argues that K_2 should become large, with $K_2\tau$ nearly constant. Several experiments confirm de Gennes' predictions, insofar as the relative behavior of the moduli are concerned. Cheung et al. [*60*] confirm the behavior of K_1 and K_3 for CBOOA (*p*-cyanobenzylidene *p'*-octyloxyaniline). Leger [*134*] also confirms the behavior of K_3. Delaye et al. [*135*] check K_2. If we stopped here, the value of $\nu \approx \frac{2}{3}$ would look good. The behavior of ν for BBMA (*p*-butoxybenzylidene-*p*-*d*-methylbutylaniline) is somewhat anomalous according to Cheung and Meyer [*136*]. For it, they find $\nu \approx 1$. According to Torza and Cladis [*137*], the transition of CBOOA is actually weakly of first order; variations in measured values of ν occur which they attribute to differences in purity.

A more classical value of $v \approx \frac{1}{2}$ attains for a clean sample, v being larger for dirtier samples. In any event, it is clear that, in various cases, we can pass rather smoothly from typical nematics to smectics A, in the process getting some moduli which are much larger than normal. Clearly, this has important ramifications with respect to static mechanical phenomena, rather independent of the details of growth. For a cholesteric, Pindak et al. [138] find the behavior suggested by Alben, K_2 and K_3 increasing as for a nematic, with $v \approx \frac{2}{3}$. Pretransition anomalies are not restricted to smectic A–nematic transitions. For further discussion of other possibilities, we refer the reader to de Gennes [139].

There is a rather extensive experimental and theoretical study by Cladis and Torza [140] of nematic–smectic A phase mixtures. If, as with CBOOA, the phase transition involves small latent heat and if the nematic is held firmly in a bent state, then some cooling can generate mixtures having a striped, virgule, or honeycomb texture. In the above, we assumed, rather tacitly, that the nematic phase occurs at higher temperatures than the smectic. As a rule, this is true, but Cladis [141] notes a case where the reverse obtains.

With the strong constraints, it is somewhat tricky to ascertain which parallel surfaces correspond to extremals of the energy given by Eq. (38), and Frank does not explore this. Recently Geurst [142, 143] has attempted this, concluding that they are the Dupin cyclides, and noting that

$$(\nabla \cdot \mathbf{n})^2 = K_M^2, \qquad (\nabla \cdot \mathbf{n})^2 - \mathrm{tr}(\nabla \mathbf{n})^2 = K_G \tag{40}$$

where K_M and K_G are, respectively, the mean and gaussian curvatures of the parallel surfaces. The result might be correct, but we are unable to follow the analysis. Leaning on the classical theory of thin elastic shells, with adaptations tending to make them behave more like smectic A monolayers, Kléman [144] attempts a study of what deformations a monolayer or stack of layers might prefer. With this theory, which is not precisely equivalent, he finds that it is exceptional for the layers to be Dupin cyclides, although an exceptional layer can be. As he notes, observations favor certain types of cyclides, as do his equations. Various writers, e.g., Friedel [6, pp. 377–378] mention that moderate electromagnetic fields, etc., have little effect on what we now call smectics A. Order of magnitude estimates such as are discussed by Parodi [145] indicate that, at the least, it requires ingenuity to find any transition involving enough change to be observed. Thus Frank's proposal [Eq. (38)], seems reasonable as a prescription for calculating energies, barring those associated with imperfections such as occur in the focal conics. If one accepts this model, it is still not very evident what boundary conditions or other side conditions are physically and mathematically appropriate, so there is some difficulty in predicting which orientation pattern will occur.

The indicated tendency of the smectics to exhibit these imperfections might be of deeper significance. Let us accept these classical views. If we consider the parallel plane structure filling all space, then, clearly, there is no admissible variation of the structure which can reasonably be considered infinitesimal everywhere. Landau and Lifshitz [*146*, Chapter XIII] argue against the existence of "two-dimensional" fluids somewhat similar to this because their analysis implies that infinite samples would be unstable with respect to infinitesimal disturbances. Perhaps it is important for the survival of the smectics that they meet some changes with much more resistance than others. On the other hand, this theory has awkward features which are leading to its replacement. With the assumed rigidity of structure, it is hard to move on to develop good theories of motion or to sensibly analyze dislocations. De Gennes [*147*] proposes that we let the thickness of the layers vary a bit, which gives a much more flexible structure. The molecular direction is assumed to remain essentially normal to the layer. He also emphasizes another point, which plays no explicit role in the older work. He assumes that matter in one layer can move to another, an idea deriving from Helfrich [*148*]. Decrease in layer thickness could be accomplished by letting the molecules tilt relative to the normal. However, Ribotta et al. [*149*] find that this occurs only if there is a compressive stress exceeding a certain threshold. Linear (dynamical) equations, incorporating the indicated changes, are proposed by de Gennes [*147*]. He does acknowledge the aforementioned objection of Landau and Lifshitz, but argues that it is not too serious. Effectively, the linear theory assumes that **n** varies little from a fixed direction, which is not the case in the focal conic textures, for example, even if we avoid close approach to the singularities. In this respect, Frank's proposal is superior. Also, there is no prescription for calculating forces or moments. De Gennes [*11*, Chapter 6], Kléman [*150*], and Kléman and Parodi [*151*] have attempted to improve the theory in these respects.

There have been various other attempts to formulate theories of smectics, including some which might be applicable to some of the more complicated types of smectics now known to exist. These occur in the works of de Gennes and Sarma [*152*], Jahnig and Schmidt [*153*], Lee and Eringen [*154*], Martin et al. [*155*], Rapini [*156*], Saupe [*157*], and those well-known anonyms forming the Orsay group [*158*]. A synthesis of some of these views, in the form of a relatively crude theory, is set forth by de Gennes [*11*, Chapter 7]. Research in this area is, as yet, in a rather tentative state. After some reflection, we have decided to discuss only a few of the ideas involved. Some of the notions are of more import away from equilibrium, but it seems sensible to rephrase static theory so that it better accords with dynamical theory.

Considering the smectics as layered structures, we consider them first as composed of identical plane layers with uniform thickness $\hat{d}$, in what is commonly considered as their relaxed state. Let these now be distorted, in such a way that the layers are not destroyed, though they may be bent and of non-uniform thickness. In such a configuration, it is rather obvious that we can label surfaces representing interfaces of adjacent layers by numbers as follows

$$\psi_n = n\hat{d}, \quad n = 0, \pm 1, \pm 2, \ldots$$

Granted that the layers are identical, with tops indistinguishable from their bottoms, it makes no difference to which we assign the value $n = 0$, and the relabeling $n \to -n$ is equivalent. In the customary, somewhat vague, passage to the continuum limit, we regard $\hat{d}$ as infinitesimal, replacing ψ_n by a generally smooth function

$$\psi = \psi(\mathbf{x});$$

the surfaces $\psi = $ const. roughly represent the layers once our vision is so blurred. Here ψ is equivalent to $\bar{\psi}$ whenever the two are related by a transformation of the type

$$\bar{\psi} = \pm \psi + \text{const.} \tag{41}$$

When the aforementioned planar configuration is attained, $\psi = \hat{\psi}$, where $\hat{\psi}$ satisfies the conditions

$$\nabla\nabla\hat{\psi} = 0, \quad \nabla\hat{\psi} \cdot \nabla\hat{\psi} = 1. \tag{42}$$

If, as seems appropriate for the smectics A, the molecular direction remains normal to the layers, we can set

$$\mathbf{n} = \nabla\psi/\|\nabla\psi\|, \tag{43}$$

and

$$d = \hat{d}/\|\nabla\psi\| = d(\mathbf{x}) \tag{44}$$

gives a measure of the actual thickness of the layers. The classical view, $d = \hat{d}$, then requires that $\nabla\psi$ be a unit vector. Note that Eq. (36) then implies that $\mathbf{n}$ is the gradient of a scalar, as indicated by Eq. (43). Thus far, nothing has been said about deformation, as this term is commonly interpreted in elasticity theory. Indeed, with the aforementioned views of the smectics A, including the Helfrich self-permeation, matter can move relatively freely within and through the layers. Thus ψ and the deformation can vary independently, according to this view. As is noted by de Gennes [*147*], it does seem likely that changes in thickness of the layers will be accompanied by changes in the mass density ρ, making the assumption of incompressibility slightly dubious. Given the fluidity of the smectics A, we expect little or no resistance to shearing deformations, so a likely assumption is that the mechanical energy $\mathscr{W}$ is senstitive only to changes in the layering and ρ. Taking into account Eq. (41), we then assume that, for smectics A,

$$\mathscr{W} = \mathscr{W}(\rho, \nabla\psi, \nabla\nabla\psi) = \mathscr{W}(\rho, -\nabla\psi, -\nabla\nabla\psi)$$
$$= \mathscr{W}(\rho, \mathbf{Q}\nabla\psi, \mathbf{Q}\nabla\nabla\psi\mathbf{Q}^T), \tag{45}$$

where $\mathbf{Q}$ is an arbitrary orthogonal transformation. For other types of smectics, such a description of layering seems to remain pertinent, but ψ and ρ are inadequate to describe their configurations. The kinematical problem is then to introduce such other variables as may seem pertinent, with some hypothesis as to how their variations might be related to changes in the layering. One must then provide some prescription for the energy. References cited above contain proposals of this kind. We content ourselves with discussing only Eq. (45).

As introduced, $\nabla\psi$ is dimensionless, $\nabla\nabla\psi$ having the dimensions of reciprocal length, so it is natural to expect that some constant having the dimensions of length will enter the picture. Commonly, it is assumed to be of molecular size, say of the order of the layer thickness $\hat{d}$. Then, plausibly,

$$\nabla\nabla\psi\hat{d}$$

is small, so we might assume $\mathscr{W}$ quadratic in these variables. Also, Frank's theory seems not to be far off, suggesting that

$$2\delta = \|\nabla\psi\|^2 - 1 \tag{46}$$

is small. It then seems likely that the combination

$$\hat{d}\nabla\delta = \hat{d}(\nabla\nabla\psi)\nabla\psi$$

is unusually small. Discarding quadratic terms involving the latter, we can grind out a preliminary form of $\mathscr{W}$, satisfying Eq. (45), not yet fully exploiting the smallness of δ. Omitting detail, we get

$$\mathscr{W} = C_0 + C_1(\nabla^2\psi)^2 + C_2\,\mathrm{tr}(\nabla\nabla\psi)^2,$$

where

$$C_N = C_N(\rho, \|\nabla\psi\|^2).$$

This can be readjusted in various ways by adding terms assumed negligible. A common variant results from noting that, with Eq. (43), our assumptions imply that

$$(\nabla^2\psi)^2 \approx (\nabla\cdot\mathbf{n})^2. \tag{47}$$

These being liquids, ρ should not vary much from a constant value $\hat{\rho}$ occurring when Eq. (42) is satisfied, i.e.,

$$\rho = \hat{\rho}(1-\theta), \tag{48}$$

where θ is small. The usual formal type of estimate then gives, to within an unimportant constant,

$$2C_0 = A\theta^2 + B\delta^2 + 2C\delta\theta + 2D\delta + 2E\theta,$$
$$2C_1 = K_1 - \bar{K},$$
$$2C_2 = \bar{K},$$

where A, B, etc., are constants, two being labeled for easy comparison with Eq. (38). For the reasons discussed before, some writers discard the term multiplied by $\bar{K}$. Thus, a likely form is, say

$$2W = A\theta^2 + 3\delta^2 + 2C\delta\theta + 2E\theta + 2D\delta$$
$$+ K_1(\nabla \cdot \mathbf{n})^2 + \bar{K}[\text{tr}(\nabla\nabla\psi)^2 - (\nabla^2\psi)^2]. \tag{49}$$

This is essentially equivalent to the proposal of de Gennes [*147, 11*, Chapter 6]. As is to be discussed later, the energy to which he refers is best interpreted as a slightly different one. To account for the effects of magnetic fields, he recommends using Eqs. (21) and (22) wherein **n** is given by Eq. (43) or a compatible approximation, e.g.,

$$\mathbf{n} \approx \nabla\psi.$$

It is rather common and useful for some linearizations, etc., to introduce some measure of layer displacement. Generally these amount to setting

$$\psi = \hat{\psi} + u, \tag{50}$$

where $\hat{\psi}$ satisfies Eq. (43), at least when the deformation is infinitesimal. What is commonly referred to as "displacement normal to the layers" seems to me a concept whose meaning is not always clear, particularly if large distortions and self-permeation occur. When the direction of the normal to the layers varies appreciably, as in the focal conic textures, introduction of u will only complicate analyses. Allowing for such things as the indicated magnetic contributions, the bulk energy of smectics A is likely to be of the form

$$E = \int U(\nabla\psi, \nabla\nabla\psi, \rho, \mathbf{x}) \, dV. \tag{51}$$

As is to be discussed later, the dependence on ρ is, in a certain sense, ignorable. In brief, this describes the simplest type of theory. Experience with it is still rather limited, but it does show some promise.

Here again, interfaces tend to promote characteristic orientations. The strong anchoring condition is invoked, e.g., by de Gennes [*11*, Chapter 6], albeit with something less than complete trust. Several of the older writers describe the remarkable droplet shapes called "*bâtonnets.*" For example,

Equilibrium Theory of Liquid Crystals

Friedel [6] includes sketches of some, discusses the complicated orientation within them and speculates on the probable nature of the interfacial energies involved. There are also the Grandjean [159] terraces (*gouttes á gradins*), layers of quite uniform thickness, this being large compared to $\hat{d}$, stacked in terraced form. They are simply oriented except near their edges, where imperfections occur. Bouligand [121] discusses their morphology. Lawrence [160] has argued that these owe their existence to a small yield stress associated with sliding of the fundamental layers over each other. Their stability does seem to require some explanation. Possibly, this is effected by the action of surface energies.

While ferroelectric liquid crystals are mentioned in the literature from time to time, examples have been lacking and most common symmetry assumptions, etc., tend to exclude theories appropriate for them. Meyer *et al.* [161] report finding two, a smectic B and a smectic C, and mention some of the unusual phenomena associated with them. There is then some reason to develop appropriate theory.

III. EQUILIBRIUM EQUATIONS

A. Nematics and Cholesterics

For these fluids, considered incompressible, we first calculate the variation of the bulk energy for a fixed volume of fluid using Eq. (27). In part, we can, roughly, vary positions of the centers of mass of the little rods. For this, we introduce a one-parameter family of volume-preserving mappings; deformations described relative to fixed rectangular cartesian coordinates,

$$\mathbf{x} \to \bar{\mathbf{x}} = \bar{\mathbf{x}}(\mathbf{x}, t),$$

$$\frac{\partial(\bar{x}_1, \bar{x}_2, \bar{x}_3)}{\partial(x_1, x_2, x_3)} = 1, \tag{52}$$

$$\bar{\mathbf{x}}(\mathbf{x}, 0) = \mathbf{x}.$$

These map the putative equilibrium region $R(0)$ onto a region $R(t)$. Let $\partial R(t)$, the boundary, be described parametrically in terms of any surface coordinates, by

$$\bar{\mathbf{x}} = \hat{\mathbf{x}}(u^\alpha, t), \qquad \alpha = 1, 2.$$

Also introduce a one-parameter family of unit vector fields

$$\mathbf{n} = \mathbf{n}(\bar{\mathbf{x}}, t),$$

$$\mathbf{n} \cdot \mathbf{n} = 1, \tag{53}$$

and let variations be given by the usual prescriptions

$$\delta(\ ) = \partial/\partial t(\ )|_{t=0},$$

where, in the differentiations, the remaining arguments indicated above are held constant. For simplicity, we write $\delta \mathbf{x}$ in place of $\delta \bar{\mathbf{x}}$. It follows that, on the boundary, at least the normal components of $\delta \mathbf{x}$ and $\delta \hat{\mathbf{x}}$ must agree. Then, calculating the variation of Eq. (27), we get, in component form,

$$\delta E = \int_{R(0)} h_i \, \delta n_i \, dV + \int_{\partial R(0)} \left(\frac{\partial U}{\partial n_{i,k}} \delta n_i + U \, \delta \hat{x}_k \right) dS_k, \tag{54}$$

where $\mathbf{h}$, the *molecular field*, is given by

$$h_i = \frac{\partial U}{\partial n_i} - \left(\frac{\partial U}{\partial n_{i,k}} \right)_{,k}, \tag{55}$$

and $d\mathbf{S}$ denotes the usual vector element of area. Here $\delta \mathbf{n}$, representing the change in $\mathbf{n}$ occurring at a fixed point in space, must satisfy

$$\mathbf{n} \cdot \delta \mathbf{n} = 0.$$

In the interior, it is otherwise arbitrary. On the boundary, $\mathbf{n}$ and $\hat{\mathbf{x}}$ may be restricted by assumptions of rigid containers, strong anchoring, etc. The inherent restriction that volumes remain unchanged imposes the requirement that

$$\oint \delta \hat{\mathbf{x}} \cdot d\mathbf{S} = 0. \tag{56}$$

If only the inherent restrictions obtain, the surface integral in Eq. (54) is to be balanced by the variation in surface energy. Of course, similar statements apply to intermediate cases, say a rigid container, $\delta \hat{\mathbf{x}} = 0$, not accompanied by strong anchoring. In all such cases, we get the interior equilibrium equations

$$\mathbf{h} = \mu \mathbf{n}, \tag{57}$$

where the scalar μ, a Lagrange multiplier, is arbitrary. For problem solving, it is convenient to rearrange these equations as follows: introduce curvilinear coordinates

$$\mathbf{y} = \mathbf{y}(\mathbf{x})$$

and express cartesian components of $\mathbf{n}$ in the form

$$n_i = f_i(y^j, \alpha, \beta), \tag{58}$$

where α and β, which may be visualized as surface coordinates on a unit sphere, are chosen so that

$$f_i f_i \equiv 1,$$

Equilibrium Theory of Liquid Crystals

and so that essentially all unit vectors can be so represented. Since every coordinatization of the unit sphere is somewhere singular, it is inherent that one or more vectors will fail to be well represented, and one must exercise some care in treating such exceptions. Now set

$$\overline{U} = \left|\frac{\partial(x_1, x_2, x_3)}{\partial(y^1, y^2, y^3)}\right| U = \overline{U}(y^i, \alpha, \beta, \alpha_{,k}, \beta_{,m}), \tag{59}$$

where commas now denote partial derivatives with respect to the curvilinear coordinates y^i. Then, Eq. (57) can be shown to be equivalent to the system

$$\frac{\partial \overline{U}}{\partial \alpha} - \left(\frac{\partial \overline{U}}{\partial \alpha_{,k}}\right)_{,k} = 0,$$

$$\frac{\partial \overline{U}}{\alpha \beta} - \left(\frac{\partial \overline{U}}{\partial \beta_{,k}}\right)_{,k} = 0. \tag{60}$$

A consequence of these is that

$$\left(\overline{U}\,\delta_l^k - \frac{\partial \overline{U}}{\partial \alpha_{,k}}\alpha_{,l} - \frac{\partial \overline{U}}{\partial \beta_{,k}}\beta_{,l}\right)_{,k} = \frac{\partial \overline{U}}{\partial y^l}. \tag{61}$$

In cases where $\overline{U}$ does not depend explicitly on **y** and Eq. (60) reduces to a system of ordinary differential equations, Eq. (61) integrates trivially to give the so-called "energy integral." Here the transformations are assumed real. Provided one properly accounts for the reality of **n**, etc., the same apparatus applies when they are complex. It can be helpful to note that an additive contribution to $\overline{U}$, depending only on the coordinates **y**, cancels out of Eqs. (60) and (61). If the boundary is known and if the orientation boundary condition is of the strong anchoring type, we will have pertinent information concerning boundary values of α and β. We then have a boundary value problem of the classical (Dirichlet) type.

It remains to consider the surface energy given by Eq. (28), say, to generate boundary conditions guaranteeing that

$$\delta(E + \mathscr{E}) = 0. \tag{62}$$

Very commonly, it is assumed, at least tacitly, that the boundary is immobilized, so we are left only with the variations of **n** which may be envisaged in weak anchoring. For any energy of the form

$$\mathscr{E} = \int f(\hat{\mathbf{x}}_{,\alpha}, \hat{\mathbf{x}}, \mathbf{n}, u^\beta)\, dS, \tag{63}$$

in particular those of the form given by Eq. (28), an easy calculation, with $\delta \hat{\mathbf{x}} = 0$, gives the orientation boundary conditions

$$\mathbf{l}\mathbf{v} = \mathbf{n} \wedge (\partial f/\partial \mathbf{n}),$$

$$l_{ij} \equiv \varepsilon_{ipq} n_p (\partial U/\partial n_{q,j}). \tag{64}$$

If Eq. (28) applies, we have

$$\partial f/\partial \mathbf{n} = 2f'\mathbf{n}\cdot\mathbf{vv}, \tag{65}$$

implying that

$$\mathbf{lv} = 0 \quad \text{if} \quad \mathbf{n}\|\mathbf{v}, \quad \mathbf{n}\perp\mathbf{v} \quad \text{or} \quad f' = 0. \tag{66}$$

Jenkins and Barratt [162] derive equations equivalent to these. For a flexible interface, the remaining conditions obtaining can be rather elaborate in detail, but a schematic form is fairly simple. We assume adherence, so that $\delta\hat{\mathbf{x}} = \delta\mathbf{x}$ on the boundary. With Eq. (63), write

$$f\,dS = \bar{f}\,du^1\,du^2.$$

Using the geometrical relations

$$\begin{aligned} dS &= \sqrt{g}\,du^1\,du^2, \\ g &= \det \hat{\mathbf{x}}_{,\alpha}\cdot\hat{\mathbf{x}}_{,\beta}, \\ \sqrt{g}\,\mathbf{v} &= \hat{\mathbf{x}}_{,1}\wedge\hat{\mathbf{x}}_{,2}, \\ \bar{f} &= f\sqrt{g}, \end{aligned} \tag{67}$$

It is clear that f and $\bar{f}$ depend on the same list of variables. To calculate the variation of $\mathscr{E}$, we should express $\mathbf{n}$ as a function of u^α, t or, alternatively, note that

$$\left.\frac{\partial\mathbf{n}(u^\alpha,t)}{\partial t}\right|_{t=0} = \delta\mathbf{n} + \delta\mathbf{x}\cdot\nabla\mathbf{n} \equiv \Delta\mathbf{n}.$$

Then, assuming smoothness and that the interface has no boundary, we get, after using the divergence theorem,

$$\delta\mathscr{E} = \int\left\{\frac{\partial\bar{f}}{\partial n_i}\Delta n_i + \left[\frac{\partial\bar{f}}{\partial\hat{x}_i} - \left(\frac{\partial\bar{f}}{\partial\hat{x}_{i,\alpha}}\right)_{,\alpha}\right]\delta x_i\right\}du^1\,du^2.$$

With the variations arbitrary except for the inherent restrictions we get, in addition to Eq. (64), the conditions

$$\sqrt{g}\left[(a+U)\delta_{ik} - \frac{\partial U}{\partial n_{p,k}}n_{p,i}\right]v_k = \left(\frac{\partial\bar{f}}{\partial\hat{x}_{i,\alpha}}\right)_{,\alpha} - \frac{\partial\bar{f}}{\partial\hat{x}_i}, \tag{68}$$

where a, an arbitrary constant, is a Lagrange multiplier associated with the constraint given by Eq. (56). For cases where Eq. (28) applies, Jenkins and Barratt [162] discuss in some detail analyses of this type, including contributions occurring when the interface does not cover the entire boundary of the liquid crystal.

Formulations of Eq. (64) appropriate for use with Eq. (60) are, of course, derivable. We express $\bar{f}$ in terms of **y** and its surface derivatives, as well as α and β. Omitting detail, we find that

$$\frac{\partial \overline{U}}{\partial \alpha_{,k}} N_k = -\frac{\partial \bar{f}}{\partial \alpha},$$

$$\frac{\partial \overline{U}}{\partial \beta_{,k}} N_k = -\frac{\partial \bar{f}}{\partial \beta}, \qquad (69)$$

$$N_k \equiv \varepsilon_{kij} \frac{\partial y^i}{\partial u^1} \frac{\partial y^j}{\partial u^2}.$$

For some interpretations, it is important to know how to calculate forces and moments and, on occasion, one does specify something about them in setting a problem. Extrapolating from experience based on more conventional continuum theories, Ericksen [*163*] proposed prescriptions for these, which now seem to be rather generally accepted. We do not reproduce the reasoning involved, but only record some of the conclusions. For the mechanical stress tensor, we use

$$t_{ij} = -p\delta_{ij} - \frac{\partial W}{\partial n_{k,j}} n_{k,i}, \qquad (70)$$

$$t_{ij} v_j = t_i,$$

the vector **t** measuring the force per unit area exerted on a surface, in the manner which has become familiar since it was introduced by Cauchy. Here p is an arbitrary scalar, a Lagrange multiplier corresponding to the constraint of incompressibility. It is then a small step to introduce a total stress tensor by replacing $\mathscr{W}$ by U, i.e.,

$$T_{ij} = -p\delta_{ij} - \frac{\partial U}{\partial n_{k,j}} n_{k,i} \qquad (71)$$

With the most common elementary estimate, $U - \mathscr{W}$ does not depend on $\nabla \mathbf{n}$, so the two coincide. Using Eqs. (54) and (56), we calculate that

$$T_{ij,j} - \frac{\partial U}{\partial x_i} = -(p+U)_{,i}.$$

If we then set

$$p + U = -a = \text{const.}$$

we then have equilibrium equations of conventional form, with $-\partial U/\partial x_i$ interpreted as a body force. Then Eq. (68) becomes

$$\sqrt{g}\, T_{ij} v_j = \left(\frac{\partial \bar{f}}{\partial \hat{x}_{i,\alpha}}\right)_{,\alpha} - \frac{\partial \bar{f}}{\partial \hat{x}_i}, \qquad (72)$$

balancing stress against surface tension. Surface couples are only partly accounted for as moments of stress. For the remainder, the proposal is that they are measured by a couple stress which, again extrapolating by a small step, is the tensor **l** occurring in Eq. (64).

In general terms, this completes the basic theory of nematics and cholesterics now in common usage.

B. Smectics

Here, we are concerned with energies of the form indicated by Eq. (51). Proceeding as we did to get Eq. (57), we write

$$\psi = \psi(\bar{\mathbf{x}}, t), \qquad \rho = \rho(\bar{\mathbf{x}}, t),$$

with the constraint corresponding to conservation of mass,

$$\delta\rho + \nabla \cdot (\rho \delta \mathbf{x}) = 0.$$

Then, with the traditional integration by parts, we get

$$\delta \int U \, dV = \int \left[\rho \left(\frac{\partial U}{\partial \rho} \right)_{,k} \delta x_k + h \, \delta \psi \right] dV$$
$$\oint \left[\left(U - \rho \frac{\partial U}{\partial \rho} \right) \delta x_k + p_k \, \delta \psi + \tau_{ik} \, \delta \psi_{,i} \right] dS_k, \qquad (73)$$

where h is the molecular field here appropriate,

$$h = \tau_{jk,jk} - \left(\frac{\partial U}{\partial \psi_{,k}} \right)_{,k} = -p_{k,k},$$

$$p_k = \frac{\partial U}{\partial \psi_{,k}} - \tau_{kj,j}, \qquad (74)$$

$$\tau_{ik} = \frac{\partial U}{\partial \psi_{,ik}} = \tau_{ki}.$$

If we forbid matter to move from one layer to another, we might then require the surfaces $\psi = $ const. to be material surfaces, giving the constraint

$$\Delta \psi = \delta \psi + \delta \mathbf{x} \cdot \nabla \psi = 0. \qquad (75)$$

One might impose the constraint of incompressibility,

$$\nabla \cdot \delta \mathbf{x} = 0. \qquad (76)$$

Kléman [*150*] employs a more complicated constraint, intended to represent the condition that "the component of the displacement parallel to the layers

Equilibrium Theory of Liquid Crystals

corresponds to an incompressible deformation," writing

$$\nabla \cdot (\mathbf{u} - u\mathbf{n}) = 0, \tag{77}$$

where $\mathbf{u}$ is the displacement and u measures the displacement of the layers along its normal which is $\mathbf{n}$. Normally, the divergence of a displacement correctly measures volume change only if the displacement gradients are small, so some such restriction seems to be involved. To be definite, following de Gennes [147], we exclude such constraints, assuming that the variations $\delta\psi$ and $\delta\mathbf{x}$ are arbitrary, at least in the interior. This yields the equations

$$\begin{aligned} h &= 0 \\ (\partial U/\partial \rho)_{,k} &= 0 \Rightarrow \partial U/\partial \rho = a = \text{const.} \end{aligned} \tag{78}$$

Assuming that the latter can be solved for ρ in terms of the remaining variables, we can introduce the Legendre transform

$$\hat{U} = U - \rho(\partial U/\partial \rho) = \hat{U}(a, \nabla\psi, \nabla\nabla\psi, \mathbf{x}), \tag{79}$$

with

$$\begin{aligned} \partial \hat{U}/\partial a &= -\rho, \\ \partial \hat{U}/\partial \psi_{,k} &= \partial U/\partial \psi_{,k}, \text{ etc.} \end{aligned} \tag{80}$$

For ordinary hydrostatics, $-\hat{U}$ reduces to the usual hydrostatic pressure, if $U = \mathscr{W}$. Here the remaining equilibrium equation becomes

$$\left(\frac{\partial \hat{U}}{\partial \psi_{,ij}}\right)_{,ij} = \left(\frac{\partial \hat{U}}{\partial \psi_{,k}}\right)_{,k}. \tag{81}$$

A calculation shows that this implies that

$$(\hat{U}\delta_{ik} - p_i\psi_{,k} - \tau_{pi}\psi_{,pk})_{,i} - \partial \hat{U}/\partial x_i = 0. \tag{82}$$

In brief, the basic equations are of essentially the same form as they would have been had we not introduced ρ as a variable, except for the dependence on the parameter a. In practice, effects of variations of a are assumed negligible. More precisely, no allowance is made for dependence on such a parameter.

If one takes $U = \mathscr{W}$, given by Eq. (49), and assumes that

i. The parameter a has the value occurring in the ground state where $\theta = \delta = 0$, etc.
ii. $\mathscr{W} = \hat{U}$ is a minimum in this ground state,
iii. $\bar{K} = 0$,

one obtains the energy introduced by de Gennes [147, 11, Chapter 7] viz.

$$\begin{aligned} 2\mathscr{W} &= \bar{B}\delta^2 + K_1(\nabla\cdot\mathbf{n})^2, \\ \bar{B} &= B - C^2/A > 0, \qquad K_1 > 0. \end{aligned} \tag{83}$$

The question of how surface forces and couples are to be calculated seems to involve some unusual subtleties. For different, but mathematically similar theories, Toupin [*164*] has made a proposal which can be adapted. For theories of smectics A, Kléman [*150*] has made a different proposal. His version of theory differs somewhat from ours; he employs the constraint given by Eq. (77), for example. Even by making some allowance for such factors, we are not able to reconcile the two proposals or to see precisely what is the relation between them. His version has since been revised by Kléman and Parodi [*151*]. Their work, done quite independently of ours, covers much the same ground, albeit in a rather different way. Their proposal again differs from Toupin's, but we can at least see how each fits into the general scheme of things. We present the facts, as we see them, leaving it to the reader to make his own assessment.

Starting with Toupin's thoughts, we first replace $\delta\psi$ by

$$\Delta\psi = \delta\psi + \delta\mathbf{x} \cdot \nabla\psi,$$

$\Delta\psi$ measuring the change in ψ as we ride on a fixed particle. With this change, we calculate an equation entirely similar to Eq. (73), with U replaced by $\mathscr{W}$ viz.,

$$\delta \int \mathscr{W} \, dV = \oint [t_{ik}\, \delta x_i + \bar{p}_k\, \Delta\psi + \bar{\tau}_{ik}(\Delta\psi)_{,i} \\ - \bar{\tau}_{pk}\psi_{,i}\, \delta x_{i,p}] \, dS_k \qquad (84) \\ + \int (f_i\, \delta x_i + \bar{h}\, \Delta\psi) \, dV,$$

where barred quantities are defined as were the unbarred, with U replaced by $\mathscr{W}$, and

$$\begin{aligned} t_{ik} &= [\mathscr{W} - \rho(\partial\mathscr{W}/\partial\rho)]\, \delta_{ik} - \bar{p}_k \psi_{,i} - \bar{\tau}_{pk}\psi_{,ip}, \\ f_k &= \rho(\partial\mathscr{W}/\partial\rho)_{,k} - \bar{h}\psi_{,k}, \end{aligned} \qquad (85)$$

Following Toupin [*164*], we now compose gradients of variations into derivatives normal and tangent to the surface, using the rules

$$Df = v_i f_{,i},$$
$$D_i f = f_{,i} - v_i Df.$$

Toupin allows for some lack of smoothness, but we assume the surface and pertinent functions smooth to avoid "edge" contributions. We then have a simplified and corrected version of the integral identity which he uses, viz.

$$\oint D_i(fv_j) \, dS = -\oint (b_{kk} v_i v_j) f \, dS,$$
$$b_{ij} = b_{ji} = -D_i v_j.$$

Equilibrium Theory of Liquid Crystals

Here **b** is the second fundamental tensor of the surface, measuring its curvatures. The aim is to get a linear functional of the variations which can vanish only if all the coefficients vanish, a property not exhibited by the right member of Eq. (84). With the indicated manipulation, Eq. (84) becomes

$$\delta \int \mathscr{W} dV = \oint [F_i^1 \delta x_i + F_i^2 D(\delta x_i) + F^3 \Delta \psi$$
$$+ F^4 D(\Delta \psi)] dS$$
$$+ \int (f_i \delta x_i + \bar{h} \Delta \psi) dV, \tag{86}$$

where

$$\begin{aligned}
F_i^1 &= t_{ik} v_k - \bar{\tau}_{pk} v_p v_k b_{mm} \psi_{,i} + D_p(\bar{\tau}_{pk} \psi_{,i} v_k), \\
F_i^2 &= -\bar{\tau}_{pk} v_p v_k \psi_{,i}, \\
F^3 &= \bar{p}_k v_k + b_{mm} \bar{\tau}_{ik} v_i v_k - D_i(\bar{\tau}_{ik} v_k), \\
F^4 &= \bar{\tau}_{ik} v_i v_k.
\end{aligned} \tag{87}$$

For a virtual displacement corresponding to a rigid translation,

$$\delta \mathbf{x} = \text{const.}, \quad \Delta \psi = 0,$$

the left side of Eq. (86) vanishes, because of Galilean invariance. The right-hand side yields the equation

$$\oint \mathbf{F}^1 dS + \int \mathbf{f} dV = 0, \tag{88}$$

and normally, such a condition of translational invariance is reasonably interpreted as a balance of forces. Similar consideration of virtual displacements corresponding to rigid rotations gives

$$\oint (\mathbf{x} \wedge \mathbf{F}^1 + v \wedge \mathbf{F}^2) dS + \int \mathbf{x} \wedge \mathbf{f} dV = 0, \tag{89}$$

which, experience suggests, represents balance of moments. Thus, Toupin suggests, $\mathbf{F}^1$ should measure the surface force, $\mathbf{f}$ the body force, etc. From Eq. (87), this surface force depends on the curvatures of the surface. The classical proof of existence of a mechanical stress tensor, due to Cauchy and reproduced in various texts, assumes, usually tacitly, that the surface force depends on the normal, but not on curvatures. Here this existence breaks down, as we can see quite explicitly. There is a similar situation with the couple stress tensor. Mathematically, the situation is clear enough. What is lacking are intuitive arguments or perhaps mechanistic calculations which might support or overturn the appropriateness of the indicated interpretations.

If we apply similar analyses to Eq. (84), we find that translational invariance of $\mathscr{W}$ gives rise to the condition

$$\oint \mathbf{t}\, dS + \int \mathbf{f}\, dV = 0, \tag{90}$$

where

$$t_i = t_{ij} v_j. \tag{91}$$

Using Eqs. (74) and (85), one can easily show that the differential equivalent

$$t_{ij,j} + f_i = 0 \tag{92}$$

reduces to an identity. When the equilibrium Eq. (78) is satisfied, one can of course write $\mathbf{f}$ in terms of derivative of the difference potential $U - \mathscr{W}$ to make it appear as a force due to external actions. After doing this, one can write $\mathbf{f}$, in part, as the divergence of a tensor which, when added to t_{ij} gives the tensor whose divergence occurs in Eq. (82). Further, from Eq. (87), the tensor t_{ij} delivers the first term in $\mathbf{F}^1$, a part which is independent of curvatures and linear in $\mathbf{v}$. Naturally, there are cases where the remainder is negligible. In brief, if we insist upon having a stress tensor of the classical variety, t_{ij} emerges as the likely candidate. In essence, the proposal of Kléman and Parodi [*151*] is that we adopt it.

Similarly, by considering invariance with respect to rigid rotations, one can derive from Eq. (84) the identity

$$\oint (\mathbf{x} \wedge \mathbf{t} + \mathbf{l})\, dS + \int \mathbf{x} \wedge \mathbf{f}\, dV = 0, \tag{93}$$

where

$$l_i = \varepsilon_{imn} \bar{\tau}_{mk} v_k \psi_{,m}. \tag{94}$$

This, then, generates the analogous version of balance of moments, $\mathbf{l}$ being a couple stress vector. After using Eq. (92), one finds that the differential equivalent reduces to the statement that

$$A_{ij} = t_{ij} - (\bar{\tau}_{jk} \psi_{,i})_{,k} = A_{ji}. \tag{95}$$

Using Eqs. (74) and (85), one finds that Eq. (95) reduces to the statement that

$$B_{ij} = \frac{\partial \mathscr{W}}{\partial \psi_{,i}} \psi_{,j} + 2 \frac{\partial \mathscr{W}}{\partial \psi_{,ik}} \psi_{,jk} = B_{ji}. \tag{96}$$

If in Eq. (45) we put in an infinitesimal rotation

$$\mathbf{Q} = \mathbf{1} + \mathbf{\Omega}, \qquad \mathbf{\Omega} = -\mathbf{\Omega}^T,$$

expanding to get terms of first order in $\mathbf{\Omega}$, we find that Eq. (96) is an identity.

Again, in essence, this covers the proposal of Kléman and Parodi [*151*] for describing moments.

In detail, the Kléman–Parodi treatment differs from ours in that they consider $\mathcal{W}$ to depend both on **n** and gradients of ψ, employing a method of Lagrange multipliers to treat constraints generated by Eq. (43). It seems fair to warn that the method is not always reliable. Using it, Ericksen [*163*] drew a conclusion which Leslie [*38*] later found to be unsound. In the calculus of variations, it is known that the method is not always reliable, as is discussed by Young [*165*], Vol. II]. Here we can cross-check, since our analyses do not employ multipliers, and we have found no obvious error. It is less easy to cross-check the earlier analysis of Kléman [*150*], and we have not done so.

From the above analyses, it is rather plain what is the difference between the two proposals for calculating forces and moments. Since it is subtle, it is fortunate that not all analyses require their calculation.

IV. LINEAR THEORY

At least for the nematic and cholesteric types, the mildly nonlinear equations associated with the Oseen–Frank theory are moderately tractable, though less tractable than linearized versions. Study of linear theory is a relatively recent development, beginning with the work of Davison [*166*], who discussed the simplest case: equations governing nematics are linearized about a configuration of minimum energy, **n** = const. The resulting equations are much like the well-studied equations of linear elasticity theory, so it is relatively easy to adapt general theorems, techniques for solution, etc. Results of this kind are reported by Davison [*166*] and by Warren [*45*], who discusses in some detail the use of complex variable methods. Jenkins [*167*] similarly treats the linearized theory of cholesterics. For the solution of particular problems, linear theory is now in common usage. Numerous applications can be found in the book of de Gennes [*11*].

The common linearizations are straightforward and, since readers are well acquainted with the general mathematical features of linear theories, we shall be brief. For example, we might start with some solution of Eq. (59), with $\alpha = \hat{\alpha}_1$, $\beta = \hat{\alpha}_2$, say. Then, writing

$$\alpha = \hat{\alpha}_1 + v_1, \qquad \beta = \hat{\alpha}_2 + v_2,$$

with v_N assumed small, we approximate the energy $\overline{U}$ by a quadratic

$$\overline{U} \approx \hat{U} + (\partial \hat{U}/\partial \alpha_M) v_M + (\partial \hat{U}/\partial \alpha_{M,i}) v_{M,i} + \chi, \tag{97}$$

$$2\chi = \frac{\partial^2 \hat{U}}{\partial \alpha_M \, \partial \alpha_N} v_M v_N + 2 \frac{\partial^2 \hat{U}}{\partial \alpha_M \, \partial \alpha_{N,i}} v_M v_{N,i} + \frac{\partial^2 \hat{U}}{\partial \alpha_{M,i} \, \partial \alpha_{N,j}} v_{M,i} v_{N,j}, \tag{98}$$

where carets denote evaluation at $v_M = 0$ and repeated indices are summed as usual. A calculation shows that v_M satisfies

$$\left(\frac{\partial \chi}{\partial v_{M,i}}\right)_{,i} - \frac{\partial \chi}{\partial v_M} = 0, \qquad (99)$$

again of the familiar Euler–Lagrange form. If one linearizes not about a solution, but, say, a guess, there is of course an additional "source" term depending on it. In cases where the reference configuration is so stable that $\chi \geqslant 0$, the equality holding only when v_M corresponds to an infinitesimal rigid motion, uniqueness of solutions for a variety of boundary value problems is easily established by classical methods. Also, if v_M and v_M' are two smooth solutions, defined in the same region, it follows easily that

$$\frac{\partial \chi}{\partial v_M} v_M' + \frac{\partial \chi}{\partial v_{M,i}} v_{M,i}' = \left(\frac{\partial \chi}{\partial v_M}\right)' v_M + \left(\frac{\partial \chi}{\partial v_{M,i}}\right)' v_{M,i}. \qquad (100)$$

From Eqs. (99) and (100) follows a reciprocal theorem of rather classical type, viz.

$$\oint \left[\frac{\partial \chi}{\partial v_{M,i}} v_M' - \left(\frac{\partial \chi}{\partial v_{M,i}}\right)' v_M\right] N_i \, du^1 \, du^2 = 0, \qquad (101)$$

where **N** is defined as in Eq. (69).

Linearization of the rectangular cartesian equation (57) for nematics, about

$$\mathbf{n} = \hat{\mathbf{n}} = (0, 0, 1)$$

is particularly simple. Introducing the infinitesimal vector **w**,

$$\mathbf{n} = \hat{\mathbf{n}} + \mathbf{w},$$

we have

$$0 = \mathbf{n} \cdot \mathbf{n} - 1 \approx 2w_3,$$

so **w** is planar, but may depend on all three coordinates. From Eq. (10),

$$2\mathscr{W} \approx K_1 (w_{M,M})^2 + K_2 (w_{1,2} - w_{2,1})^2$$
$$+ K_3 w_{M,3} w_{M,3}$$
$$- 2(K_2 + K_4)(w_{1,1} w_{2,2} - w_{1,2} w_{2,1}), \qquad (102)$$

where capital indices take on values 1 and 2. Similarly, the magnetic energy, Eqs. (21) and (22) is, to within a term independent of **w**,

$$2\mathscr{W}_M \approx -\chi_a [2H_3 H_M w_M + (H_M w_M)^2]. \qquad (103)$$

Corresponding equilibrium equations are, if no other bulk energies occur,

$$\chi = \mathscr{W}_M + \mathscr{W},$$

$$\left(\frac{\partial \chi}{\partial w_{M,i}}\right)_{,i} - \frac{\partial \chi}{\partial w_M} = 0. \tag{104}$$

Clearly, when **H** is constant, these are linear equations with constant coefficients. That no Lagrange multiplier shows up here stems from the fact that we have (approximately) satisfied the constraint $\mathbf{n} \cdot \mathbf{n} = 1$ by setting $w_3 = 0$. In other linearizations, error could arise from mishandling this constraint, but one only needs to exercise reasonable care.

For smectics A, there is a subtlety of more interest to mathematicians than physicists, we think. The simplest case emerges from linearizing about the ground state, say

$$\psi = \hat{\psi} = x_3.$$

Writing, as in Eq. (50),

$$\psi = \hat{\psi} + u$$

and employing Eq. (83), we infer that

$$2\mathscr{W} = \bar{B} u_{,3}^2 + K_1 (\nabla_2^2 u)^2, \tag{105}$$
$$\nabla_2^2 (\ ) = (\ )_{,11} + (\ )_{,22}.$$

If no other energies are involved, Eq. (81) yields what are becoming the standard equations, viz.

$$u_{,33} = \lambda^2 \nabla_2^4 u$$
$$\lambda^2 = K_1/\bar{B}. \tag{106}$$

Within the context of the linear theory, a presumption made in getting these equations is that the derivatives

$$u_{,33}, \quad u_{,31}, \quad u_{,32}$$

are small, relative to other second derivatives. In this spirit, $\nabla \cdot \mathbf{n}$ might have been replaced by $\nabla^2 \psi$ in Eq. (83), for example, which would modify Eq. (106). Formally, the classical theory results for $\bar{B} \to \infty$, K_1 remaining finite, or $\lambda \to 0$. As a matter of mathematics, it does not seem entirely obvious how shifting the differential operator as indicated will affect solutions of these equations, but partial answers to such questions may well lie within the capabilities of analysts. The limit behavior as $\lambda \to 0$ lies within the province of boundary layer theory. In the limit, we lose some ability to specify boundary conditions. With λ small, the main effects of its not being zero are most likely to occur near walls or near singularities, e.g., near the edges of dislocations. Since the solution of Eq. (106) with $\lambda = 0$ is easy, there might be some merit in better developing appropriate boundary layer theory. This linearization is valid only if ∇u be small, among other things. For a finite rigid rotation of the layers, it is generally not. For example, with a 90° rotation, we can have

$$\psi = x_2 \Rightarrow \|\nabla u\|^2 = 2,$$

and this linear theory incorrectly predicts a large change of energy. Not infrequently, the normal to the layers varies substantially from one part of the sample to another, as in the focal conic textures, in which case this type of linearization can lead to serious error. Kléman and Parodi [*151*] study equations linearized about configurations where the layers are circular cylinders. As they note, there are important differences between their linear equation and Eq. (106).

It would expend more ink than we can afford to cover the linear theories, in detail, so we content ourselves with these brief remarks and, later, with comments on a few predictions.

V. NONLINEAR PROBLEMS

For nematics and cholesterics, the common approach is to find cases where the governing equations reduce to ordinary differential equations, particularly the subset of those which are easily integrated once, giving the solution to within quadratures. A considerable variety of physically interesting problems have been so analyzed. It seems likely that many such solutions remain to be discovered.

Ordinary Differential Equations

1. PLANAR VARIATION

For nematics and cholesterics, one broad class of solutions can be obtained as follows. Write

$$\mathbf{n} = (\cos\alpha \cos\beta, \cos\alpha \sin\beta, \sin\alpha). \tag{107}$$

Assume that α and β depend only on the one coordinate

$$x_3 = z$$

and let primes denote derivatives with respect to this. Note that β is ill defined for $2\alpha = \pi$, say, so normal caution is here dictated. If $2\alpha \equiv \pi$, we have the trivial case of uniform orientation. Employing Eq. (33), we include cholesterics and nematics ($\tau = 0$) and a routine calculation gives

$$2\mathscr{W} = (K_1 \cos^2 \alpha + K_3 \sin^2 \alpha) \alpha'^2$$
$$+ K_3 \cos^2 \alpha \sin^2 \alpha \beta'^2$$
$$+ K_2 (\tau - \beta' \cos^2 \alpha)^2. \tag{108}$$

If a magnetic field is to be accounted for, it can reasonably depend only on z, in which case Eq. (20) requires it to be constant. Using Eqs. (21) and (22), we then calculate

$$2\mathscr{W}_M = -\chi_a (H_1 \cos \alpha \cos \beta + H_2 \cos \alpha \sin \beta + H_3 \sin \alpha)^2 + \text{const.} \tag{109}$$

With the most elementary estimate, taking **E** as a fixed vacuum field, one makes the substitutions

$$\chi_a \to \varepsilon_a/4\pi, \quad \mathbf{H} \to \mathbf{E},$$

to treat analogous dielectric effects. In this approximation, it is not difficult to show, and is noted by Zwetkoff [*168*], that the effect of parallel electric and magnetic fields is mathematically equivalent to applying a single field in the same direction. Most studies employ a single field, so we consider only the two energies given by Eq. (108) and (109). Deuling *et al.* [*169*] discuss a more general case, involving orthogonal electric and magnetic fields, with a more sophisticated estimate of the electric field. From Eqs. (59) and (60), we must then satisfy the ordinary differential equations

$$\partial \overline{U}/\partial \alpha - (\partial \overline{U}/\partial \alpha')' = 0,$$
$$\partial \overline{U}/\partial \beta - (\partial \overline{U}/\partial \beta')' = 0, \tag{110}$$
$$\overline{U} = U = \mathscr{W} + \mathscr{W}_M.$$

One integral follows from Eq. (61), viz.

$$U - \alpha'(\partial U/\partial \alpha') - \beta'(\partial U/\partial \beta') = \text{const.} \tag{111}$$

In special cases, another is available. For example,

$$H_1 = H_2 = 0 \Rightarrow \partial U/\partial \beta = 0 \Rightarrow \partial U/\partial \beta' = \text{const.} \tag{112}$$

There are solutions with $\beta = \text{const.}$, reducible by rotation of coordinates to the form

$$\tau = 0, \quad \beta = 0, \quad H_2 = 0 \Rightarrow \partial U/\partial \beta = (\partial U/\partial \beta')' = 0. \tag{113}$$

Nontrivial solutions with $\alpha = \text{const.}$ are of the form

$$\alpha = H_3 = 0 \Rightarrow \partial U/\partial \alpha = (\partial U/\partial \alpha')' = 0. \tag{114}$$

Generally, workers have been content to treat cases falling in one of these three categories, with the assumption that χ_a (or ε_a) is positive. Early studies of special cases were made by Zocher [20, 21] and Oseen [170], numerous later workers following in their footsteps.

One of the simplest solutions included among those obtained by Oseen is of the form

$$\alpha = \mathbf{H} = 0, \qquad \beta' = \text{const.}$$

entirely similar to the twisted state which is characteristic of cholesterics, except that we need not have $\beta' = \tau$. It provides a simplistic description of old observations of twisted patterns in nematics, first correctly interpreted by Mauguin [171], the "*plage tordue.*" They are discussed in some detail by Friedel [6, pp. 350–352] and de Gennes [11, Chapters 3 and 4]. Idealistically, we prepare two infinite parallel plates, say by rubbing, to uniformly orient **n** tangent to them. Not knowing the technique, older workers observed a similar effect which sometimes occurs without such treatment. We start twisting one plate relative to the other, about its normal, through an angle β_0. The appropriate strong anchoring boundary condition can be stated as follows:

$$\beta(0) = 0, \qquad \beta(L) = \beta_0 + n\pi, \qquad n = 0, \pm 1, \pm 2, \ldots$$

L being the obvious gap width, giving infinitely many possible values for β'. The total energy of the infinite sample is infinite, but one can compare energy densities and these are least when $|\beta'|$ is least, and these "least energy" solutions reasonably accord with what is observed. For $2\beta_0 = \pi$, there are two, one twisted to the left, the other to the right. If one varies β_0 a little around this value, one should see the twist flip from right to left or vice versa, depending on which will give the least value of $|\beta'|$. Friedel reports observations of this kind. According to Turner [172], one can, with care, exceed the critical value by quite a bit; the overly twisted state is stable with respect to small enough disturbances. In practice, other phenomena complicate the picture, and there is a fair literature on these, partly summarized in the above references. For example, regions twisted to the left may occur in part of a sample, the remainder being twisted to the right, and disclinations may intrude to help relax the twist. The latter possibility is analyzed by Friedel and de Gennes [173], using Eq. (16). Observations are discussed by Spruijt [174], among others. Transition from right to left twists can occur through a twist wall or transition layer. Work on this topic is summarized by Turner [172], who gives an approximate solution for them and finds that they are unstable unless

$$2K_2 \leqslant K_1 + K_3.$$

A disadvantage of Eq. (16) is, of course, that one may miss seeing conditions like this. Besides this case of twist, Oseen [170] analyzes in some detail the various solutions occurring when $\mathbf{H} = 0$. Forces and moments associated with twisting are discussed by Ericksen [175].

A "Fréedericksz transition" now commonly means a transition of the following type: a sample, usually nematic but occasionally cholesteric, with $\chi_a > 0$, is somehow constrained by boundaries, being uniformly or at least simply oriented in the absence of fields. Commonly, if one imposes a magnetic field at right angles to $\mathbf{n}$, there is no effect if the field strength is smaller than a critical value, depending on the material and sample dimensions. Above this critical value, the orientation begins to distort.

Early observations of such transitions are summarized by Fréedericksz and Zolina [176], who found the empirical rule that, in simpler cases, the critical field strength is inversely proportional to the gap width. Early analyses of this type were made by Zocher [21], giving the empirical rule a theoretical basis. Further, if χ_a be known, a measurement of the critical field gives an experimental value of some combination of moduli occurring in $\mathscr{W}$, the combination depending on the choice of boundary conditions. Typically, above a Fréedericksz transition, there are different possible patterns, related by simple symmetry operations, and one may see two or more like this in a single sample, connected by an inversion wall, wherein the orientation smoothly adjusts from one pattern to the other. Approximate solutions for some such walls are worked out by Brochard [177], who also discusses their motion. Leger [178] also discusses these, reporting observations of them. As might be obvious, they are too complex to be included among the solutions here discussed.

Numerous workers have followed Zocher's lead, using solutions of the type here discussed to analyze similar transitions. Saupe [179] analyzes three cases in some detail, including optical analyses and corresponding experimentation for PAA. Strong anchoring is assumed, the three cases being

$$\beta = \tau = 0, \quad H_1 = H_2 = 0, \quad H_3 = H, \quad \alpha(0) = \alpha(L) = 0, \quad (115)$$

$$\alpha = \tau = 0, \quad H_1 = H_3 = 0, \quad H_2 = H, \quad \beta(0) = \beta(L) = 0, \quad (116)$$

$$\beta = \tau = 0, \quad H_1 = H, \quad H_2 = H_3 = 0, \quad \alpha(0) = \alpha(L) = \pi/2. \quad (117)$$

All give Fréedericksz transitions, the critical field strengths H_c being, respectively,

$$H_c L = \pi(K_1/\chi_a)^{1/2}, \quad H_c L = \pi(K_2/\chi_a)^{1/2}, \quad H_c L = \pi(K_3/\chi_a)^{1/2}, \quad (118)$$

so we here have what might well be called the classical methods for obtaining experimental values for K_1, K_2, and K_3. Rapini et al. [180] discuss the (rather unrealistic) cases where K_1 or K_3 vanishes. For $H > H_c$, Saupe calculates the solutions, finding some discrepancies between theory and experiment. He suggests that the strong anchoring assumption might be at fault. Much earlier, Zocher [21] had remarked that, with surfaces prepared by rubbing, departures from strong anchoring occur, with strong fields. Pieranski et al. [181] employ patterns occurring with $H > H_c$, in conjunction with thermal measurements. They discuss discrepancies in measurements of K_1/χ_a and K_3/χ_a for MBBA, obtained by different techniques and obtain some data concerning the thermal conductivity tensor. Calculations similar to those of Saupe for the effect of an electric field on MBBA indicate a similar discrepancy, according to Deuling [182]. Actually, each of these physical problems has infinitely many solutions of the one-dimensional type here discussed. For the case covered by Eq. (115), Dafermos [183] has established that the remaining solutions have higher energy per unit surface area and that, for $H > H_c$, the uniform pattern has higher energy than the distorted one commonly considered. He and Zocher [21], earlier, note that, if **H** is very slightly misaligned, these two solutions merge into one. Then, the orientation is affected by the smallest fields. However, the effect is slight until $H \approx H_c$, when it becomes more dramatic, so the transition is still recognizable. Explicit, approximate calculations of the effect are given by Malraison et al. [184], together with corresponding experimentation. Gruler and Meier [48] similarly treat the analogous dielectric effect.

Using a surface energy of the type given by Eq. (29), Zocher [21] attempts to estimate error resulting from the strong anchoring assumption, for the special case where $K_1 = K_3$. Rapini and Papoular [185] make a similar study without assuming $K_1 = K_3$. The difference in boundary conditions affects the critical field somewhat; it is no longer simply inversely proportional to L, but a more complex function. However, the authors agree that there is no significant difference, at least for $H \approx H_c$. Rapini and Papoular [185] do estimate that small misalignments can have a more serious effect, as is also noted by Pieranski et al. [181]. The possibility suggested by Saupe, that error resulting from the strong anchoring assumption is more serious for H well above H_c, seems not to have been analyzed. For this regime, then, discrepancies between theory and experiment are not fully explained. Barratt and Jenkins [186] as well as Jenkins and Barratt [162] treat cases where strong anchoring obtains at $z = 0$, considered as a solid wall and weak anchoring obtains at $z = L$, visualized as an interface with an isotropic fluid, using energy comparisons to sort out the more relevant solutions. They leave the form of the interfacial energy densities arbitrary. In cases where anchoring is weak enough to let the boundary orientation shift appreciably, such calculations might be used to obtain experimental determinations of the forms of these densities. These studies all deal with situations described by Eqs. (115)–(117), except for the noted replacement of strong by weak anchoring conditions. With MBBA and the experiments of Bouchiat and Langevin-Cruchon [65] in mind, Prost and Gasparoux [187] study the case of strong anchoring where

$$\beta = \tau = 0, \quad H_1 = H, \quad H_2 = H_3 = 0, \quad \alpha(0) = \alpha_0, \quad \alpha(L) = 0, \tag{119}$$

α_0 being an arbitrary constant. Here $z = 0$ is considered as a free surface, $z = L$ being a solid wall. The aforementioned experiments give estimates of α_0. They describe their optical measurements which should, it seems, give experimental values of $\eta = (K_1 - K_3)/K_3$, but not without additional calculation. They report that the assumption that $\eta = 0$ gives results not in accord with experiment.

Physically quite different are the alignment inversion walls of Helfrich [188]. Unlike other walls mentioned earlier, these are simple enough to fall within the analysis here considered. As yet, there seem to be no clear-cut observations of these, presumably because their energy is higher than that of competitors. One, a twist wall, is of the type

$$\alpha = \tau = 0, \quad H_1 = H, \quad H_2 = H_3 = 0, \tag{120}$$

most of the variation in β taking place in a zone whose thickness is of the order of $(K_2/\chi_a)^{1/2} H^{-1}$. The others are of the type

$$\beta = \tau = 0, \quad H_1 = H, \quad H_2 = H_3 = 0. \tag{121}$$

Helfrich remarks that, of the three, the first is, on energetic grounds, favored. For all, **n** undergoes a 180° rotation as we move from $z = -\infty$ to $z = \infty$. Most of the variation occurs in a zone whose thickness is of the order of the coherence length L_c, where L_c is commonly taken as

$$L_c = (K_c/\chi_a)^{1/2} H^{-1},$$

K_c being an appropriate combination of moduli occurring in $\mathscr{W}$. Roughly, it obtains from equating magnetic and orientational energies. From Eq. (118), those Fréedericksz transitions occur when the gap width is of the order of the coherence length and, more generally, this sets a natural scale of length for nematics in magnetic fields. It is then useful for rough estimation, particularly in the common cases where the moduli are all of the same order of magnitude. Clearly, near nematic–smectic A transitions, where large differences occur, more art is involved in judicious choice of K_c.

Various other cases involving twist have been studied. Though they are unusual, there are cholesterics with $\chi_a > 0$. If a magnetic field is applied perpendicular to the twist axis, it then tends to diminish the twist. Rather surprisingly, there are cases where the twist can be reduced to zero with a field of reasonable strength. The effect was first inferred from NMR measurements by Sackmann et al. [189]. The first analysis of this, for an infinite sample, given by de Gennes [190] and, independently, by Meyer [191], employs a solution of the type

$$\alpha = 0, \quad H_1 = H_3 = 0, \quad H_2 = H, \tag{122}$$

with β periodic. Integration constants are fixed by energy arguments. As H increases, the calculated twist, more precisely the spatial period, smoothly decreases, reaching zero at a critical field calculated by de Gennes to be

$$2H_c = \pi\tau(K_2/\chi_a)^{1/2} = (2\pi^2/P)(K_2/\chi_a)^{1/2},$$
$$P = 2\pi/\tau, \tag{123}$$

P being the pitch at $H = 0$. Said differently, the critical condition occurs when the magnetic coherence length L_c is of the order of the pitch. Of course, P sets a natural scale of length for the cholesterics. Dreher [192] allows for the effect of walls, to which the twist axis is normal. Here the pitch increases with H by steps, rather than smoothly. As is discussed by de Gennes [11, Chapter 6], a variety of experiments confirm his prediction in some detail. Two examples, employing magnetic fields, are the work of Durand et al. [193] and Meyer [194]. Employing electric fields, with the dielectric analog, Baessler and Labes [195] and Baessler et al. [196] verify the inverse proportionality of field strength to pitch. Gerritsma and Van Zanten [197] verify the dependence on ε_a, exploiting the fact that it varies with frequency. As is discussed by Kahn [198], there is a blue to red color shift associated with the untwisting. Using this, he checks the dependence of pitch on electric field. Williams and Cladis [199] dope MBBA to give it a slight twist, then untwist it to obtain experimental values of K_2/χ_a. They also employ Eq. (117), with an interference technique, to measure K_3/χ_a. Luckhurst and Smith [96] employ the effect, with an ESR technique, to measure K_2/χ_a at several temperatures. Values obtained this way tend to be higher and more strongly dependent on temperature than some obtaining from observations of Fréedericksz transitions.

A variation of the case covered by Eq. (116) is discussed by Ericksen [200]. Replace $z = L$ by a free surface, for a nematic where **n** tends to align tangent to the free surface, no tangent direction being preferred. One possibility is PAA. In such cases, a strong anchoring assumption must be supplemented by another boundary condition. A reasonable one is that the normal component of the couple stress vanish. This might be viewed as an approximate form of Eq. (63), f being a minimum when $\mathbf{n} \cdot \mathbf{v} = 0$, changing rapidly when its argument strays from this value. This leads to the case

$$\alpha = \tau = 0, \quad H_1 = H_3 = 0, \quad H_2 = H, \quad \beta(0) = 0, \quad \beta'(L) = 0. \tag{124}$$

A Fréedericksz transition is predicted, with

$$2H_c L = \pi (K_2/\chi_a)^{1/2}. \tag{125}$$

In this and other situations, more complicated instabilities involving domain structures may occur instead, as is discussed by de Gennes [*201*, *11*, pp. 134–135].

A different situation is treated by Leslie [*202*]. We first twist a nematic, as discussed in the first example. We then apply a magnetic field parallel to the twist axis. With strong anchoring, a solution is of the form

$$\tau = 0, \quad H_1 = H_2 = 0, \quad H_3 = H, \quad \alpha(0) = \alpha(L) = 0,$$

$$\beta(0) = -\beta_0, \quad \beta(L) = \beta_0, \tag{126}$$

β_0 being chosen to give the proper twist. A Fréedericksz transition occurs to a more distorted configuration calculated to be more stable. For small twists, the critical field is given by

$$\chi_a L^2 H_c^2 = K_1 \pi^2 + 4(K_3 - 2K_2)\beta_0^2. \tag{127}$$

Corresponding experiments of Gerritsma *et al.* [*203*] fix $4\beta_0 = \pi$ and employ five gap widths, checking the dependence of H_c on L. There is satisfactory argument, except for the smallest ($L = 7.2$ μm) for which $H_c L$ is observed to be much smaller. They do note a significant difference between nominal optical and capacitance thresholds. Van Doorn [*119*] gives results of numerical calculations of the director field and at least partially explains the difference. By the more refined optical calculations, the "optical threshold" does not occur at H_c, but at a field varying differently with L. The analogous dielectric experiment is discussed by Schadt and Helfrich [*204*], who find that it is feasible to remove most of the twist with a strong field. Baise and Labes [*205*], employing electric fields and varying ε_a, find reasonably good agreement between theory and experiment, at least for lower values of ε_a.

Leslie [*202*] also discusses a similar problem for cholesterics, adjoined by plates forcing **n** to be tangent to them, but not favoring one tangent direction over another. Again, the normal component of couple stress is required to vanish. These solutions are of the type

$$H_1 = H_2 = 0, \quad H_3 = H, \quad \alpha(0) = \alpha(L) = 0,$$

$$\beta'(0) = \beta'(L) = \tau. \tag{128}$$

With this, he again predicts a Fréedericksz transition provided either

$$2L < [K_2/3(K_3 - K_2)]^{1/2} P, \tag{129}$$

or

$$K_2 > K_3. \tag{130}$$

Here again, the calculations apply only to the relatively rare cholesterics with $\chi_a > 0$. When $\chi_a < 0$, all indications are that the field has no influence. The critical field strength is given by

$$\chi_a H_c^2 = K_1 (\pi/L)^2 + K_3 \tau^2. \tag{131}$$

This seems to be at least roughly consistent with what is observed by Rondelez and Hulin [206] when $L \ll P$, judging from their descriptions; for L/P larger, other things occur, as will be covered in Section VI.

When Eq. (130) applies, the transition may occur with L infinite, and one has a transition which is superficially like that briefly described earlier by Meyer [191]. According to him, the orientation changes smoothly with field strength above the critical field given by Eq. (131), with L infinite, the pitch being inversely proportional to field strength, until one approaches another critical value

$$\bar{H}_c = \tau (K_2^2/K_3)^{1/2}. \tag{132}$$

Here, complete unwinding occurs. Involved in the earlier stages is a pitch contraction, and a (blue) color shift suggesting this has been observed. However, Gerritsma and Van Zanten [207] conclude that the observed shift is explainable without contraction, in imperfectly aligned samples and, more probably, it is the latter which produces the effect. The assumption that the effects of boundaries are insignificant rather naturally attends the assumption that L is infinite. Then, as is discussed by de Gennes [11, p. 240], the first effect of a weak field should be to rotate the chiral axis so that it becomes perpendicular to **H**. Unwinding might then occur as described earlier or, conceivably, a stronger field might induce a transition more like that discussed by Meyer. As an alternative, we might begin with the nematic texture induced by a strong field, then decrease the field strength. If the chiral axis first appears perpendicular to **H**, we might well simply reverse the untwisting described above, beginning at the critical field given by Eq. (123). If the critical field given by Eq. (132) is higher, i.e., if

$$4K_2 > \pi^2 K_3, \tag{133}$$

then, as is discussed by de Gennes [208, 11, p. 244], the chiral axis might first emerge parallel to **H**, at the critical field given by Eq. (132). If some other instability does not intervene, one might see Meyer's pattern evolving as the field strength drops. As is hinted at here, there is some lack of stability analyses in the intermediate field regions which might eliminate ambiguities in prediction. Using electric fields and presuming the dielectric analogy, Baessler and Labes [209] as well as Gerritsma and van Zanten [210] observe an effect resembling that predicted by Meyer, but there are discrepancies between Meyer's prediction and experiment. Neither have independent data to establish that Eq. (130) or Eq. (133) holds and, as is mentioned by de Gennes [11, p. 244], the inequality $K_3 > K_2$ is more common. Perhaps similar but different phenomena are being confused. If Eq. (129) applies, it can only apply for L/P sufficiently small. To describe what commonly occurs for L/P larger and at lower field strengths, we cannot use our "one-dimensional" solutions. We return to this question later.

Greubel [211] considers a somewhat similar problem. He considers a cholesteric, untwisted by a sufficiently strong electric field, strongly anchored with **n** normal to the walls; **E** must also be normal. He then studies what happens as the field is decreased. Again, the dielectric analog is employed, with $\varepsilon_a > 0$. This involves solutions of the type

$$E_1 = E_2 = 0, \quad E_3 = E, \quad \alpha(0) = \alpha(L) = \pi/2. \tag{134}$$

The analysis is much like that employed by Leslie for Eq. (128); the boundary conditions differ. It indicates that nothing happens until the field drops below the critical value given by

$$\varepsilon_a E_c^2 = 4\pi(K_2^2 \tau^2 / K_3 - K_3 \pi^2 / L^2). \tag{135}$$

He estimates that below this, a simple conical configuration may be stable if

$$\pi^2 K_1 K_3^2 > 3L^2 \tau^2 (K_3 - K_2) K_2^2. \tag{136}$$

He reports experimentation on MBBA, strongly anchored to walls coated with a lecithin–ethanol solution. In these, Eq. (136) is violated. He observes a transition to a focal conic texture. By arranging L so that the right member of Eq. (135) is negative, he finds that the nematic texture persists when the field is turned off, in accord with theory.

In brief, study of these simple solutions illuminates a variety of basic physical phenomena. There are shadowy areas remaining to be studied.

2. Radial Variation

Other similarly tractable equations obtain when we introduce the transformations

$$x_1 = e^\xi \cos\theta, \quad x_2 = e^\xi \sin\theta, \quad x_3 = z,$$

$$n_\xi = \sin\alpha, \quad n_\theta = \cos\alpha \cos\beta, \quad n_z = \cos\alpha \sin\beta, \tag{137}$$

where n_ξ, n_θ, and n_z are (physical) components in this curvilinear coordinate system. Clearly, the coordinate curves are the same as for the familiar polar coordinate system with $\xi = \ln r$, so n_ξ, etc., are also physical components relative to the latter. We assume that

$$\alpha = \alpha(\xi), \qquad \beta = \beta(\xi).$$

Again using Eqs. (33) and (59), we calculate that

$$\begin{aligned}
2\overline{\mathscr{W}} = 2e^{2\xi}\mathscr{W} = \; & K_1(\sin\alpha + \alpha'\cos\alpha)^2 \\
& + K_2[\cos^2\alpha(\sin\beta\cos\beta - \beta') + \tau e^\xi]^2 \\
& + K_3[(\alpha'^2 + \beta'^2\cos^2\alpha)\sin^2\alpha \\
& - \sin 2\alpha\cos\beta(\alpha'\cos\beta + \beta'\cos\alpha\sin\alpha\sin\beta) \\
& + \cos^2\alpha\cos^2\beta(1 - \sin^2\beta\cos^2\alpha)] + 2\mathscr{W}_1,
\end{aligned} \qquad (138)$$

where

$$\mathscr{W}_1 = -(K_2 + K_4)(\cos^2\alpha\cos^2\beta)' \qquad (139)$$

will not influence the equilibrium equations, but, in some cases, does influence stability. If the components of **H** depend only on ξ, they must be of the form

$$H_\xi = H_r = ae^{-\xi}, \qquad H_\theta = be^{-\xi}, \qquad H_2 = c, \qquad (140)$$

where a, b, and c are constants. The corresponding magnetic energy is of the form

$$2\mathscr{W}_M = -\chi_a(a\sin\alpha + b\cos\alpha\cos\beta + ce^\xi\cos\alpha\sin\beta)^2 + 2\mathscr{W}_2, \qquad (141)$$

where

$$\mathscr{W}_2 = -\chi_\perp(a^2 + b^2 + c^2 e^{2\xi}) \qquad (142)$$

will not influence the equilibrium equations. Ignoring other possible contributions to the energy, the ordinary differential equations to be satisfied can then be taken as

$$\begin{aligned}
\overline{\overline{U}} = \overline{\mathscr{W}} + \overline{\mathscr{W}}_M - \mathscr{W}_1 - \mathscr{W}_2 & = \overline{\overline{U}} - \mathscr{W}_1 - \mathscr{W}_2, \\
(\partial\overline{\overline{U}}/\partial\alpha')' - \partial\overline{\overline{U}}/\partial\alpha & = 0, \\
(\partial\overline{\overline{U}}/\partial\beta')' - \partial\overline{\overline{U}}/\partial\beta & = 0.
\end{aligned} \qquad (143)$$

U will not depend explicitly on ξ in the cases where

$$c = \tau = 0,$$

which we now assume. We then have the energy integral

$$\overline{\overline{U}} - \alpha'(\partial\overline{\overline{U}}/\partial\alpha') - \beta'(\partial\overline{\overline{U}}/\partial\beta') = \text{const.} \qquad (144)$$

Equilibrium Theory of Liquid Crystals

Also, calculations show that

$$\alpha = ab = 0 \Rightarrow (\partial \overline{\overline{U}}/\partial \alpha) = (\partial \overline{\overline{U}}/\partial \alpha')' = 0, \tag{145}$$

$$\beta = 0 \Rightarrow \partial \overline{\overline{U}}/\partial \beta = \partial \overline{\overline{U}}/\partial \beta' = 0, \tag{146}$$

$$\beta = \pi/2, \; ab = 0 \Rightarrow \partial \overline{\overline{U}}/\partial \beta = \partial \overline{\overline{U}}/\partial \beta' = 0. \tag{147}$$

When $2\alpha = \pi$, β is not well defined, so cases involving this require special attention. When $ab = 0$, there is a degenerate solution of this type,

$$n_\xi = 1, \quad n_\theta = n_z = 0. \tag{148}$$

As might be expected, it is these more tractable cases which are encountered in the literature, and there are examples of the main types. Equation (148), singular on the z-axis, is one of the classical disclination solutions discussed by Frank [18]. The singularity causes the energy integral to diverge. Not uncommonly, these are viewed along the z-axis, making it easy to confuse them with solutions covered by Eq. (147), which may be similarly singular, singular only at a point, or involve no singularity. We here begin to encounter questions which, only rather recently, have received serious attention. With disclinations of the above and certain other types, it is topologically feasible for the orientation pattern to adjust to rid itself of the singular line or curve, perhaps retaining point singularities. To sensibly compare energies of singular vs. nonsingular patterns, one needs to adopt some hypothesis concerning core energies, likely candidates having been discussed in Section II,A,3. One finds that the smoother solutions are sometimes, but not always favored, depending on values of moduli, sample size, etc. Theoretical studies illuminating this have been made by Anisomov and Dzyaloskinskiĭ [212], Barratt [213], Cladis and Kléman [99], Meyer [214] and Williams et al. [215]; there are general discussions by de Gennes [11, Chapter 4] and Saupe [105]. Some, but not all patterns occurring in capillaries conform to solutions of the simple type here considered. The stabler patterns do not always exhibit the symmetry which naive reasoning might lead one to expect. This should not be surprising; naive symmetry arguments would exclude most of the familiar Fréedericksz transitions. Experimentally, it is found that, as is indicated by theory, singular or nonsingular patterns may occur. As is discussed by Williams and Bouligand [216] and made clear in the photographs of Nehring [217], there are optical differences between the singular and nonsingular patterns. Pertinent observations are reported by them, Cladis [218], Meyer [214], Rault [219], and Williams et al. [220]. By varying the temperature near a nematic–smectic A transition, Cladis [218] is able to induce transitions from a planar (disclination) texture to a three-dimensional texture involving point defects. Both she and Williams et al. [221] emphasize the fact that point defects seem to play an important role in flow processes. Later, we will say a bit more about this general topic.

Currie [222] suggests that Eq. (147) includes orientations resembling those occurring in very slow capillary flow, with compatible strong anchoring conditions at the wall, analyzing the various possibilities occurring when $H = 0$. Some involve disclinations, some not. When they do, he assumes that there is a change of phase in the core. In these studies, for example, the value of K_4 does influence stability, as does the critical energy occurring at the phase boundary. Hints as to what occurs in flow might be gleaned from the observations of Cladis [218], for example. There is further discussion of textures occurring in capillaries by Saupe [105]. Some such calculations would be needed to soundly design a capillary method for measuring surface tension, such as is used by Ferguson and Kennedy [82].

An interesting plane pattern given by Eq. (146) with $H = 0$ obtains for the region between two concentric cylinders. One is prepared to give normal, the other parallel orientation, with strong anchoring. This is the "magic spiral" which de Gennes [11, p. 120] attributes to Parodi. De Gennes uses it to illustrate the forces and torques acting on the boundaries.

Leslie [223] and Atkin and Barratt [224] treat cases involving Fréedericksz transitions, including calculations of critical conditions, for samples contained between coaxial cylinders with strong anchoring. Cases treated are azimuthal fields opposing axial orientation and radial field opposing azimuthal or axial orientation. In some cases, the indicated pattern is less stable than another when the field is absent. Critical conditions depend on what ranges moduli lie in and on the ratio of radii of the two cylinders. Conditions under which the elementary patterns become unstable are sorted out by these writers, with some limitations on range of parameters. Here, as in essentially all other studies of a one-dimensional nature, the energy comparisons involve only a limited set of patterns, so a judgment of stability is at best provisional. To show instability, we need only produce one other pattern with less energy so, if one such is included among the set used, the question is settled. At least as yet, there seem not to be experimental checks of critical conditions here calculated.

As we get into these and other nonuniform patterns, it seems worth bearing in mind that they give rise to stresses which may sometimes become appreciable. This is most likely to occur near nematic–smectic A transitions, where some moduli grow larger. One likely consequence is that free surfaces may take unusual shapes. From such idle thoughts, it seems possible that we have not yet fully exploited the solutions here considered.

3. Azimuthal Variation

Again using the coordinates given by Eq. (137), but setting

$$n_\xi = \cos\alpha \cos\beta, \quad n_\theta = \cos\alpha \sin\beta, \quad n_z = \sin\alpha, \tag{149}$$

we can obtain ordinary differential equations by setting

$$\alpha = \alpha(\theta), \qquad \beta = \beta(\theta).$$

A calculation gives

$$\begin{aligned}2\overline{\overline{\mathscr{W}}} = &\; K_1 [\alpha' \sin\alpha \sin\beta - (\beta'+1)\cos\alpha\cos\beta]^2 \\ &+ K_2 [\alpha' \cos\beta + (\beta'+1)\sin\alpha\cos\alpha\sin\beta]^2 \\ &+ K_3 [(\alpha')^2 + (\beta'+1)^2 \cos^2\alpha]\cos^2\alpha\sin^2\beta,\end{aligned} \qquad (150)$$

the term involving K_4 again vanishing. We here consider only nematics, setting $\tau = 0$. With magnetic fields of the form

$$H_\xi = ae^{-\xi}, \qquad H_\theta = be^{-\xi}, \qquad H_z = 0, \qquad (151)$$

the corresponding magnetic energy is, to within an inconsequential term,

$$2\overline{\overline{\mathscr{W}}}_M = -\chi_a(a\cos\beta + b\sin\beta)^2 \cos^2\alpha. \qquad (152)$$

The corresponding equilibrium equations, entirely similar to Eq. (143), admit the energy integral analogous to Eq. (144), with $\overline{\overline{U}}$ merely replaced by $\overline{\overline{U}} = \overline{\overline{\mathscr{W}}} + \overline{\overline{\mathscr{W}}}_M$, as calculated here. Cases considered to date are the plane patterns, with

$$\alpha = 0 \Rightarrow \partial\overline{\overline{U}}/\partial\alpha = \partial\overline{\overline{U}}/\partial\alpha' = 0, \qquad (153)$$

so the energy integral suffices to reduce the solution to quadratures. One such, due to Leslie [223], occurs for a wedge-shaped region, **n** being tangent to the walls, with an azimuthal field. This involves a Fréedericksz transition, and he calculates the critical condition. Actually, in the experiments of Fréedericksz and Zolina [176], the gap width varies, the sample lying between a plate and planoconcave glass. Judging from what they observe, it might appear that, near transition, the narrower part would remain practically undistorted, the wider part distorting appreciably. However, with the azimuthal field, there is a change in field strength with a contrary bias, so it would take a detailed calculation to see if such an effect occurs.

With $\alpha = a = b = 0$, we get a variety of simple disclination solutions. The easiest occur when $K_1 = K_3$, as assumed by Frank [18]. The energy integral then reduces to $\beta' = \text{const.}$ His assumption, generally accepted, is that **n** is continuous, or merely reverses direction as we follow a circuit around $r = 0$ ($\xi = -\infty$), so

$$2\beta = m\theta + \text{const.}, \qquad m = \pm 1, \pm 2, \ldots \qquad (154)$$

As is discussed by Frank [18], observed patterns correspond to smaller values of m; the energy goes as m^2. Here m is the topological index of these plane vector fields, a quantity which has meaning for more complex patterns.

According to the old empirical rules set down by Friedel [6], these occur in sets whose indices sum to zero; two attract or repel each other according as their indices are of opposite or like sign. Two can coalesce or one can break up to form two or more. In such cases, the total topological index, the sum of the individual contributions, is unchanged. Ericksen [97] suggests a rationale for such rules. Roughly, they accord with the view that lower energy configurations are more probable. As is discussed by him and, in more detail, by Dzyaloskinskiĭ [225], similar solutions are obtainable when $K_1 \neq K_3$. The patterns are qualitatively similar, with one exception. For $m = 1$, some of the patterns previously admitted by assigning the arbitrary constant are excluded; only the radial and circular patterns survive. This fault stems from the fact that, with $K_1 = K_3$ and planar orientations, the basic equations are invariant under independent rotations of **n** and the coordinates. For any pattern, this is true of the energy given by Eq. (16). It would not be implausible to expect this degeneracy to produce atypical predictions on occasion, and we here have an example. Anisomov and Dzyaloskinskiĭ [212] analyze, in some detail, stability of such simpler disclinations.

As was mentioned earlier, it can be topologically feasible for such plane patterns to adjust to rid themselves of the singularity, forming a three-dimensional pattern. As is discussed by de Gennes [11, Chapter 4], the primary condition for this is that the topological index m be even. As was mentioned before, the smoothing need not occur if m is even; what happens depends on where the energy advantage lies.

More complex "umbilical" defects, or apparent defects are sometimes found in patterns associated with Fréedericksz transitions, when one sees **n** projected on a suitable plane. The projected vector is then somewhat analogous to **n** for a plane disclination. Rapini [226] presents an approximate theory useful for analyzing these.

4. OTHER CASES

There are other cases, very likely some undiscovered, where the equilibrium equations are reducible to ordinary differential equations. If one introduces spherical coordinates, and assumes that corresponding components of **n** depend only on the latitude angle, this occurs if the magnetic field is suitably restricted. Corresponding equations are recorded by Atkin and Barratt [224], who find them not very tractable. With some approximations, they treat a sample bounded by a cone and a plate, subject to fields directed along latitude or longitude lines, finding phenomena akin to Fréedericksz transitions.

Using the special form of $\mathscr{W}$ given by Eq. (16), Meyer [214] notes some curious cases where, in polar coordinates, components of **n** depend in particular ways on θ, otherwise depending on ζ or r in a way determined by solving equilibrium equations, reducible to ordinary differential equations in r. Such cases are also discussed by Saupe [*105*]; they generate some disclination or pseudo-disclination solutions of interest for capillaries.

For rough calculations concerning nematics, it is sometimes useful to note that, with $\mathscr{W}$ of the form given by Eq. (16) taken as the only volume energy, there are solutions of the form

$$\mathbf{n} = (\cos\alpha, \sin\alpha, 0), \tag{155}$$

in cartesian coordinates, where α may depend on all three coordinates. The equilibrium equations then take the very simple form

$$\nabla^2 \alpha = 0. \tag{156}$$

Normally, α is treated as a scalar in transforming to curvilinear coordinates. Thus, without linearizing, but with a special relation between moduli, we get a familiar linear equation. As already noted, this form has degenerate invariance features which may induce qualitative error. Also, it is worth bearing in mind that, even if a solution is not bad, one may be misled if one uses the same form to estimate its stability. This point is emphasized by Turner [*172*], who gives an illustrative example of some importance. However, it is only to be expected that that tractability will induce workers to use it in the future, as they have in the past.

The disclination solutions given by Eq. (154) can also be regarded as solutions of Eq. (156), and we can superpose to get solutions involving sets of disclinations. By solving a problem in potential theory, one can then adjust this, within some limits, to meet desired boundary conditions. This, and the calculation of forces acting on disclinations is discussed in some detail by Dafermos [*227*] as well as Nehring and Saupe [*228*]. This lays the groundwork for calculation of forces exerted on a disclination by boundaries or other disclinations. Some of the simpler examples are treated by Imura and Okano [*229*] and de Gennes [*11*, Chapter 4]. Cladis [*218*] analyzes the problem of two disclinations in a capillary, with normal orientation at the walls, with corresponding observations. Nehring [*230*] treats curved disclinations associated with oppositely twisted regions occurring in the *plage tordue*. He also treats cases where the disclination is smoothed out, becoming tangent to the apparent line. The latter do not have the form given by Eq. (137). He presents photographic evidence of the existence of both types. Turner [*172*] presents a different approximate analysis, evading use of the special form of $\mathscr{W}$ which, he indicates, accords better with observations. In discussing their observations of surface singularities associated with twist, Williams *et al.* [*230*] employ simple solutions, given in polar coordinates by

$$\alpha = ar^b \sin b\theta, \tag{157}$$

where *a* and *b* are constants.

For cholesterics, the analogous Eq. (35), inserted into the equilibrium Eq. (57), yields two equations for α, reducible to

$$K\nabla^2 \alpha + M\alpha_{,33} = 0, \tag{158}$$

$$((|\tau - \alpha_{,3}|)^{1/2} \sin \alpha)_{,1} = ((|\tau - \alpha_{,3}|)^{1/2} \cos \alpha)_{,2}, \tag{159}$$

if the x_3-axis is the twist axis. Here the second equation derives from $h_3 = 0$, a condition trivially satisfied for nematics. In cases like this, where the basic representation of **n** does not cover all possibilities, one must take some care to see that all equilibrium equations are satisfied. Clearly, Eqs. (158) and (159) admit some solutions of the form

$$\alpha = \tau x_3 + \beta(x_1, x_2), \quad \nabla^2 \beta = 0, \tag{160}$$

a slightly more general linear dependence on x_3 being feasible if α depends only on x_3. Solutions of this type are briefly mentioned by Frank [*18*], in an early sketch of a treatment of some disclinations in cholesterics. Rather obviously, we can merely superpose the twist on the disclination solutions obtaining for nematics, to obtain such solutions. Cladis and Kléman [*232*] seem to imply that β must be constant. If our algebra is correct, they have slipped in their calculation of the molecular field, their Eq. (16). Of course, the alternative, $\alpha = \alpha(x_3)$, is one of the planar cases described earlier. We then conclude that, excepting the latter case, n_3 must become nonzero whenever we try to change the natural twist. For this reason, de Gennes' [*233*] analysis of the Grandjean–Cano wedge is incomplete, for he satisfies only the first equation. The same is true of Scheffer's [*234*] analysis, which puts the solution in neater form and otherwise gives a more general analysis of the basic problem. Here, a specimen is anchored, more or less strongly, to the walls of a (small angle) wedge-shaped opening. It is observed that orientation singularities occur periodically to help the natural twist adjust to the boundary conditions. By knowing the wedge angle and measuring the distance between singularities, one obtains one of the older estimates of the natural twist. De Gennes' solution exhibits such a singularity. Cladis and Kléman [*99, 232*] discuss somewhat similar patterns, assessing the error involved in violating Eq. (159). Similar reasoning might be used in de Gennes' case. Caroli and Dubois-Violette [*235*] use perturbation methods to remove the restriction on moduli inherent in de Gennes' work, but again ignore the second equation. De Gennes [*236*] does mention that Caroli and Dubois-Violette had become aware of the discrepancy and, apparently, had made some error estimate. It is noted by the Orsay group [*237*], for example, that different types of singularities are seen and, as is discussed by Friedel and Kléman [*101*], there is a rather different topological model of the wedge, in better accord with some of the observations. It would seem feasible to further improve analysis of the

Equilibrium Theory of Liquid Crystals

Grandjean–Cano wedge, but a neat, simple solution seems not to exist. As is observed and roughly analyzed by Rault [*238*], patterns somewhat similar in nature occur in drops with a free surface in the presence of magnetic fields.

To accommodate more realistic forms of $\mathcal{W}$, when **n** can be expected to be consistent with Eq. (155), one can work out a perturbation procedure. As a first guess, take the appropriate solution of Laplace's equation. This will not satisfy the correct equations. However, if the two energy functions involved are not too different, one might sensibly get a correction by linearizing about the first guess, proceeding by iteration to get successive approximations. The linear equations to be solved are Poisson equations, as is illustrated by the example of Caroli and Dubois-Violette [*235*]. In fact, it would seem wise to use this or some other method for a check more often, since the special case sometimes yields unrepresentative results.

Suppose a glass plate is rubbed and that this produces more or less periodic undulations in the glass, perpendicular to the direction of rubbing. A nematic sample, say PAA, is placed in contact with it. An intuitive argument suggests what should happen, neglecting other influences. Assume an interfacial energy of the form given by Eq. (28). Often, with PAA, **n** is tangent to glass, suggesting that

$$f[(\mathbf{n}\cdot\mathbf{v})^2] \geq f(0).$$

Then, the bulk and surface energies can both be minimized only if **n** is constant and parallel to the surface, requiring it to line up parallel to the grooves. We could minimize the surface energy by keeping **n** tangent but, say, perpendicular to the grooves. Then, the finer the grooves, the more rapidly **n** must change as we traverse the grooves. This will produce large gradients driving up the bulk energy near the surface, at least within the limits where it can be estimated by the Oseen–Frank theory. It might seek relief by giving up the strong anchoring condition and probably would, if it had no alternative. There is then at least this mechanism for the observed alignment produced by rubbing, discussed in Section II,A,2.

To illustrate the effect, Berreman [*69*] uses a simple solution of Eq. (156), viz.

$$\alpha = abe^{-bx_2}\sin bx_1, \tag{161}$$

where a and b are constant. For ab small, a matching undulating surface is, approximately,

$$x_2 = -a\cos bx_1.$$

One can of course fix the value of the small number

$$\varepsilon = ab,$$

and increase b to get more rapid oscillations. With

$$bx_2 = \varepsilon x_2/a,$$

we see that x_2 must be quite large compared to the amplitude a of the surface oscillation before the exponential factor gets small. Wolff et al. [70] present an approximate analysis, not employing special relations between moduli, with similar conclusions. With the rubbing effect, we exploit desirable effects of controlled surface roughness. At the same time, we might well heed the warning; small surface roughness can produce significant effects on orientation. In this light, it is perhaps surprising that theory and experiment agree as well as they do. Of course, there is a moderating factor; where the effect is big enough, the careful observer will see it and react accordingly.

Since this is a matter of some importance, and variations of the argument are useful elsewhere, it is perhaps worthwhile to consider a scaling analysis. With $\mathscr{W}$ given by Eq. (10) as the only bulk energy, the transformation

$$\tilde{x}_i = cx_i, \qquad \tilde{n}_i = n_i(\tilde{\mathbf{x}}),$$

c any positive constant, will take equilibrium solutions into equilibrium solutions. Were there a magnetic field, we would have to divide the field strength by c. In obvious notation, we calculate that

$$\tilde{\mathscr{W}} = c^{-2}\mathscr{W}.$$

The transformation preserves angles, hence typical boundary conditions of the strong anchoring type. To fix our ideas, suppose we have a solution defined above a surface oscillating about the plane $x_3 = 0$, described by a periodic function

$$x_3 = g(x_1, x_2) = g(x_1 + T_1, x_2) = g(x_1, x_2 + T_2),$$

T_1 and T_2 being nonzero constants. The transformation will yield a different surface, again periodic, with periods

$$\tilde{T}_1 = cT_1, \qquad \tilde{T}_2 = cT_2;$$

with $c < 1$ it will oscillate more rapidly. Then the energy E_p associated with a period cell, assumed finite, changes by the rule

$$\tilde{E}_p = \int_0^{T_2}\int_0^{T_1}\int_{\tilde{g}}^{\infty} \tilde{\mathscr{W}}\, dx_3\, dx_1\, dx_2 = cE_p,$$

where

$$\tilde{g} = cg(\tilde{x}_1/c, \tilde{x}_2/c).$$

The volume of the period cell is infinite. However, the finite cross sectional area goes as

$$\tilde{A} = c^2 A,$$

whether one takes a plane cross section or the undulating cap of a period cell.

Thus the energy per unit surface area goes as

$$\tilde{E}_p/\tilde{A} = c^{-1}E_p/A. \tag{162}$$

Typical volumes go as c^3, so energy per unit volume goes as c^{-2}. If $E_p \neq 0$, i.e., if there is nonuniformity in the orientation, this energy per unit area will, for c sufficiently small, exceed the maximum surface energy per unit area, roughly the maximum value of f in Eq. (28), granted that this is finite. Thus, for a surface undulating very rapidly, the sample would be better off to uniformly orient, even if it maximizes the surface energy in so doing this. If no other influences are of importance, the best uniform **n** would, of course, be that which minimizes the surface energy for the undulating surface. Given the form of f and the surface, one could estimate this. With indications that controlled irregularities can be advantageous, the converse problem might be worthwhile; for what shapes of surfaces can we make **n** align in an assigned direction?

An amusing problem seemingly analyzed using Eq. (156) is mentioned by de Gennes [239], who does not give details. It is the "Archimedes problem" of a cylinder floating above a flat surface in a nematic fluid, with **n** normal to the cylinder and wall. Roughly, the resulting distortion gives rise to a stress, which can be calculated using Eq. (70). It helps buoy up the cylinder. He estimates that, for a cylinder of density 2 gm/cm^3, and a radius of the order of a micron or less, the cylinder would float near the top, being close to the bottom if the radius is 20 times larger; with moduli of the order of 10^{-6} dynes, it is hard to generate much stress. Again, such effects are likely to be enhanced near the nematic–smectic A transitions.

VI. LINEAR PROBLEMS

Of course, analysis of the various linearizations of the nonlinear equations is much more a matter of routine. Along the way, we have mentioned some linear analyses, and a collection of physically interesting examples is available in the book by de Gennes [11]. Thus we will treat this area only sketchily.

In Section V, we discussed what happens to a cholesteric, in a few cases, when it is subject to a uniform magnetic field. In cases not covered there, important advances have been made, using linear theory. The Fréedericksz transition, with critical condition given by Eq. (131), applies only for L/P relatively small, and nothing else there discussed covers what should occur if the field strength is not too high. In the regime where L/P is large, we encounter the Helfrich–Hurault transitions first roughly analyzed by Helfrich [240], with Hurault [241] correcting and improving the analysis. Again, **H** is parallel to the twist axis, $\chi_a > 0$, etc. Briefly, for H below the critical value H_c given by

$$\chi_a H_c^2 = 2\pi^2 (6K_2 K_3)^{1/2}/PL, \tag{163}$$

the naturally twisted pattern is unaffected. For fields slightly above this, a square periodic pattern is predicted, the "layers" perpendicular to the twist axis now undulating with a period σ given by

$$2\sigma^2 = (3K_3/2K_2)^{1/2}PL. \tag{164}$$

Here one must be particularly cautious in employing the usual dielectric analogy. As is discussed by Helfrich [59] and de Gennes [11, Chapter 6], there is a rather similar pattern originating from electrohydrodynamic effects, not predictable from equilibrium theory. The naive analogy yields different critical fields, etc. According to Gerritsma and Van Zanten [242], similar periodic patterns can be induced by temperature gradients or mechanical actions; de Gennes [11, Fig. 7.9] presents a photograph of a very regular pattern induced by tension. It is well to bear in mind that theory underlying two patterns with some common features may be quite different. It is not yet very certain how far we have come in understanding such subtleties.

There are experiments employing magnetic fields of the Helfrich–Hurault transitions by Scheffer [243] or Rondelez and Hulin [206]. Scheffer uses Helfrich's equation for H_c rather than Hurault's correction [Eq. (163)]. He finds some discrepancy between the value of $K_2 K_3$, calculated from the critical field and that obtaining from other measurements. Otherwise, the measurements accord well with theory. The reader might like to calculate for himself how the fit is altered by using Eq. (163). Experimentally, if one continues to increase H above H_c, one encounters another transition to a "fingerprint" texture. Qualitatively, it seems that the twist axis is attempting to become perpendicular to the field, the helix to untwist; we are beginning the untwisting process described earlier.

Linear analyses and experiments covering a variety of similar problems are presented by Hurvet et al. [244]. These treat cases where χ_a may be positive or negative, **H** parallel or perpendicular to the twist axis, with L/P small, near one, or large. Among other things, the Helfrich–Hurault transitions can give way to a striped pattern if L/P is near one.

Typically, in such calculations, sensibly linearized equations are used, H occurring as a parameter. At a critical value H_c, bifurcation occurs, there being some multiplicity of solutions satisfying appropriate boundary conditions. Even if we ignore nonequilibrium effects, such as the electrohydrodynamic, this time-honored procedure is not foolproof. Abrupt transitions, such as the left-right transitions in the *plage-tordue*, can well be lost in the linearization. In the case at hand, the justification for the linearization is supplied more by the experimentalists.

Not infrequently, foreign matter is introduced into nematics. This may be by design, as in cases where we add chiral matter to give it a twist, or add impurities to decorate disclinations, or add elongate dye particles to make a shutter. Of course, impurities also occur accidentally. Linear analyses of such problems are feasible, at least in the "dilute suspension" approximation, where interactions between two such bodies can be ignored. De Gennes [*11*, p. 236] treats the twisting induced by chiral agents, for example. There is a relatively long range interaction; one body disturbs the nematic orientation, influencing the environment of its neighbors. Thus the "dilute suspension" assumption is somewhat restrictive. Our knowledge of how such bodies interact with walls, imperfections, other bodies, etc., is still very sketchy.

De Gennes [*11*, Chapter IV] also gives a linear analysis of domain structures which may form in a nematic sample with a free surface in the presence of a magnetic field.

The linear equation (106) for smectics A is quite simple, and it is easy to generate a great variety of particular solutions, a bit more difficult to match them to situations of physical interest. Among the more interesting possibilities is the analysis of dislocations similar to those occurring in crystalline solids. Roughly, one can disrupt the near planar layering by removing part of a layer or, if you prefer, by adding a piece of a layer, as is discussed by Bouligand [*121*]. Optical observations of edge dislocations in smectics A are reported by Williams and Kléman [*245*]. A simple solution of this type is given by de Gennes [*246*, *11*, pp. 300–301]. Kléman [*150*] gives a simple solution for a straight screw dislocation and for a circular defect. He also gives a general discussion of interactions between such defects. In some such analyses, one sees the need for reliable prescriptions for calculating forces and moments. Kléman and Williams [*247*] analyze forces exerted on each other by two parallel edge dislocations. Under certain conditions discussed by de Gennes [*248*], one may expect formation of phases containing rather regular arrays of dislocations. As he explains, this is based on an interesting analogy with superconductors. Scheffer *et al*. [*249*] observe a phase more or less like this.

In principle, smectics A should, theoretically, admit instabilities analogous to the Helfrich–Hurault instabilities for cholesterics. Estimates such as are given by de Gennes [*11*, p. 291] indicate that relatively high field strengths and larger samples would be needed to make the effect observable and, as yet, these seem not to have been observed. Similarly, theory predicts analogs of the Fréedericksz transitions for nematics but, as is discussed by Rapini [*156*], for example, requirements for their observability are not easily met. Parodi [*145*] discusses the possibility of magnetically induced transitions involving defects; some might be observable. There is an interesting instability, missed by the indicated linear theory, but predictable from a very slightly nonlinear theory. This occurs when a sample is between parallel plates; a tension tending to pull the plates apart can cause the layers to buckle in an undulating mode. Theory and corresponding experiments are presented by Delaye *et al*. [*250*].

This yields an experimental value for the modulus λ occurring in Eq. (106). It is close to 20 Å for CBOOA. Even here, the experimental part of the study is not strictly within the realm of static equilibrium. A transient undulation is observed, induced by touching the sample. The authors expect that, on a longer time scale, dislocations will come in to relax the distortion. There is a similar study by Clark and Meyer [251]. They find no distortion at very low strain rates presumably because of such dislocation motion. Also, they briefly discuss similar phenomena in cholesterics. In such geometries, small irregularities in the plates can produce sizable changes in interior orientations. Because λ has length as its physical dimension, solutions of Eq. (106), or nonlinear versions, do not scale as for nematics, so a scaling argument is trickier. However, a linear analysis is feasible, de Gennes [11, pp. 287–288] discussing one attributed to Durand and Clark. For a wavy plate, of wavelength l, the distortion is significant to a depth of the order of l^2/λ. With $\lambda \approx 20$ Å, this is easily large compared to l. It seems likely that some such wall irregularity produces the static distortion observed by Ribotta et al. [252], and at least some of those observed by Clark and Pershan [253], who describe an optical method for measuring λ. It does appear that a common presumption—that λ is of the order of the layer thickness (20–30 Å)—is not far wrong.

Kléman and Parodi [151] discuss some simple solutions for a different linearization, perturbing about textures where the layers have the form of circular cylinders. They cover a spiral dislocation, solutions oscillating with the polar coordinate z, and solutions depending exponentially on z. They suggest that the latter might correlate with orientations observed in some of the focal conic textures.

The older theory of Frank [18] does not lend itself well to the analysis of problems such as are here mentioned; the presumed structure has too much rigidity, and this will, no doubt, promote its replacement by some theory better adapted to the treatment of boundary value problems, etc. It would be rash to say that any such theory is well established, even for the smectics A, but the approach initiated by de Gennes [147] does show promise. As yet, for smectics A, there seem to be no gross inconsistencies between theory and experiment.

ACKNOWLEDGMENTS

The research reported herein was supported by a grant from the National Science Foundation. I am greatly indebted to numerous workers for supplying me with preprints, reprints, and helpful advice. Especial thanks go to P. G. de Gennes, J. T. Jenkins, and F. M. Leslie. Mrs. Mary K. Thuma has done yeoman's work in library searches and in converting my bad handwriting into legible form. Such faults as remain are rightfully attributed to the author.

REFERENCES

1. G. H. Brown and W. G. Shaw, *Chem. Rev.* **57**, 1049–1157 (1957).
2. G. H. Brown, J. W. Doane, and V. D. Neff, *CRC Crit. Rev. Solid State Sci.* **1**, 303–379 (1970).
3. P. Chatelaine, *Bull. Soc. Fr. Mineral.* **77**, 323–352 (1954).
4. I. G. Chistyakov, *Usp. Fiz. Nauk* **89**, 563–602 (1966); *Sov. Phys.—Usp.* **9**, 551–573 (1967).
5. G. Durand and J. D. Litster, *Annu. Rev. Mat. Sci.* **3**, 269–292 (1973).
6. G. Friedel, *Ann. Phys. (Paris)* **18**, 273–474 (1922).
7. G. Friedel and E. Friedel, *Z. Kristallogr.* **79**, 1–60 (1931).
8. Ch. Maugin, *in* "Traité de Chimie Organique" (de V. Grignard, ed.), Vol. 1, pp. 81–119. Masson, Paris (1934).
9. A. Saupe, *Angew. Chem., Int. Ed. Engl.* **7**, 97–112 (1968).
10. A. Saupe, *Annu. Rev. Phys. Chem.* **24**, 441–471 (1973).
11. P. G. de Gennes, "The Physics of Liquid Crystals." Oxford Univ. Press (Clarendon), London and New York, 1974.
12. G. W. Gray, "Molecular Structure and the Properties of Liquid Crystals." Academic Press, New York, 1962.
13. M. J. Stephen and J. P. Straley, *Red. Mod. Phys.* **46**, 617–704 (1974).
14. J. L. Fergason, T. R. Taylor, and T. B. Harsch, *Electro-Technol. (New York)* **85**, 41–50 (1970).
15. G. H. Heilmeier, *Sci. Amer.* **222**, 100–106 (1970).
16. H. Kelker, *Mol. Cryst. Liquid Cryst.* **21**, 1–48 (1973).
17. Э. Л. Аэро, А. Н. Бчлыпцн, "Гидромеханика," том7, Всесоюзный институт научной технической, информаии, Moscow, 1973.
18. F. C. Frank, *Discuss. Faraday Soc.* **25**, 19–28 (1958).
19. C. W. Oseen, *Ark. Mat., Astr. Fys. A* **19**, 1–19 (1925).
20. H. Zocher, *Phys. Z.* **28**, 790–796 (1927).
21. H. Zocher, *Trans. Faraday Soc.* **29**, 945–957 (1933).
22. H. Zocher, *Mol. Cryst. Liquid Cryst.* **7**, 165–180 (1969).
23. J. L. Ericksen, *Quart. Appl. Math.* **25**, 474–479 (1968).
24. P. G. de Gennes, *Phys. Lett. A* **30**, 454–455 (1969).
25. P. G. de Gennes, *Mol. Cryst. Liquid Cryst.* **12**, 193–214 (1971).
26. C. Fan and M. J. Stephen, *Phys. Rev. Lett.* **25**, 500–503 (1970).
27. J. A. Fisher, Doctoral Dissertation, University of Minnesota (1969).
28. J. T. Sullivan, Doctoral Dissertation, University of Minnesota (1966).
29. G. L. Hand, *J. Fluid Mech.* **13**, 33–46 (1962).
30. T. C. Lubensky, *Phys. Rev. A* **2**, 2497–2514 (1970).
31. M. Miesowicz, *Bull. Int. Akad. Polon. Cl. Sci. Math. Nat. Ser. A* pp. 228–247 (1936).
32. M. Miesowicz, *Nature (London)* **158**, 27 (1946).
33. B. D. Coleman, *Arch. Ration. Mech. Anal.* **20**, 41–58 (1965).
34. C.-C. Wang, *Arch. Ration. Mech. Anal.* **20**, 1–40 (1965).
35. P. C. Martin, R. S. Pershan, and J. Swift, *Phys. Rev. Lett.* **25**, 844–848 (1970).
36. J. D. Lee and A. C. Eringen, *J. Chem. Phys.* **54**, 5027–5034 (1971).
37. M. Shahinpoor, *J. Chem. Phys.* **63**, 1319–1320 (1975).
38. F. M. Leslie, *Arch. Ration. Mech. Anal.* **28**, 265–283 (1968).
39. F. M. Leslie, *Proc. Roy. Soc., Ser. A* **307**, 359–372 (1968).
40. M. E. Gurtin, *Arch. Ration. Mech. Anal.* **52**, 93–103 (1973).

41. J. T. Jenkins, *Mol. Cryst. Liquid Cryst.* **18**, 309–312 (1972).
42. J. Nehring and A. Saupe, *J. Chem. Phys.* **54**, 337–343 (1971).
43. J. L. Ericksen, *Phys. Fluids* **9**, 1205–1207 (1966).
44. J. L. Ericksen, *Arch. Ration. Mech. Anal.* **10**, 189–196 (1962).
45. W. E. Warren, *Quart. J. Mech. Appl. Math.* **23**, 525–547 (1970).
46. C. W. Oseen, *Trans. Faraday Soc.* **29**, 883–899 (1933).
47. H. J. Deuling and W. Helfrich, *Appl. Phys. Lett.* **25**, 129–130 (1974).
48. H. Gruler and G. Meier, *Mol. Cryst. Liquid Cryst.* **16**, 299–310 (1972).
49. M. Schadt, *J. Chem. Phys.* **56**, 1494–1497 (1972).
50. E. F. Carr, *Mol. Cryst. Liquid Cryst.* **7**, 253–268 (1969).
51. R. B. Meyer, *Phys. Rev. Lett.* **22**, 918–921 (1969).
52. W. Helfrich, *Phys. Lett. A* **35**, 393–394 (1971).
53. W. Haas, J. Adams, and J. B. Flannery, *Phys. Rev. Lett.* **25**, 1326–1327 (1970).
54. D. Schmidt, M. Schadt, and W. Helfrich, *Z. Naturforsch. A* **27**, 277–280 (1972).
55. H. J. Deuling, *Solid State Commun.* **14**, 1073–1074 (1974).
56. W. Helfrich, *Appl. Phys. Lett.* **24**, 451–452 (1974).
57. C. Fan, *Mol. Cryst. Liquid Cryst.* **13**, 9–15 (1971).
58. A. Derzhanski and A. G. Petrov, *Phys. Lett. A* **34**, 427–428 (1971).
59. W. Helfrich, *Mol. Cryst. Liquid Cryst.* **21**, 187–209 (1973).
60. L. Cheung, R. B. Meyer, and H. Gruler, *Phys. Rev. Lett.* **31**, 349–352 (1973).
61. A. F. Schenz, V. D. Neff, and T. W. Schenz, *Mol. Cryst. Liquid Cryst.* **23**, 59–67 (1973).
62. P. G. de Gennes, Gravitational instabilities of liquid crystals, *in* "Proceedings of the Erice Summer School on Quantum Electronics," 1974. North-Holland Publ., Amsterdam.
63. F. Grandjean, *Bul.. Soc. Fr. Mineral.* **29**, 164–213 (1916).
64. Ch. Mauguin, *C. R. Acad. Sci., Paris* **156**, 1246–1247 (1913).
65. M. Bouchiat and D. Langevin-Cruchon, *Phys. Lett. A* **34**, 331–332 (1971).
66. V. Naggiar, *C. R. Acad. Sci., Paris* **208**, 1916–1918 (1939).
67. P. Chatelaine, *Bull. Soc. Fr. Mineral.* **66**, 105–130 (1943).
68. H. Zocher and K. Coper, *Z. Phys. Chem.* **132**, 295–302 (1928).
69. D. W. Berreman, *Phys. Rev. Lett.* **28**, 1683–1686 (1972).
70. U. Wolff, W. Greubel, and H. Krüger, *Mol. Cryst. Liquid Cryst.* **3**, 187–196 (1973).
71. J. E. Proust, L. Ter-Minassian-Saraga, and E. Guyon, *Solid State Commun.* **11**, 1227–1230 (1972).
72. J. L. Janning, *Appl. Phys. Lett.* **21**, 173–174 (1972).
73. E. Guyon, P. Pieranski, and M. Boix, *Lett. Appl. Engr. Sci.* **1**, 19–24 (1973).
74. G. D. Dixon, T. P. Brody, and W. A. Hester, *Appl. Phys. Lett.* **24**, 47–49 (1974).
75. F. J. Kahn, *Appl. Phys. Lett.* **20**, 199–201 (1972).
76. T. Uchida, H. Watanabe, and M. Wada, *Jap. J. Appl. Phys.* **11**, 1559–1565 (1972).
77. A. Rapini, Thèse 3e cycle, Orsay (1970).
78. I. Haller, *J. Chem. Phys.* **57**, 1400–1405 (1972).
79. C. W. Oseen, *Z. Kristallogr.* **79**, 173–185 (1931).
80. S. Chandrasekhar, *Mol. Cryst.* **2**, 71–80 (1966).
81. M. J. Press and A. S. Arrott, *Phys. Rev. Lett.* **33**, 403–406 (1974).
82. A. Ferguson and S. J. Kennedy, *Phil. Mag.* **26**, 41–49 (1938).
83. V. Naggiar, *Ann. Phys. (Paris)* **18**, 5–55 (1943).
84. W. M. Schwartz and H. W. Moseley, *J. Phys. Colloid Chem.* **51**, 826–837 (1947).
85. D. Langevin and M. A. Bouchiat, *Mol. Cryst. Liquid Cryst.* **22**, 317–331 (1973).
86. J. L. Ericksen, *Quart. J. Mech. Appl. Math.* **27**, 213–219 (1974).
87. E. Dubois-Violette and O. Parodi, *J. Phys. (Paris)* **30**, C4, 57–64 (1969).

88. P. M. Naghdi, *in* "Handbuch der Physik" (C. Truesdell, ed.), Vol. VIa/2, pp. 424–640. Springer Verlag, Berlin and New York, 1972.
89. J. T. Jenkins, "The Equations of Mechanical Equilibrium of a Model Membrane." Submitted for publication.
90. R. B. Meyer and P. S. Pershan, *Solid State Commun.* **13**, 989–992 (1973).
91. W. Helfrich, *Z. Naturforsch. C* **28**, 693–703 (1973).
92. C.-K. Yun, Doctoral Dissertation, University of Minnesota (1970).
93. J. Rault, *J. Phys. (Paris)* **33**, 383–395 (1972).
94. A. Saupe, *Z. Naturforsch. A* **15**, 810–814 (1960).
95. J. Nehring and A. Saupe, *J. Chem. Phys.* **56**, 5527–5528 (1972).
96. G. R. Luckhurst and H. J. Smith, *Mol. Cryst. Liquid Cryst.* **20**, 319–341 (1973).
97. J. L. Ericksen, "Liquid Crystals and Ordered Fluids" (J. F. Johnson and R. S. Porter, eds.), pp. 181–193. Plenum, New York, 1970.
98. O. Lehmann, "Flüssige Kristalle und ihre Scheinbares Leben." Voss, Leipzig, 1921.
99. P. E. Cladis and M. Kléman, *J. Phys. (Paris)* **33**, 591–598 (1972).
100. M. Kléman and J. Friedel, *J. Phys. (Paris)* **30**, C4, 43–53 (1969).
101. J. Friedel and M. Kléman, *Nat. Bur. Standards Spec. Publ.* **317**, 607–636 (1970).
102. Y. Bouligand and M. Kléman, *J. Phys. (Paris)* **31**, 1041–1054 (1970).
103. J. Rault, *Solid State Commun.* **9**, 1965–1969 (1971).
104. Orsay Group, *J. Phys. (Paris)* **30**, C4, 38–42 (1969).
105. A. Saupe, *Mol. Cryst. Liquid Cryst.* **21**, 211–238 (1973).
106. R. B. Meyer, *Mol. Cryst. Liquid Cryst.* **16**, 355–369 (1972).
107. M. Kléman, *J. Phys. (Paris)* **34**, 931–935 (1973).
108. J. L. Ericksen, *Trans. Soc. Rheol.* **11**, 5–14 (1967).
109. J. T. Jenkins, *J. Fluid Mech.* **45**, 465–475 (1971).
110. E. F. Carr, J. H. Parker, and D. P. McLemore, "Liquid Crystals and Ordered Fluids" (J. F. Johnson and R. S. Porter, eds.), pp. 201–213. Plenum, New York, 1970.
111. J. S. Prasad and M. S. Madhava, *Mol. Cryst. Liquid Cryst.* **22**, 165–174 (1973).
112. S. C. Chou, L. Cheung, and R. B. Meyer, *Solid State Commun.* **11**, 977–981 (1972).
113. D. Coates and G. W. Gray, *Phys. Lett. A* **45**, 115–116 (1973).
114. R. Dreher and G. Meier, *Solid State Commun.* **13**, 607–610 (1973).
115. R. D. Ennulat, *Mol. Cryst. Liquid Cryst.* **13**, 337–355 (1971).
116. J. L. Fergason, *Mol. Cryst.* **1**, 293–307 (1966).
117. I. Teucher, K. Ko, and M. M. Labes, *J. Chem. Phys.* **56**, 3308–3311 (1972).
118. C. Z. van Doorn and J. L. A. M. Heldens, *Phys. Lett. A* **47**, 135–136 (1974).
119. C. Z. van Doorn, *Phys. Lett. A* **42**, 537–539 (1973).
120. H. Sackmann and D. Demus, *Mol. Cryst. Liquid Cryst.* **2**, 81–102 (1966).
121. Y. Bouligand, *J. Phys. (Paris)* **33**, 525–547 and 715–736 (1972).
122. Y. Bouligand, *J. Phys. (Paris)* **34**, 603–614 and 1011–1020 (1973).
123. Y. Bouligand, *J. Phys. (Paris)* **35**, 215–236 (1974).
124. L. Eisenhart, "An Introduction to Differential Geometry." Princeton Univ. Press, Princeton, New Jersey, 1947.
125. V. Hlavatý, "Differentialgeometrie der Kurven und Flächen." Noordhoff, Gröningen-Batavia, 1939.
126. G. Friedel and F. Grandjean, *Bull. Soc. Fr. Mineral.* **33**, 409–465 (1910).
127. W. Bragg, *Trans. Faraday Soc.* **29**, 1056–1060 (1933).
128. W. Bragg, *Proc. Roy. Inst. G. Brit.* **28**, 57–92 (1933).
129. K. K. Kobayashi, *Mol. Cryst. Liquid Cryst.* **13**, 137–148 (1971).
130. W. L. McMillan, *Phys. Rev. A* **4**, 1238–1246 (1971).
131. P. G. de Gennes, *Solid State Commun.* **10**, 753–756 (1972).

132. P. E. Cladis, *Phys. Rev. Lett.* **31**, 1200–1203 (1973).
133. R. Alben, *Mol. Cryst. Liquid Cryst.* **20**, 231–238 (1973).
134. L. Leger, *Phys. Lett. A* **44**, 535–536 (1973).
135. M. Delaye, R. Ribotta, and G. Durand, *Phys. Rev. Lett.* **31**, 443–445 (1973).
136. L. Cheung and R. B. Meyer, *Phys. Lett. A* **43**, 261–262 (1973).
137. S. Torza and P. E. Cladis, *Phys. Rev. Lett.* **32**, 1406–1409 (1974).
138. R. S. Pindak, C.-C. Huang, and J. T. Ho, *Phys. Rev. Lett.* **32**, 43–46 (1974).
139. P. G. de Gennes, *Mol. Cryst. Liquid Cryst.* **21**, 49–76 (1973).
140. P. E. Cladis and S. Torza, *J. Appl. Phys.* **46**, 584–599 (1975).
141. P. E. Cladis, *Phys. Rev. Lett.* **35**, 48–51 (1975).
142. J. A. Guerst, *Phys. Lett. A* **34**, 283–284 (1971).
143. J. A. Guerst, *Phys. Lett. A* **37**, 279–280 (1971).
144. M. Kléman, *Proc. Roy. Soc., Ser. A* **347**, 387–404 (1976).
145. O. Parodi, *Solid State Commun.* **11**, 1503–1507 (1972).
146. L. D. Landau and E. M. Lifshitz, "Statistical Physics." Pergamon, London, 1958.
147. P. G. de Gennes, *J. Phys. (Paris)* **30**, C4, 65–71 (1969).
148. W. Helfrich, *Phys. Rev. Lett.* **23**, 372–374 (1969).
149. R. Ribotta, R. B. Meyer, and G. Durand, *J. Phys. (Paris) Lett.* **35**, 161–164 (1974).
150. M. Kléman, *J. Phys. (Paris)* **35**, 595–600 (1974).
151. M. Kléman and O. Parodi, *J. Phys. (Paris)* **36**, 671–681 (1975).
152. P. G. de Gennes and G. Sarma, *Phys. Lett. A* **38**, 219–220 (1972).
153. F. Jahnig and H. Schmidt, *Ann. Phys.* **71**, 129–166 (1972).
154. J. D. Lee and A. C. Eringen, *J. Chem. Phys.* **58**, 4203–4211 (1973).
155. P. Martin, O. Parodi, and P. Pershan, *Phys. Rev. A* **6**, 2401–2429 (1972).
156. A. Rapini, *J. Phys. (Paris)* **33**, 237–247 (1972).
157. A. Saupe, *Mol. Cryst. Liquid Cryst.* **7**, 59–74 (1969).
158. Orsay Liquid Crystal Group, *Solid State Commun.* **9**, 653–655 (1971).
159. F. Grandjean, *C. R. Acad. Sci., Paris* **166**, 165–167 (1918).
160. A. S. C. Lawrence, *J. Roy. Microsc. Soc.* **58**, 30–48 (1938).
161. R. B. Meyer, L. Liebert, L. Strzelecki, and P. Keller, *J. Phys. (Paris) Lett.* **36**, 69–71 (1975).
162. J. T. Jenkins and P. J. Barratt, *Quart. J. Mech. Appl. Math.* **27**, 111–127 (1974).
163. J. L. Ericksen, *Arch. Ration. Mech. Anal.* **9**, 371–378 (1962).
164. R. A. Toupin, *Arch. Ration. Mech. Anal.* **11**, 387–414 (1962).
165. L. C. Young, "Lectures on the Calculus of Variations and Optional Control Theory." Saunders, Philadelphia, Pennsylvania, 1969.
166. L. Davison, *Phys. Fluids* **10**, 2333–2338 (1967).
167. J. T. Jenkins, *SIAM J. Appl. Math.* **25**, 603–612 (1973).
168. V. Zwetkoff, *Acta Physiocochim (USSR)* **6**, 865–894 (1937).
169. H. J. Deuling, E. Guyon, and P. Pieranski, *Solid State Commun.* **15**, 277–279 (1974).
170. C. W. Oseen, *Ark. Mat. Astr. Fys. A* **21**, 1–35 (1928).
171. Ch. Mauguin, *Bull. Soc. Fr. Mineral.* **34**, 71–117 (1911).
172. R. Turner, *Phil. Mag.* **30**, 13–20 (1974).
173. J. Friedel and P. G. de Gennes, *C. R. Acad. Sci. Paris* **268**, 257–259 (1969).
174. A. M. J. Spruijt, *Solid State Commun.* **13**, 1919–1922 (1973).
175. J. L. Ericksen, *J. Fluid Mech.* **27**, 59–64 (1967).
176. F. Fréedericksz and V. Zolina, *Trans. Faraday Soc.* **29**, 919–930 (1933).
177. F. Brochard, *J. Phys. (Paris)* **33**, 607–611 (1972).
178. L. Leger, *Solid State Commun.* **11**, 1499–1501 (1972).
179. A. Saupe, *Z. Naturforsch. A* **15**, 815–822 (1960).

180. A. Rapini, M. Papoular, and P. Pincus, *C. R. Acad. Sci., Paris* **267**, 1230–1233 (1968).
181. P. Pieranski, F. Brochard, and E. Guyon, *J. Phys. (Paris)* **33**, 681–689 (1972).
182. H. J. Deuling, *Mol. Cryst. Liquid Cryst.* **19**, 123–131 (1972).
183. C. M. Dafermos, *SIAM J. Appl. Math.* **16**, 1305–1318 (1968).
184. B. Malraison, P. Pieranski, and E. Guyon, *J. Phys. Lett. (Paris)* L-9-10 (1974).
185. A. Rapini and M. Papoular, *J. Phys. (Paris)* **30**, C-4, 54–56 (1969).
186. P. J. Barratt and J. T. Jenkins, *J. Phys. A. Math. Nucl. Gen.* **6**, 756–769 (1973).
187. J. Prost and H. Gasparoux, *C. R. Acad. Sci., Paris* **273**, 335–338 (1971).
188. W. Helfrich, *Phys. Rev. Lett.* **21**, 1518–1521 (1968).
189. E. Sackmann, S. Meiboom, and L. C. Snyder, *J. Amer. Chem. Soc.* **89**, 5981–5982 (1967).
190. P. G. de Gennes, *Solid State Commun.* **6**, 163–165 (1968).
191. R. B. Meyer, *Appl. Phys. Lett.* **12**, 281–282 (1968).
192. R. Dreher, *Solid State Commun.* **13**, 1571–1574 (1973).
193. G. Durand, L. Leger, F. Rondelez, and M. Veyssie, *Phys. Rev. Lett.* **22**, 227–228 (1969).
194. R. B. Meyer, *Appl. Phys. Lett.* **14**, 208–209 (1969).
195. H. Baessler and M. M. Labes, *Phys. Rev. Lett.* **21**, 1791–1793 (1968).
196. H. Baessler, T. M. Larouge, and M. M. Labes, *J. Chem. Phys.* **51**, 3213–3219 (1969).
197. C. J. Gerritsma and P. Van Zanten, *Phys. Lett. A* **42**, 127–128 (1972).
198. F. J. Kahn, *Phys. Rev. Lett.* **24**, 209–212 (1970).
199. C. Williams and P. E. Cladis, *Solid State Commun.* **10**, 357–359 (1972).
200. J. L. Ericksen, *ZAMP* **20**, 383–388 (1969).
201. P. G. de Gennes, *Solid State Commun.* **8**, 213–216 (1970).
202. F. M. Leslie, *Mol. Cryst. Liquid Cryst.* **12**, 57–72 (1970).
203. C. J. Gerritsma, W. H. de Jeu, and P. Van Zanten, *Phys. Lett. A* **36**, 389–390 (1971).
204. M. Schadt and W. Helfrich, *Appl. Phys. Lett.* **18**, 127–128 (1971).
205. A. I. Baise and M. M. Labes, *Appl. Phys. Lett.* **24**, 298–300 (1974).
206. F. Rondelez and J. P. Hulin, *Solid State Commun.* **10**, 1009–1012 (1972).
207. C. J. Gerritsma and P. Van Zanten, *Phys. Lett. A* **42**, 329–330 (1972).
208. P. G. de Gennes, *Mol. Cryst. Liquid Cryst.* **7**, 325–345 (1969).
209. H. Baessler and M. M. Labes, *J. Chem. Phys.* **51**, 1846–1852 (1969).
210. C. J. Gerritsma and P. Van Zanten, *Mol. Cryst. Liquid Cryst.* **15**, 257–258 (1971).
211. W. Greubel, *Appl. Phys. Lett.* **25**, 5–7 (1974).
212. S. I. Anisimov and I. E. Dzyaloskinskiĭ, *Zh. Eksp. Teor. Fiz.* **63**, 1460–1471 (1972); *Sov. Phys.—JETP* **36**, 774–779 (1973).
213. P. J. Barratt, *Quart. J. Mech. Appl. Math.* **27**, 505–522 (1974).
214. R. B. Meyer, *Phil. Mag.* **27**, 405–424 (1973).
215. C. Williams, V. Vitek, and M. Kléman, *Solid State Commun.* **12**, 581–584 (1973).
216. C. Williams and Y. Bouligand, *J. Phys. (Paris)* **35**, 589–593 (1974).
217. J. Nehring, *Phys. Rev., A* **7**, 1737–1748 (1973).
218. P. E. Cladis, *Phil. Mag.* **29**, 641–663 (1974).
219. J. Rault, Thèse, Orsay (1971).
220. C. Williams, P. Pieranski, and P. E. Cladis, *Phys. Rev. Lett.* **29**, 90–92 (1972).
221. C. E. Williams, P. E. Cladis, and M. Kléman, *Mol. Cryst. Liquid Cryst.* **21**, 355–373 (1973).
222. P. K. Currie, *Trans. Soc. Rheol.* **17**, 197–208 (1973).
223. F. M. Leslie, *J. Phys. D: Appl. Phys.* **3**, 889–897 (1970).
224. R. J. Atkin and P. J. Barratt, *Quart. J. Mech. Appl. Math.* **26**, 109–128 (1973).
225. I. E. Dzyaloskinskiĭ, *Zh. Eksp. Teor. Fiz.* **58**, 1443–1452 (1970); *Sov. Phys.—JETP* **31**, 773–777 (1970).
226. A. Rapini, *J. Phys. (Paris)* **34**, 629–633 (1973).

227. C. Dafermos, *Quart. J. Mech. Appl. Math.* **23**, S 49–64 (1970).
228. J. Nehring and A. Saupe, *J. Chem. Soc.* **68**, II, 1–15 (1972).
229. H. Imura and K. Okano, *Phys. Lett. A* **42**, 405–406 (1973).
230. J. Nehring, *Phys. Rev. A* **7**, 1737–1748 (1973).
231. C. Williams, V. Vitek, and M. Kléman, *Solid State Commun.* **12**, 581–584 (1973).
232. P. E. Cladis and M. Kléman, *Mol. Cryst. Liquid Cryst.* **16**, 1–20 (1972).
233. P. G. de Gennes, *C. R. Acad. Sci., Paris* **266**, 571–573 (1968).
234. T. J. Scheffer, *Phys. Rev. A* **5**, 1327–1336 (1972).
235. C. Caroli and E. Dubois-Violette, *Solid State Commun.* **7**, 799–802 (1969).
236. P. G. de Gennes, *Mol. Cryst. Liquid Cryst.* **7**, 325–345 (1969).
237. Orsay Liquid Crystal Group, *Phys. Lett. A* **28**, 687–688 (1969).
238. J. Rault, *Mol. Cryst. Liquid Cryst.* **16**, 143–152 (1972).
239. P. G. de Gennes, "Nematodynamics," Les Houches Lectures (1973). Gordon & Breach, New York. To be published.
240. W. Helfrich, *Appl. Phys. Lett.* **17**, 531–532 (1970).
241. J. P. Hurault, *J. Chem. Phys.* **59**, 2068–2075 (1973).
242. C. J. Gerritsma and P. Van Zanten, *Phys. Lett. A* **32**, 47–48 (1971).
243. T. J. Scheffer, *Phys. Rev. Lett.* **28**, 593–596 (1972).
244. H. Hurvet, J. P. Hurault, and F. Rondelez, *Phys. Rev. A* **8**, 3055–3064 (1973).
245. C. E. Williams and M. Kléman, *J. Phys. Lett. (Paris)* **35**, L-33-35 (1974).
246. P. G. de Gennes, *C. R. Acad. Sci., Paris* **275**, 939–941 (1972).
247. M. Kléman and C. Williams, *J. Phys. Lett. (Paris)* **35**, L-49-51 (1974).
248. P. G. de Gennes, *Solid State Commun.* **10**, 753–756 (1972).
249. T. J. Scheffer, H. Gruler, and G. Meier, *Solid State Commun.* **11**, 253–257 (1972).
250. M. Delaye, R. Ribotta, and G. Durand, *Phys. Lett. A* **44**, 139–140 (1973).
251. N. A. Clark and R. B. Meyer, *Appl. Phys. Lett.* **22**, 493–494 (1973).
252. R. Ribotta, G. Durand, and J. D. Litster, *Solid State Commun.* **12**, 27–29 (1973).
253. N. A. Clark and P. S. Pershan, *Phys. Rev. Lett.* **30**, 3–6 (1973).

SUBJECT INDEX

A

Absorption modes, 206
Accordion mode, 204, 206, 223, 228
Accordionlike vibration, 207
Achiral compounds, 79
Aeronautical applications, 154
Aerosol OT, 2
Alicyclic rings, 13
Alignment, 80
  perfect, 235
Alka-2,4-dienoic acids, 9
Alkoxy group, 13
Alkoxyazoxy benzines, 218, 223
4'-n-Alkoxybenzylidene-2-methyl-4-n-alkylanilines, 42
4-(p-n-Alkoxybenzylideneamino), 41
4-(p-n-Alkoxybenzylideneamino) acetophenones, 29
2-(p-n-Alkoxybenzylideneamino) fluorenones, 29
4'-n-Alkoxybiphenyl-4-carboxylic acids, 40
4'-n-Alkoxy-4-cyanobiphenyls, 12, 29, 97, 98
Alkyl chains, odd, 28
4'-n-Alkyl-4-cyanobiphenyls, 12, 24, 29, 57, 58
4''-n-Alkyl-4-cyano-p-terphenyl, 57
Alkyl terminal substituents, 11
Alkyl thoicarbonates, 63
Alkylalkoxy compounds, 13
  dipole moments, 13
4-n-Alkylbenzoic acids, 14
4-n-Alkylcyclohexane-1-carboxylic acids, 14
Amines, 181
1-Amino-octane, 190
1-Amino-octane acid, 140
Amphiphile bilayer, 177
Amphiphiles, 2
Androst-5-en-3β-ol, 115
Androst-5-en-3β-ol-17-one, 115

Androst-5,6-en-3β-ol, 67
Androst-5,6-en-3β-ol-17-one, 67
5α-Androstan-3β-ol, 115
Angle of incidence, 92
Angular resolution, 80
Anharmonic force theory, 95
Anisoles, p-substituted, 12
Anisotropic, 6
Anisotropy, 6
  polarizability, 6
  dipolarity, 6
APAPA, 211
Applications
  aeronautical, 154
  medical, 154
  space, 154
Aromatic rings
  p-phenylene, 10
  planar, 10
  polarizable, 10
  rigid, 10
Aryl benzoate esters, 10
Attraction, lateral, 33, 78
Attraction potential, 193
Axial polarizability, 28, 59
Axis
  chiral, 247
  helical, 247
  twist, 247
20-Azacholest-5-en-3β-ol, 115
Azimuthal variation, 282
Azoxy linking, 42
p-Azoxyanisole, 242
4,4'-Azoxyphenol, 29

B

Band shape, 201
Bâtonnets, 76, 256
BBCA, 210
Bending, 239

Subject Index

Bend/splay elastic constant, 209
Bent state, 252
p-Benzoyloxybenzoates, 42
Benzylideneanilines, 54
1,4-Bicyclo[2,2,2]octylene rings, 13, 53
Biphenyl nucleus, 46
4,4'-Biphenylene ring, 11
"Blue phase," 76
Boltzmann factors, 205, 225
Branching, 48, 51
   chain, 51
   terminal alkyl chains, 48
Brassicasterol, *see* (22E,24R)-Argosta- 5,22-dien-3β-ol
Brownian movement, 193
Bulk energies, 237
n-*p*-Butoxybenzylidene, *p*'-cyanoaniline (BBCA), 210
p-Butoxybenxylidene-*p*-*d*-methylbutylaniline, 251

C

Campesterol, *see* (24R)-Ergost-5-en-3β-ol
17β-Carbomethyxyandrosta-3,5-dien-3β-ol, 115
17β-Carbomethoxyandrost-5-en-3β-ol, 115
17β-Carbomethoxyandrost-5,6-en-3β-ol, 67
20β-Carbomethoxy-pregn-5-en-3β-ol, 115
Carbon substituents
   p-CH$_3$, 25
   p-Cl, 25
Chain
   axis, 316
   branching, 51
   flexing, 29
   fluidity, 206, 208
   molecules, 6
Chalinasterol, *see* 24-Methylcholesta-5,24(28)-dien-3β-ol
Chemical stability, 80
Chiral axis, 247
Chiral mesogens, 57, 58, 62
Chirality, 227
4-Chloro-ω-phenylalkyl esters, 36
Chol-5-en-3β-ol, 116
Chol-5-en-3β-ol-22-one, 116
Cholest-5-en-3β-ol-22-one, 111
Cholest-5-en-3β-ol-24-one, 112

(Cholest-5,6-en-3β-ol)-structure, 62
Cholest-5-enes, 3β-substituted, 100
5α-Cholest-8,14-en-3β-ol, 64
5α-Cholest-8(14)-en-3β-ol, 64
5α-Cholest-6,7-en-3β-yl, 66
5α-Cholest-7,8-en-3β-yl *n*-alkanoates, 64
5α-Cholest-2,3-en-3β-yl 4'-*n*-alkoxybenzoates, 64
Cholesta-3,4;5,6-dien-3β-ol, 64
Cholesta-5,6;20,21-dien-3β-ol, 66
Cholesta-5,6;22,23-dien-3β-ol, 66
Cholesta-5,6;25,26-dien-3β-ol, 66
Cholesta-5,20-dien-3β-ol, 114
Cholesta-5,22-dien-3β-ol, 114
Cholesta-5,24-dien-3β-ol, 114
Cholesta-5,25-dien-3β-ol, 114
5α-Cholesta-7,22-dien-3β-ol, 119
5α-Cholesta-8,24-dien-3β-ol, 117
5α-Cholesta-14,25-dien-3β-ol, 119
5α-Cholesta-7,8;9,11-dien-3β-yl benzoate, 66
Cholesta-6,7;8,14;9,11-trien-3β-yl octadecanoate, 65
Cholestadienes, derivatives of, 117
Cholestadienols, 65
5α-Cholestan-3β-ol, homologous series, 99, 103
5α-Cholestan-3β-*n*-alkanoates, 63
5α-Cholestanes, 3β-substituted, 99, 100
Cholestanol, homologous series, 63, 65, 103
Cholestatrienes, derivatives of, 117
Cholestenes, derivatives of, 118
Cholestanol, 65
Cholesteric focal conics, 76
Cholesteric mesophase, 75
Cholesteric–nematic mixtures, 136
Cholesteric phase, 3, 75, 225
   vibrational spectra, 225
Cholesterics, 234, 246
Cholesterogens, 4, 31, 62
Cholesterol, 17, 63, 100
   homologous series, 103
   phosphate, 181
   sulfate, 181
Cholesteryl acetate, 226
Cholesteryl 1-admantanecarboxylate, 101
Cholesteryl *n*-alkanoates, 31, 63
   Cinnamates, substituted, 101
   2,2-Dimethylpropionate (pivalate), 101
Cholesteryl carbonates, 106ff

Cholesteryl esters, 225, 226, 227
Cholesteryl 2-(2'-butoxyethoxy)ethyl carbonate, 79
  ω-ferrocenylalkanoates, 101
  ω-phenylalkanoates, 36, 103
Cholesteryl formate, 226
Cholesteryl iodide, 79
Cholesteryl 2-methylpentanoate, 79
Cholesteryl propionate, 225 226
Cholesteryl propionate (V), 4
Cholesteryl S-alkyl thiocarbonate, 107
Cholesteryl tetradecanoate, 76
Circular defect, 291
Circular dichroism, 79, 91, 249
Clearing point, 235
Clionasterol, see (24S)-Stigmast-5-en-3β-ol
Coalescence, 193
Color band, 127
"Complex" formation, 190, 186
Composite structures, 153
Compositions eutectic, 55
Conjugated system, 24
Conjugative interactions, 11, 24, 53, 61
Continuous films, 141
Continuum
  limit, 250
  theory, 233
Coplanar, 53
Coprostanol, 63
Core energies, 246
Correlation functions, 212
Crinosterol, see (22E,24S)-Ergosta-5,22-dien-3β-ol
Cubic phases, 7
  symmetry, 178
p-Cyanobenzylidene-p'-octyloxyaniline, 251
α-Cyanostilbenes, 48
Cycloartanol, see 9,19-Cyclo-9β-lanostan-3β-ol
Cycloartenol, see 9,19-Cyclo-9β-lanost-24-en-3β-ol
Cycloeucalanol, see 4α,14α,24ζ-Trimethyl-9,19-cyclo-5α,9β-cholestan-3β-ol
Cycloeucalenol, see 4α,14α-Dimethyl-9,19-cyclo-5α,9β-ergost-24(28)-en-3β-ol
Cycloartanyl palmitate, 65, 66
Cycloeucalenyl esters, 66
Cyclohexanone, 17
Cyclohexyl ring, 16
1,4-Cyclohexylene ring, 14

9,19-Cyclo-9β-lanostan-3β-ol, 120
9,19-Cyclo-9β-lanost-24-en-3β-ol, 120

D

Decanol/potassium oleate, 188
Desmosterol, see Cholesta-5,24-dien-3β-ol
Degree of motional freedom, 211
Degree of orientation, 235, 245
Depolarization ratio, 210
Deuterated lipids, 228
Devices
  electrical, 146
  mechanical, 146
  optical, 146
  thermal, 146
Dialkoxy
  compounds, 13
  dipole moments, 13
  series, 21
4,4'-Dialkoxyazoxy benzenes, 211
4,4'-Di-(p-n-alkoxybenxylidenamino) benzidines, 43
Diamantane, 8
p,p'-Diaminophenyl, 211
20,25-Diazacholest-5-en-3β-ol, 115
Dicholesteryl dicarboxylates, 101
Dielectric anisotropy, 55, 57, 91
Diethylazoxybenzoate, 223
Diethyl 4,4'-azoxybenzoate (III), 4
Diethylazoxycinnamate, 223
4,4''-Diethyl-p-terphenyl, 17
4,4'-Diheptyloxy azoxybenzene, 212
22,23-Dihydrobrassicasterol, see (24S)-Ergost-5-en-3β-ol
Dihydrolanosterol, see Lanost-8-en-3β-ol
"Dilute suspension," 291
4,4-Dimethyl-5α-cholest-7-en-3β-ol, 119
4α,14α-Dimethyl-5-cholest-8-en-3β-ol, 119
4,4-Dimethyl-5α-cholest-8(14)-en-3β-ol, 120
4α,14α-Dimethyl-5α-cholest-9(11)-en-3β-ol, 120
4,4-Dimethyl-5α-cholesta-8,14-dien-3β-ol, 120
4α,14α-Dimethyl-9,19-cyclo-5α,9β-cholestan-3β-ol, 120
4α,14α-Dimethyl-9,19-cyclo-5α,9β-ergost-24(28)-en-3β-ol, 121
4α,14α-Dimethyl-5α-cholest-9,11-en-3β-yl acetate, 64, 66

17,17β-Dimethyl-D-homo-18-nor-5α-androsta-13,15,17-trion-3β-ol, 115
Dimers, lathlike, 20
Dimodan LS, 185
4,4″-Di-n-pentyl-p-terphenyl, 17
Dipalmitoyl phosphatidyl choline, 228
Dipalmitoyl phosphatidylethanolamine, 228
1,2-Diphenylethanes, 9
Dipolar properties, 61
Dipolarity, 6, 40
Dipole–dipole forces, 6
Dipole–induced dipole forces, 6
Dipole moments, 13
  alkylalkoxy, 13
  dialkoxy, 13
  off-axial, 24
Director, 235
Disclination solutions, 285
Disclinations, 245
Dispersion forces, 6, 28, 40
  attractive, 28
Dispersive reflection, 74
Distribution function, 210
4,4′-Disubstituted biphenyls, 12, 56
4,4″-Disubstituted p-terphenyls, 12
Divergence theorem, 238
Dupin cyclides, 89, 250
Dynamic scattering, 55, 57
Dynamic scattering mode, 211
Dynamical theory, 237

E

EBBA, 20, 211
Edge energies, 245
Effective rotary power, 99
Elastic dislocation theory, 245
Elastic constants, 209
Elasticity, 236
Electrical devices, 146
Electrohydrodynamic effects, 241
π-Electron system, 13, 14
Electron withdrawing, 12
Electrohydrodynamic theory, 240
Emulsifier/emulsifier interaction, 186
Emulsifier–water systems, 179
Emulsion stability, 190
Emulsions, 173
Enantiotropic
  nematic phase, 35

smectic phase, 62
S_A, 36
Encapsulated liquid crystals, 142
Energetics, 236
Energy coupling, 144
Enthalpy of fusion, 57
Entropy of transition, 8
Epicholesterol, 63
Equilibrium
  equations, 257
  theory, 233, 234, 240
(24R)-Ergost-5-en-3β-ol, 111
(24S)-Ergost-5-en-3β-ol, 111
(22E,24R)-Ergosta-5,22-dien-3β-ol, 111
(22E,24S)-Ergosta-5,22-dien-3β-ol, 111
(22E,24R)-Ergosta-5,7,22-trian-3β-ol, 113
Erythrocyte ghosts, 229
Esters of 4-(p-cyanobenzylideneamino)cinnamic acid, 36
p-(p-Ethoxyphenyl)heptonoate, 206
p-(p-Ethoxyphenyl azo)phenylundecylenate, 206
Ergosterol, see (22E,24R)-Ergosta-5,7,22-trien-3β-ol
Ethyl-(p-methoxybenzylidene amino)cinnamate, 223
Eutectic compositions, 55

F

Faraday rotation, 241
Fermi resonance, 202
Ferroelectric liquid crystals, 257
Ferroelectric phase, 215
Flat molecules, 7
Flexing motion, 28
Flexo-electric effect, 240
Flocculated droplets, 191
Flocculation, 191, 193
Focal conic texture, 31, 76, 250
Fourth-order moments, 235
Free energy difference, 28
Fréedericksz transition, 273
Frequency maximum, 201, 206
Fucosterol, see (24E)-Stigmasta-5,24(28)-dien-3β-ol

G

Gaussian curvatures, 252
Gel phase, 176, 181, 228

Subject Index

I, 182
II, 182
Glassy liquid crystals, 101
  cholesteric, 136
Globular molecules, 7
Goossens' theory, 79
Grandjean terraces, 257
Gravitational field, 241
  potential, 241
Group efficiency order, 60

H

Handedness, 79
Helical
  arrangement, 89
  axis, 247
  structure, 7, 77
  twisting power, 78
Helix, 75
Helmholtz free energy, 237
Hemoglobin, 229
Heptyloxyazoxybenzene, 223
Heterocyclic rings, 17, 21
Heteronitrogen, 19, 21
Hexagonal liquid crystals, 177
Homeotropic texture, 76
Honeycomb texture, 252
Hydrocarbon/emulsifier interaction, 189
Hydrophilic amphiphiles, 2
Hydrophilic surface properties, 176
Hydrophobic amphiphiles, 2, 187
2-Hydroxy Schiff's bases, 41
Hysteresis effect, 228
Hysteretic mixtures, 136

I

Imbedded liquid crystals, 143
Imperfection, 245
  theory, 246
Induced dipoles, 204
Induced dipole–induced dipole forces, 6
Inelastic scattering, 222
Infrared
  dichroism, 204
  radiation, 148
Intensity, 201
  integrated, 206
Interactions, steric, 28

Interchain distance, 187
Interdigitated
  bilayer, 25
  lamellar smectic, 59
Interfacial
  effects, 241
  energy, 244
Intermolecular
  forces, 6, 206
    dipole–dipole, 6
    dipole–induced dipole, 6
    induced dipole–induced dipole, 6
  modes, 213
Internal vibrations, 201, 204, 222
Inversed hexagonal liquid crystals, 177
Ionizing radiation, 147
Isotropic liquid
  form, 239
  phase, 7
  viscous, 7

K

22-Ketocholesterol, see Cholest-5-en-3β-ol-22-one
24-Ketocholesterol, see Cholest-5-en-3β-ol-24-one

L

L-shape molecule, 34, 36
Lamellar
  arrangement, 176
  smectic order, 40
Lanost-8-en-3β-ol, 119
Lanosta-8,24-dien-3β-ol, 120
13α,14β,17β$_H$-Lanosta-8,24-dien-3β-ol, 119
Lanosterol, see Lanosta-8,24-dien-3β-ol
Lathosterol, see 5β-Cholest-7-en-3β-ol
Lattice
  layer crystal, 5
  modes, 213, 216, 220
  vibrations, 204, 213, 216, 226
Lauryl alcohol, 186
Lauryl sulfate, 186
Layer crystal lattices, 5, 59
Librating, 226
Linear
  problems, 289
  theory, 236, 267

Liposomes, 179
Liquid-condensed phase, 183
Lophenol, see 4α-Methyl-5α-cholest-7-en-3β-ol
Liquid crystals
  hexagonal, 177
  imbedded, 143
  mixed, 132
Liquid crystal phase
  cholesteric, 3
  nematic, 3
  smectic, 3
LS phase, 183
Long chain lipids, 227
Low frequency region, 226
Lyotropic, 2
  mesophase, 227
  vibrational spectra, 227

M

Macroscopic motion, 236
Magic spiral, 282
Maier–Saupe theory, 13, 40
Marker's acid, 101
Mass density, 241
MBBA, 20, 211, 213, 218
Mean curvature, 252
Mechanical bulk energy, 237
Mechanical coupling forces, 201
Mechanical
  devices, 146
  response, 235
  theory, 235
Medical applications, 154
  cutaneous thermography, 156
  dentistry, 157
  gynecology, 158
  neurology, 158
  oncology, 158
  ophthalmology, 159
  pediatrics, 159
  radiology, 160
  surgery, 160
  urology, 160
5-Membered rings, 17
6-Membered rings, 17
Membrane mesicules, 229
Memory displays, 57
Mesogens, 3, 39

chiral, 62
optically active, 61
steryl, 67
Mesophase, 74
  cholesteric, 75
  lyotropic, 2
  nematic, 75
  nonamphiphilic, 3
  smectic, 75
  thermotropic, 2
Metallographical applications, 153
4′-Methoxybenxylidene-4-n-butylaniline (II), 4
(p-Methoxybenzylidene amino)cinnamate, 223
17β-Methyl-5α-androstan-3β-ol, 115
Methyl 3β-chlorochol-5-en-24-oate, 115
Methyl 3β-hydroxychol-5-en-24-oate, 115
Methyl 3β-hydroxy-24-norchol-5-en-23-oate, 115
4α-Methyl-5α-cholest-7-en-3β-ol, 119
14α-Methyl-9,19-cyclo-5α,3β-cholestan-3β-ol, 120
24ξ-Methyl-9,19-cyclolanostan-3β-ol, 121
p-Methoxybenzylidene-p'-n-butylaniline, 240
4′-(2″-Methylbutyloxy)-4-cyanobiphenyl, 98
24-Methylchol-5-en-3β-ol, 116
24-Methylchol-5-en-3β-ol-22-one, 116
24α-Methylcholesta-5,6;7,8;22,23-trien-3β-yl alkanoates, 66
24-Methylcholesta-5,24(28)-dien-3β-ol, 111
24-Methylenecycloartanol, see 4,4,14α-Trimethyl-9,19-cyclo-5α,9β-ergost-24(28)-en-3β-ol
24-Methylenecycloartanyl esters, 66
17β-(1-Methylheptyl) androst-5-en-3β-ol, 116
2-Methylhydroquinone, 42
17β-(1-Methyloctyl) androst-5-en-3β-ol, 116
4′Methyl-ω-phenylalkyl esters, 36
17β-(1-Methyl-5-phenylpentyl) androst-5-en-3β-ol, 114
20β-Methylpregn-5-en-3β-ol, 116
5-Methylthienyl groups, 23
Micellar solubilization, 189
Microemulsions, 187
Microwave radiation, 149
Miscibility, 24
Mixed liquid crystals, 132

Modes, 201
 absorption, 206
 accordion, 204, 223, 228
 C—C, 228
 dynamic scattering, 211
 intermolecular, 213
 lattice, 213, 216, 220
 Raman, 223
 soft, 217
 stretching, 208
Molecular
 arrangement, 74
 broadening, 40, 45
 conformation, 206
 field, 258
 geometry, 1
 length/breadth ratio, 29
 optical activity, 75
 polarizability, 28, 34, 36, 41, 53
 rigidity, 11
 structure, 58
 vector, 235
Molecules
 chiral, 6
 globular, 7
 lathlike, 5, 7, 11, 53, 99
Monolaurin, 176
Monotropic phase, 224
 nematic, 35
 smectic, 62
Mono-substituted di-anils, 47
Monotropic S_A, 35, 36

N

Nematic–chiralic mixtures, 137
Nematic phase, 3, 75
 enantiotropic, 35
 internal vibrations, 201
 monotropic, 35
Nematics, 234, 237, 246
Nematic–isotropic liquid transition, 235
Nonamphiphilic liquid crystals, 1ff
 molecules, spherical, 7
  lathlike, 7
Noncholesterogens, 5, 62
Noncoplanarity, 41
Nondestructive testing, 152
 aeronautical and space applications, 154
 composite structures, 153

 metallographical applications, 153
 quality control of components, 153
Nonequilibrium phenomena, 237
Nonlinear problems, 270
Nonoriented chains, 187
Nonsingular patterns, 281
Nonsteroidal cholesteric liquid crystals, 122
24-Norchol-5-en-3β-ol, 116
24-Norchol-5-en-3β-ol-22-one, 116
21-Norcholest-5-en-3β-ol, 114
21-Norcholest-5-en-3β-ol-20-one, 114
27-Norcholest-5-en-3β-ol, 116
27-Norcholest-5-en-3β-ol-22-one, 116
27-Norcholest-5-en-3β-ol-25-one, 114
31-Norcycloartanol, see 4α,14α-Dimethyl-9,19-cyclo-5α,9β-cholestan-3β-ol

O

O-Alkyl carbonates, 63
Octadecanoic acids, 182
1-Octanic acid, 190
4'-n-Octyloxybiphenyl-4-carboxylic acid, 46
4'-n-Octyloxy-4-cyanobiphenyl, 24
Odd alkyl chains, 28
Oil/emulsifier interaction, 186
Oil/water interphase, 176, 183
"On-the-fly" effluents, 227
Optical
 activity, 77
 devices, 146
 rotation, 138
 thresholds, 249
Optically active
 mesogens, 61
 Schiff bases, 122
Order parameter, 210, 224
Ordered association, 190
Orientation, 236
 dispersion, 245
 singularity, 245
Orientational entropy, 28
Oscillatory shearing motion, 236
Oseen–Frank theory, 237

P

PAA, 213, 216, 220, 242, 287
Palmitic acid, 228
Parallel surfaces, 252

Peak wavelength shift, 138
p-Pentaphenyl, 12
p-Pentoxy benzoic acid, 211
4′-n-Pentyl-4-cyanobiphenyl, 4, 8
Penultimate carbon, 48, 50, 60
Perfect alignment, 235
Phase
  isotropic, 7
  plastic crystal, 7
  viscous isotropic cubic, 7
Phenomenological theories, 89
ω-Phenylalkanethioates, 63
ω-Phenylalkanoates, 36, 63
  cholestanyl, 36
  cholesteryl, 36
  S-cholesteryl, 36
ω-Phenylalkyl, 25
  groups, 36
ω-Phenylalkyl 4-(p-cyano-benzylideneamino)cinnamates, 34
ω-Phenylalkyl 4-(p-nitro-benzylideneamino)cinnamates, 34
ω-Phenylalkyl 4-(p-substituted-benzylideneamino)cinnamates, 44
ω-Phenylalkyl esters, 48
p-Phenylene, 13
Phenylene rings, 18
p-Phenylene rings, 10, 11, 13
Phosphtatidyl choline–phosphatidyl serine, 228
Phospholipid–water gels, 227
Photochemical stability, 56, 57
Pitch, 79
Planar
  layers, 185
  rings, 10
  texture, 76, 77
  variation, 270
Plastic crystal, 6
  lattice, 7
  mesophase, 6
  phase, 7
Point defects, 246
Polar head groups, 228
Polarizability, 6, 9, 24, 33
  molecular, 36
Polarizable rings, 10
Polarization, 209, 240
Pollinastanol, see 14α-Methyl-9,19-cyclo-5α,9β-cholestan-3β-ol

α,ω-Polymethylene-bis-cholesteryl carbonates, 63
Poriferasterol see (24R)-Stigmasta-5,22-dien-3β-ol
Posttransition effects, 206
PPU, 206
5α-Pregnan-3β-ol, 115
Pregn-5-en-3β-ol, 115
Pregn-5-en-3β-ol-20-one, 115
Pretransition effects, 205, 206, 279
17β-Propyl-5α-androstan-3β-ol, 115
"Pseudo-lattice" vibrations, 213
Pyrazinyl ring, 19
Pyridazinyl
  derivaties, 17
  ring, 19
Pyridyl compounds, 20
  3-pyridyl, 22
  4-pyridyl, 21, 22
4-Pyridyl nitrogen, 21
Pyrroyl groups, 23

Q

Quality control of components, 153
Quasi-planar arrangement, 63, 64
p-Quaterphenyl, 12

R

Racemic mixture, 75
Radial variation, 279
Radiation
  infrared, 148
  ionizing, 147
  microwave, 149
  sensors, 147
  ultrasonic, 150
  ultraviolet, 147
Raman depolarization ratios, 201, 204
Raman mode, 223
Random alignment, 235
Rayleigh scattering, 213
Reduced attraction potential, 196
Reflected light, 76
Reflection
  dispersive, 74
  selective, 74
Region
  nonpolar (hydrophobic), 2

Subject Index

polar (hydrophilic), 2
Repulsive interactions, 21, 28
Rigid rings, 10
Rotary power, 78
Rotational
  barrier, 221
  conformers, 44
  disorder, 183
  motion, 28, 41
Rotatory oscillation, 223

S

S-Cholestanyl-ω-phenylalkanotes, 36
S-Cholesteryl alkanethioates, 63
Schiff's base, 23, 218
  2-hydroxy, 41
  linkage, 21
Screw dislocation, 291
Second order, 251
  tensor, 235
Secondary perturbation, 206
Selective reflection, 73ff, 77, 80, 85, 125, 138, 249
Singular patterns, 281
β-Sitosterol, see (24R)-Stigmast-5-en-3β-ol
Smectic phase, 3, 75, 222
  A, 262
  B, 224
  C, 224
  internal vibrations, 222
  lamellae, 33
  vibrational spectra, 222
Smectics, 249, 262
Sodium docosyl sulfate, 180
Sodium dodecyl sulfate, 180
Soft mode, 217
Space applications, 154
Spatially kinked, 64
Spectra, temperature dependence, 204
Sphericity, 8
Splay, 239
Steric effects, 45, 50, 61
Steric intermolecular interactions, 28
Sterols, 62
Steryl mesogens, 67
(24R)-Stigmast-5-en-3β-ol, 112
(24S)-Stigmast-5-en-3β-ol, 112
(24R)-Stigmasta-5,22-dien-3β-ol, 112
(24S)-Stigmasta-5,22-dien-3β-ol, 112

(24E)-Stigmasta-5,24(28)-dien-3-ol, 112
(24S)-5α-Stigmastan-3β-ol, 113
Stigmasterol, see (24S)-Stigmasta-5,22-dien-3β-ol
Stilbazoles, 20
Stilbenes, 9
  cyano, 22
  planar, 22
Stretching modes, 208
  C—C, 228
  C—H, 227, 228
Striped pattern, 290
Structure
  dextro, 77
  helical, 77
  levo, 77
Substituents
  alkyl terminal, 11
  n-alkyl terminal, 25
  lateral, 39, 61
  terminal, 59
p-Substituted anisoles, 13
5-Substituted 6-n-alkoxy-2-naphthoic acids, 41, 61
4-(p-Substituted-benzylideneamino)cinnamates, 44, 50
4-(p-Substituted-benzylideneamino)cinnamic acid, 36
2-Substituted biphenyls, 45
3β-Substituted cholest-5-enes, 99, 100
3β-Substituted-5α-cholestanes, 99, 100
p-Substituted-ω-phenylalkyl 4-(p-phenylbenzylideneamino)cinnamates, 25
p-Substituted toluenes, 13
Surface
  derivatives, 261
  energy, 243, 246
  films, 182
  integral, 238
  thermography, 62, 66
Surface properties, 176
  hydrophilic, 176
  hydrophobic, 176
Symmetry, 201

T

TBBA, 224
Temperature, sensing, 137
Terephthalic esters, 12

Subject Index

Terminal substituent, 25, 56, 59
p-Terphenyl compounds, 12, 15, 17, 53
Terpthal-bis-butyl aniline, 224
Tetrazinyl ring, 19
Textures, 76
  "blue phase," 76
  focal conic, 31, 76, 250
  "homeotropic," 76
  honeycomb, 252
  planar, 76
Thermal devices, 146
Thermal stability, 13, 21
Thermography, surface, 62
Thermotropic mesophase, 2
Thiocholesterol, 63, *see also* Cholest-5-en-3β-thiol
Tirucallol, *see* 13α,14β,17β_H-Lanosta-8,24-dien-3β-ol
Toluenes, *p*-substituted, 13
Torsional vibrations, 220
Translational motions, 8
Transmission coefficient, 92
Trans-stilbenes, 47, 53
4,4,14α-Trimethyl-5α-cholest-7-en-3β-ol, 119
4α,14α,24ξ-Trimethyl-9,19-cyclo-5α,9β-cholestan-3β-ol, 121
4,4,14α-Trimethyl-9,19-cyclo-5α9β-ergost-24(28)-en-3β-ol, 121
Twist, 239
Twist axis, 247
Twisted nematic phase, 75

U

Ultrasonic radiation, 150
Ultraviolet radiation, 147
Uniaxial materials, 234
Unit vector, 234

V

Vacuum equations, 239
van der Waals attraction potential, 187, 193
van der Waals Forces, 193
Vibration frequencies, 217
Vibrations
  accordionlike, 207
  internal, 201, 204, 222
  lattice, 204, 213, 216
Vibrational spectra, 222
  cholesteric phase, 225
  lyotropic mesophase, 227
  smectic phases, 222
Vibrational spectroscopy, 199
Viscosity, 196
Viscous isotropic, 7

W

Water/emulsifier interaction, 184
Water interface, 176
Water–lecithin system, 229

Z

Zymosterol, *see* 5α-Cholesta-8,24-dien-3β-ol

QD
923
A35
v.2

NOV 10 1976